Advanced Eco-friendly Wood-Based Composites

Advanced Eco-friendly Wood-Based Composites

Editors

Roman Reh
Lubos Kristak
Petar Antov

MDPI • Basel • Beijing • Wuhan • Barcelona • Belgrade • Manchester • Tokyo • Cluj • Tianjin

Editors

Roman Reh
Faculty of Wood Sciences
and Technology
Technical University
in Zvolen
Zvolen
Slovakia

Lubos Kristak
Faculty of Wood Sciences
and Technology
Technical University
in Zvolen
Zvolen
Slovakia

Petar Antov
Faculty of Forest Industry
University of Forestry
Sofia
Bulgaria

Editorial Office
MDPI
St. Alban-Anlage 66
4052 Basel, Switzerland

This is a reprint of articles from the Special Issue published online in the open access journal *Materials* (ISSN 1996-1944) (available at: www.mdpi.com/journal/materials/special_issues/Eco_friendly_Wood_Based_Composites).

For citation purposes, cite each article independently as indicated on the article page online and as indicated below:

LastName, A.A.; LastName, B.B.; LastName, C.C. Article Title. *Journal Name* **Year**, *Volume Number*, Page Range.

ISBN 978-3-0365-6409-8 (Hbk)
ISBN 978-3-0365-6408-1 (PDF)

© 2023 by the authors. Articles in this book are Open Access and distributed under the Creative Commons Attribution (CC BY) license, which allows users to download, copy and build upon published articles, as long as the author and publisher are properly credited, which ensures maximum dissemination and a wider impact of our publications.

The book as a whole is distributed by MDPI under the terms and conditions of the Creative Commons license CC BY-NC-ND.

Contents

Roman Reh, Lubos Kristak and Petar Antov
Advanced Eco-Friendly Wood-Based Composites
Reprinted from: *Materials* **2022**, *15*, 8651, doi:10.3390/ma15238651 1

Czesław Dembiński, Zbigniew Potok, Martin Kučerka, Richard Kminiak, Alena Očkajová and Tomasz Rogoziński
The Dust Separation Efficiency of Filter Bags Used in the Wood-Based Panels Furniture Factory
Reprinted from: *Materials* **2022**, *15*, 3232, doi:10.3390/ma15093232 7

Marta Pedzik, Radosław Auriga, Lubos Kristak, Petar Antov and Tomasz Rogoziński
Physical and Mechanical Properties of Particleboard Produced with Addition of Walnut (*Juglans regia* L.) Wood Residues
Reprinted from: *Materials* **2022**, *15*, 1280, doi:10.3390/ma15041280 21

Dorota Dukarska, Tomasz Rogoziński, Petar Antov, Lubos Kristak and Jakub Kmieciak
Characterisation of Wood Particles Used in the Particleboard Production as a Function of Their Moisture Content
Reprinted from: *Materials* **2022**, *15*, 48, doi:10.3390/ma15010048 33

Anita Wronka, Eduardo Robles and Grzegorz Kowaluk
Upcycling and Recycling Potential of Selected Lignocellulosic Waste Biomass
Reprinted from: *Materials* **2021**, *14*, 7772, doi:10.3390/ma14247772 47

Sucia Okta Handika, Muhammad Adly Rahandi Lubis, Rita Kartika Sari, Raden Permana Budi Laksana, Petar Antov and Viktor Savov et al.
Enhancing Thermal and Mechanical Properties of Ramie Fiber via Impregnation by Lignin-Based Polyurethane Resin
Reprinted from: *Materials* **2021**, *14*, 6850, doi:10.3390/ma14226850 61

Vahid Nasir, Hamidreza Fathi, Arezoo Fallah, Siavash Kazemirad, Farrokh Sassani and Petar Antov
Prediction of Mechanical Properties of Artificially Weathered Wood by Color Change and Machine Learning
Reprinted from: *Materials* **2021**, *14*, 6314, doi:10.3390/ma14216314 83

Radosław Mirski, Łukasz Matwiej, Dorota Dziurka, Monika Chuda-Kowalska, Maciej Marecki and Bartosz Pałubicki et al.
Influence of the Structure of Lattice Beams on Their Strength Properties
Reprinted from: *Materials* **2021**, *14*, 5765, doi:10.3390/ma14195765 101

Vassil Jivkov, Ralitsa Simeonova, Petar Antov, Assia Marinova, Boryana Petrova and Lubos Kristak
Structural Application of Lightweight Panels Made of Waste Cardboard and Beech Veneer
Reprinted from: *Materials* **2021**, *14*, 5064, doi:10.3390/ma14175064 115

Pavlo Bekhta, Gregory Noshchenko, Roman Réh, Lubos Kristak, Ján Sedliačik and Petar Antov et al.
Properties of Eco-Friendly Particleboards Bonded with Lignosulfonate-Urea-Formaldehyde Adhesives and pMDI as a Crosslinker
Reprinted from: *Materials* **2021**, *14*, 4875, doi:10.3390/ma14174875 133

Radosław Mirski, Dorota Dukarska, Joanna Walkiewicz and Adam Derkowski
Waste Wood Particles from Primary Wood Processing as a Filler of Insulation PUR Foams
Reprinted from: *Materials* **2021**, *14*, 4781, doi:10.3390/ma14174781 . . . . **155**

Karol Tutek and Anna Masek
Hemp and Its Derivatives as a Universal Industrial Raw Material (with Particular Emphasis on the Polymer Industry)—A Review
Reprinted from: *Materials* **2022**, *15*, 2565, doi:10.3390/ma15072565 . . . . **169**

Editorial

Advanced Eco-Friendly Wood-Based Composites

Roman Reh, Lubos Kristak and Petar Antov

1. Faculty of Wood Sciences and Technology, Technical University in Zvolen, 960 01 Zvolen, Slovakia
2. Faculty of Forest Industry, University of Forestry, 1797 Sofia, Bulgaria
* Correspondence: roman.reh@tuzvo.sk

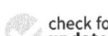

Citation: Roman Reh, Lubos Kristak and Petar Antov Advanced Eco-Friendly Wood-Based Composites. *Materials* 2022, 15, 8651. https://doi.org/10.3390/ma15238651

Received: 11 November 2022
Accepted: 18 November 2022
Published: 5 December 2022

Publisher's Note: MDPI stays neutral with regard to jurisdictional claims in published maps and institutional affiliations.

Copyright: © 2022 by the authors. Licensee MDPI, Basel, Switzerland. This article is an open access article distributed under the terms and conditions of the Creative Commons Attribution (CC BY) license (https://creativecommons.org/licenses/by/4.0/).

In collaboration with the MDPI publishing house, we are pleased to introduce the reader to our new project, the Special Issue entitled "Advanced Eco-friendly Wood-Based Composites". This Special Issue provides an opportunity to investigate the advanced eco-friendly wood-based composites from a broader perspective. The coronavirus pandemic and shutdown measures employed to contain it, as well as the ongoing war, have influenced and decelerated the world economy and adversely impacted the research activities on most levels in all countries. Surprisingly, researchers in the field of wood-based composites have continued to make progress, which is also described in this Special Issue.

The wood of forest trees is a renewable, sustainable and easily workable material and has been widely used in construction, paper making, and furniture and as a feedstock for biofuels. Wood composites are engineered wood-based materials that are fabricated from a wide variety of wood and other non-wood lignocellulosic materials, bonded with synthetic or natural bio-based adhesive systems, and designed for specific value-added applications and performance requirements [1–6]. Traditional wood-based composites are fabricated using synthetic formaldehyde-based adhesives that are commonly formed from fossil-derived constituents, such as urea, phenol, and melamine [7–9]. Along with their undisputable advantages, these adhesives are characterized by certain problems related to the emission of hazardous volatile organic compounds (VOCs), including free formaldehyde emissions from the finished wood composites, which is carcinogenic to humans and harmful to the environment [10–12]. The growing environmental concerns connected with the adoption of circular economy principles and the new, stricter legislative requirements for the emission of harmful VOCs, such as free formaldehyde, from wood composites pose new challenges for researchers and industrial practice. These challenges are related to the development of sustainable, eco-friendly wood composites [13–15], the optimization of the available lignocellulosic raw materials [16–18], and the use of alternative resources [19–23]. The harmful release of formaldehyde from wood composites can be reduced by applying formaldehyde scavengers to conventional adhesive systems [24–27], by the surface treatment of the finished wood composites, or by the application of novel bio-based wood adhesives as environmentally friendly alternatives to traditional synthetic resins [28–30]. Another alternative to the use of synthetic formaldehyde-based adhesives is the manufacturing of binderless wood composites, since wood is a natural polymer material that is rich in lignocellulosic compounds such as cellulose, hemicellulose, and lignin.

This Special Issue represents a collection of 11 high-quality original research and review papers that provide examples of the latest advancements in the development and applications of eco-friendly wood-based composites.

In their paper, Bekhta et al. investigated the potential of incorporating lignin-based additives, i.e., magnesium and sodium lignosulfonates, in urea-formaldehyde resin in order to manufacture low-toxic, eco-friendly particleboards with acceptable physical and mechanical properties and achieve reduced formaldehyde emissions [31]. The adhesive system employed by the authors also included polymeric 4,4'-diphenylmethane diisocyanate (pMDI) as a crosslinker. The authors determined that the lignosulfonate addition levels varied from 10 to 30%, resulting in particleboards with physical and mechanical properties comparable to those of panels bonded with UF resin alone. In addition, the panels bonded with lignosulfonates and pMDI exhibited a close-to-zero formaldehyde content, reaching the super E0 emission grade of ≤1.5 mg/100 g.

In another paper, Mirski et al. studied the effect of the structure of lattice beams on their strength properties [32]. Based on the results obtained from the study, it was concluded that the solutions proposed by the authors represent alternatives to wooden trusses, which are joined with flanges using punched metal plate fasteners. However, it should be noted that, at the current stage of this research, these solutions exhibited approximately 30% lower static bending strength values than trusses fabricated with metal plates.

The feasibility of employing novel lightweight panels fabricated from waste corrugated cardboard and beech veneer, as structural materials with applications in interior and furniture construction, was studied by Jivkov et al. [33]. In laboratory conditions, the authors developed two types of multi-layered panels and evaluated the bending moments and stiffness coefficients of seven different types of end corner joints (demountable joints and those fixed with an adhesive) formed from the developed composites. The authors concluded that these materials can be successfully used in the construction of furniture and other interior elements.

Following the circular economy principles, i.e., the reuse, recycling, or upcycling of materials for the purpose of the increased utilization of waste and by-products in value-added applications, Mirski et al. investigated the possibilities of using waste wood particles obtained from the primary wood processing as a filler for polyurethane foams (PUR) with an open-cell structure [34]. It was found that the addition of 10% waste wood particles resulted in 30% increased compressive strength values of the PUR foam and 10% decreased thermal conductivity, respectively. The authors concluded that the developed composite foams can be efficiently used in thermal insulation applications in the construction of prefabricated buildings.

In another interesting study, an attempt was made to predict the mechanical properties, i.e., the modulus of elasticity (MOE) and modulus of rupture (MOR), of artificially weathered fir, alder, oak, and poplar wood by investigating the variations in the color parameters of the wood samples and developing a machine learning model [35]. It was found that the deflection to failure of the wood samples increased with the weathering, which was attributed to the increased viscoelasticity of the weathered wood samples. Significantly, the experimental work was performed only on small-sized, clear wood samples without defects. Thus, the effectiveness of the developed model should be further analyzed using large-sized wood specimens.

Handika et al. reported the isolation of lignin from black liquor, used as a pre-polymer for the preparation of bio-based polyurethane resin, which was exploited for the impregnation of ramie fiber (*Boehmeria nivea* (L.) Gaudich) with the aim of improving its thermal and mechanical properties [36]. One-step fractionation of the isolated lignin was performed using methanol and acetone as solvents. Based on the experimental results, the authors concluded that the increased mechanical properties, i.e., the tensile strength and MOE, as well as the enhanced thermal stability of the impregnated ramie fiber, could expand its future potential for wider industrial application as a sustainable and functional material.

Wronka et al. studied the potential of using raspberry (*Rubus idaeus* L.) and black chokeberry (*Aronia melanocarpa* (Michx.) Elliott) lignocellulosic particles for manufacturing particleboards intended for furniture applications [37]. The authors also characterized the wooden particles, obtained from the re-milling particleboards, in order to evaluate their recycling possibilities. The authors reported the successful fabrication of particleboards from both lignocellulosic by-products. Significantly, the addition of raspberry particles should not exceed 50% in order to obtain boards with mechanical properties that fulfil the European standard requirements. In addition, it was found that the upcycling of the particles obtained from the re-milled panels is rather limited due to the significantly different fractions and shape of particles.

The study carried out by Dukarska et al. aimed to investigate and characterize the physical properties of wood particles intended for the manufacturing of particleboards according to their moisture content [38]. It was found that the increased moisture content of the wood particles resulted in an increase in their dimensions, regardless of their degree of fineness, as well as an increased slippery angle of repose. In addition, the greater moisture content of the particles resulted in an increased tapped bulk density for both types of particles evaluated, e.g., the microparticles of the outer layers of the particleboards and the particles of the core layers of the panels. The results obtained could be of great benefit in the industrial practices of the wood-based panel industry with respect to the optimization of the technological parameters and related production costs.

One of the greatest challenges for the wood composite industry is the increased demand for wood and other lignocellulosic raw materials [39–41]. This has led to significantly increased interest in the industrial and research sectors in efforts to identify alternative raw materials as natural feedstocks for the production of wood composites. In their study, Pędzik et al. evaluated the potential of using walnut (*Juglans regia* L.) wood residues as an alternative raw material for the production of particleboards [42]. The authors reported that the mechanical properties of the panels, which were produced in the laboratory with 50% walnut wood particles, fulfilled the European standard requirements for particleboards intended for load-bearing applications.

Exposure to wood dust is one of the greatest occupational hazards to the health and safety of workers in wood-processing and furniture enterprises [43–46]. The results of the study carried out by Dembiński et al. will be of great benefit for the industrial practice of furniture factories in terms of methods for predicting the separation efficiency in the long-term use of filter bags employed in the wood-based panel industry [47].

Last but not least, a comprehensive review of the possibilities of using hemp as an abundant and renewable natural raw material for the polymer industry was conducted by Tutek and Masek [48]. The authors presented and critically discussed the chemical composition and physical and mechanical properties of hemp fibers, oil, wax, and extracts and provided relevant examples of the use of hemp derivatives in polymer composites.

The ongoing transition of the wood-based panel industry toward a circular, low-carbon bio-economy is a strong prerequisite for the continuous development of sustainable and eco-friendly wood composites. The examples presented herein represent only a selection and short overview of the future research trajectories related to the development, properties, and applications of innovative, high-performance, eco-friendly wood composites with a lower environmental impact.

Author Contributions: Conceptualization, R.R., L.K. and P.A.; Writing and editing, R.R. and P.A. All authors have read and agreed to the published version of the manuscript.

Funding: This research received no external funding.

Acknowledgments: This work was supported by the Slovak Research and Development Agency under contracts No. APVV-20-004, APVV-19-0269 and No. SK-CZ-RD-21-0100 and by the project "Development, Properties, and Application of Eco-Friendly Wood-Based Composites", No. НИС-Б-1145/04.2021, carried out at the University of Forestry, Sofia, Bulgaria.

Conflicts of Interest: The authors declare no conflict of interest.

References

1. Irle, M.A.; Barbu, M.C.; Réh, R.; Bergland, L.; Rowell, R.M. Wood Composites. In *Handbook of Wood Chemistry and Wood Composites*; CRC Press: Boca Raton, FL, USA, 2012.
2. Pizzi, A.; Papadopoulos, A.N.; Policardi, F. Wood composites and their polymer binders. *Polymers* **2020**, *12*, 1115. [CrossRef] [PubMed]
3. Krišťák, Ľ.; Réh, R. Application of Wood Composites. *Appl. Sci.* **2021**, *11*, 3479. [CrossRef]
4. Papadopoulos, A.N. Advances in Wood Composites III. *Polymers* **2021**, *13*, 163. [CrossRef]
5. Mirski, R.; Derkowski, A.; Kawalerczyk, J.; Dziurka, D.; Walkiewicz, J. The Possibility of Using Pine Bark Particles in the Chipboard Manufacturing Process. *Materials* **2022**, *15*, 5731. [CrossRef] [PubMed]

6. Lee, S.H.; Lum, W.C.; Boon, J.G.; Kristak, L.; Antov, P.; Rogoziński, T.; Pędzik, M.; Taghiyari, H.R.; Lubis, M.A.R.; Fatriasari, W.; et al. Particleboard from Agricultural Biomass and Recycled Wood Waste: A Review. *J. Mater. Res. Technol.* **2022**, *20*, 4630–4658. [CrossRef]
7. Mantanis, G.I.; Athanassiadou, E.T.; Barbu, M.C.; Wijnendaele, K. Adhesive systems used in the European particleboard, MDF and OSB industries. *Wood Mater. Sci. Eng.* **2018**, *13*, 104–116. [CrossRef]
8. Dorieh, A.; Ayrilmis, N.; Pour, M.F.; Movahed, S.G.; Kiamahalleh, M.V.; Shahavi, M.H.; Hatefnia, H.; Mehdinia, M. Phenol formaldehyde resin modified by cellulose and lignin nanomaterials: Review and recent progress. *Int. J. Biol. Macromol.* **2022**, *222*, 1888–1907. [CrossRef]
9. Barbu, M.C.; Irle, M.; Réh, R. Wood Based Composites. In *Research Developments in Wood Engineering and Technology*; Aguilera, A., Davim, P., Eds.; IGI Global: Hershey, PA, USA, 2014; Chapter 1; pp. 1–45.
10. Kumar, R.N.; Pizzi, A. Environmental Aspects of Adhesives–Emission of Formaldehyde. In *Adhesives for Wood and Lignocellulosic Materials*; Wiley-Scrivener Publishing: Hoboken, NJ, USA, 2019; pp. 293–312.
11. Walkiewicz, J.; Kawalerczyk, J.; Mirski, R.; Dziurka, D.; Wieruszewski, M. The Application of Various Bark Species as a Fillers for UF Resin in Plywood Manufacturing. *Materials* **2022**, *15*, 7201. [CrossRef]
12. Bekhta, P.; Sedliačik, J.; Noshchenko, G.; Kačík, F.; Bekhta, N. Characteristics of Beech Bark and its Effect on Properties of UF Adhesive and on Bonding Strength and Formaldehyde Emission of Plywood Panels. *Eur. J. Wood Prod.* **2021**, *79*, 423–433. [CrossRef]
13. Ninikas, K.; Mitani, A.; Koutsianitis, D.; Ntalos, G.; Taghiyari, H.R.; Papadopoulos, A.N. Thermal and Mechanical Properties of Green Insulation Composites Made from Cannabis and Bark Residues. *J. Compos. Sci.* **2021**, *5*, 132. [CrossRef]
14. Antov, P.; Savov, V.; Trichkov, N.; Krišťák, Ľ.; Réh, R.; Papadopoulos, A.N.; Taghiyari, H.R.; Pizzi, A.; Kunecová, D.; Pachikova, M. Properties of High-Density Fiberboard Bonded with Urea–Formaldehyde Resin and Ammonium Lignosulfonate as a Bio-Based Additive. *Polymers* **2021**, *13*, 2775. [CrossRef] [PubMed]
15. Savov, V.; Valchev, I.; Antov, P.; Yordanov, I.; Popski, Z. Effect of the Adhesive System on the Properties of Fiberboard Panels Bonded with Hydrolysis Lignin and Phenol-Formaldehyde Resin. *Polymers* **2022**, *14*, 1768. [CrossRef] [PubMed]
16. Kminiak, R.; Orlowski, K.A.; Dzurenda, L.; Chuchala, D.; Banski, A. Effect of Thermal Treatment of Birch Wood by Saturated Water Vapor on Granulometric Composition of Chips from Sawing and Milling Processes from the Point of View of Its Processing to Composites. *Appl. Sci.* **2020**, *10*, 7545. [CrossRef]
17. Reinprecht, L.; Iždinský, J. Composites from Recycled and Modified Woods—Technology, Properties, Application. *Forests* **2022**, *13*, 6. [CrossRef]
18. Pędzik, M.; Kwidziński, Z.; Rogoziński, T. Particles from Residue Wood-Based Materials from Door Production as an Alternative Raw Material for Production of Particleboard. *Drv. Ind.* **2022**, *73*, 351–357. [CrossRef]
19. Barbu, M.C.; Sepperer, T.; Tudor, E.M.; Petutschnigg, A. Walnut and Hazelnut Shells: Untapped Industrial Resources and Their Suitability in Lignocellulosic Composites. *Appl. Sci.* **2020**, *10*, 6340. [CrossRef]
20. Kain, G.; Morandini, M.; Stamminger, A.; Granig, T.; Tudor, E.M.; Schnabel, T.; Petutschnigg, A. Production and Physical–Mechanical Characterization of Peat Moss (Sphagnum) Insulation Panels. *Materials* **2021**, *14*, 6601. [CrossRef]
21. Barbu, M.C.; Montecuccoli, Z.; Förg, J.; Barbeck, U.; Klímek, P.; Petutschnigg, A.; Tudor, E.M. Potential of Brewer's Spent Grain as a Potential Replacement of Wood in pMDI, UF or MUF Bonded Particleboard. *Polymers* **2021**, *13*, 319. [CrossRef]
22. Rammou, E.; Mitani, A.; Ntalos, G.; Koutsianitis, D.; Taghiyari, H.R.; Papadopoulos, A.N. The Potential Use of Seaweed (*Posidonia oceanica*) as an Alternative Lignocellulosic Raw Material for Wood Composites Manufacture. *Coatings* **2021**, *11*, 69. [CrossRef]
23. Pędzik, M.; Janiszewska, D.; Rogoziński, T. Alternative Lignocellulosic Raw Materials in Particleboard Production: A Review. *Ind. Crops Prod.* **2021**, *174*, 114162. [CrossRef]
24. Kristak, L.; Antov, P.; Bekhta, P.; Lubis, M.A.R.; Iswanto, A.H.; Reh, R.; Sedliacik, J.; Savov, V.; Taghiayri, H.; Papadopoulos, A.N.; et al. Recent Progress in Ultra-Low Formaldehyde Emitting Adhesive Systems and Formaldehyde Scavengers in Wood-Based Panels: A Review. *Wood Mater. Sci. Eng.* **2022**. [CrossRef]
25. Mirski, R.; Kawalerczyk, J.; Dziurka, D.; Siuda, J.; Wieruszewski, M. The Application of Oak Bark Powder as a Filler for Melamine-Urea-Formaldehyde Adhesive in Plywood Manufacturing. *Forests* **2020**, *11*, 1249. [CrossRef]
26. Medved, S.; Gajsek, U.; Tudor, E.M.; Barbu, M.C.; Antonovic, A. Efficiency of bark for reduction of formaldehyde emission from particleboards. *Wood Res.* **2019**, *64*, 307–315.
27. Kawalerczyk, J.; Walkiewicz, J.; Woźniak, M.; Dziurka, D.; Mirski, R. The effect of urea-formaldehyde adhesive modification with propylamine on the properties of manufactured plywood. *J. Adhes.* **2022**. [CrossRef]
28. Arias, A.; González-Rodríguez, S.; Vetroni Barros, M.; Salvador, R.; de Francisco, A.C.; Piekarski, C.M.; Moreira, M.T. Recent developments in bio-based adhesives from renewable natural resources. *J. Clean. Prod.* **2021**, *314*, 127892. [CrossRef]
29. Maulana, M.I.; Lubis, M.A.R.; Febrianto, F.; Hua, L.S.; Iswanto, A.H.; Antov, P.; Kristak, L.; Mardawati, E.; Sari, R.K.; Zaini, L.H.; et al. Environmentally Friendly Starch-Based Adhesives for Bonding High-Performance Wood Composites: A Review. *Forests* **2022**, *13*, 1614. [CrossRef]
30. Saud, A.S.; Maniam, G.P.; Rahim, M.H.A. Introduction of Eco-Friendly Adhesives: Source, Types, Chemistry and Characterization. In *Eco-Friendly Adhesives for Wood and Natural Fiber Composites*; Jawaid, M., Khan, T.A., Nasir, M., Asim, M., Eds.; Composites Science and Technology; Springer: Singapore, 2021.

31. Bekhta, P.; Noshchenko, G.; Réh, R.; Kristak, L.; Sedliačik, J.; Antov, P.; Mirski, R.; Savov, V. Properties of Eco-Friendly Particleboards Bonded with Lignosulfonate-Urea-Formaldehyde Adhesives and pMDI as a Crosslinker. *Materials* **2021**, *14*, 4875. [CrossRef]
32. Mirski, R.; Matwiej, Ł.; Dziurka, D.; Chuda-Kowalska, M.; Marecki, M.; Pałubicki, B.; Rogoziński, T. Influence of the Structure of Lattice Beams on Their Strength Properties. *Materials* **2021**, *14*, 5765. [CrossRef]
33. Jivkov, V.; Simeonova, R.; Antov, P.; Marinova, A.; Petrova, B.; Kristak, L. Structural Application of Lightweight Panels Made of Waste Cardboard and Beech Veneer. *Materials* **2021**, *14*, 5064. [CrossRef]
34. Mirski, R.; Dukarska, D.; Walkiewicz, J.; Derkowski, A. Waste Wood Particles from Primary Wood Processing as a Filler of Insulation PUR Foams. *Materials* **2021**, *14*, 4781. [CrossRef]
35. Nasir, V.; Fathi, H.; Fallah, A.; Kazemirad, S.; Sassani, F.; Antov, P. Prediction of Mechanical Properties of Artificially Weathered Wood by Color Change and Machine Learning. *Materials* **2021**, *14*, 6314. [CrossRef] [PubMed]
36. Handika, S.O.; Lubis, M.A.R.; Sari, R.K.; Laksana, R.P.B.; Antov, P.; Savov, V.; Gajtanska, M.; Iswanto, A.H. Enhancing Thermal and Mechanical Properties of Ramie Fiber via Impregnation by Lignin-Based Polyurethane Resin. *Materials* **2021**, *14*, 6850. [CrossRef] [PubMed]
37. Wronka, A.; Robles, E.; Kowaluk, G. Upcycling and Recycling Potential of Selected Lignocellulosic Waste Biomass. *Materials* **2021**, *14*, 7772. [CrossRef] [PubMed]
38. Dukarska, D.; Rogoziński, T.; Antov, P.; Kristak, L.; Kmieciak, J. Characterisation of Wood Particles Used in the Particleboard Production as a Function of Their Moisture Content. *Materials* **2022**, *15*, 48. [CrossRef] [PubMed]
39. Wronka, A.; Kowaluk, G. Upcycling Different Particle Sizes and Contents of Pine Branches into Particleboard. *Polymers* **2022**, *14*, 4559. [CrossRef]
40. Wronka, A.; Beer, P.; Kowaluk, G. Selected Properties of Single and Multi-Layered Particleboards with the Structure Modified by Fibers Implication. *Materials* **2022**, *15*, 8530. [CrossRef]
41. Wronka, A.; Kowaluk, G. The Influence of Multiple Mechanical Recycling of Particleboards on Their Selected Mechanical and Physical Properties. *Materials* **2022**, *15*, 8487. [CrossRef]
42. Pędzik, M.; Auriga, R.; Kristak, L.; Antov, P.; Rogoziński, T. Physical and Mechanical Properties of Particleboard Produced with Addition of Walnut (*Juglans regia* L.) Wood Residues. *Materials* **2022**, *15*, 1280. [CrossRef]
43. Očkajová, A.; Stebila, J.; Rybakowski, M.; Rogoziński, T.; Krišťák, Ľ; Ľuptáková, J. The Granularity of Dust Particles when Sanding Wood and Wood-Based Materials. *Adv. Mater. Res.* **2014**, *1001*, 432–437. [CrossRef]
44. Igaz, R.; Kminiak, R.; Krišťák, Ľ.; Němec, M.; Gergeľ, T. Methodology of Temperature Monitoring in the Process of CNC Machining of Solid Wood. *Sustainability* **2019**, *11*, 95. [CrossRef]
45. Makovicka Osvaldova, L.; Petho, M. Occupational Safety and Health During Rescue activities. *Procedia Manuf.* **2015**, *3*, 4287–4293. [CrossRef]
46. Očkajová, A.; Kučerka, M.; Kminiak, R.; Rogoziński, T. Granulometric composition of chips and dust produced from the process of working thermally modified wood. *Acta Facultatis Xylologiae Zvolen* **2020**, *62*, 103–111. [CrossRef]
47. Dembiński, C.; Potok, Z.; Kučerka, M.; Kminiak, R.; Očkajová, A.; Rogoziński, T. The Dust Separation Efficiency of Filter Bags Used in the Wood-Based Panels Furniture Factory. *Materials* **2022**, *15*, 3232. [CrossRef]
48. Tutek, K.; Masek, A. Hemp and Its Derivatives as a Universal Industrial Raw Material (with Particular Emphasis on the Polymer Industry)—A Review. *Materials* **2022**, *15*, 2565. [CrossRef] [PubMed]

Article

The Dust Separation Efficiency of Filter Bags Used in the Wood-Based Panels Furniture Factory

Czesław Dembiński [1], Zbigniew Potok [1], Martin Kučerka [2], Richard Kminiak [3], Alena Očkajová [2] and Tomasz Rogoziński [1,*]

1. Department of Furniture Design, Faculty of Forestry and Wood Technology, Poznań University of Life Sciences, 60-627 Poznań, Poland; czeslaw415@wp.pl (C.D.); zbigniew.potok@up.poznan.pl (Z.P.)
2. Department of Technology, Faculty of Natural Sciences, Matej Bel University in Banská Bystrica, Tajovského 40, 974 01 Banská Bystrica, Slovakia; martin.kucerka@umb.sk (M.K.); alena.ockajova@umb.sk (A.O.)
3. Department of Woodworking, Faculty of Wood Sciences and Technology, Technical University in Zvolen, 960 01 Zvolen, Slovakia; richard.kminiak@tuzvo.sk
* Correspondence: tomasz.rogozinski@up.poznan.pl; Tel.: +48-61-848-7483

Abstract: The relationship between the conditions of the use of filter bags made of non-woven fabric and the separation efficiency of wood dust generated in a furniture factory was experimentally determined in the conditions of pulse-jet filtration using a pilot-scale baghouse as waste during the processing of wood composites. The experiments were carried out, and we describe the results of the experiment as consisting in assembling one type of filter bag in two dust extraction installations operating under different operating conditions in the same furniture factory. The filter bags working in the assumed time intervals were then tested for their separation efficiency using a stand for testing filtration processes on a pilot scale. The test results are presented in the form of graphs and tables describing both the characteristics of the dust extraction installations and the filter fabric used, as well as the separation efficiency of bags used at different times in different industrial operating conditions for each of them. The conducted research allowed us to recognize the phenomenon of filtration in relation to a very important value, which is the separating efficiency of dust extraction in various operating conditions of dust extraction installations in a furniture factory during the long-term use of filter fabrics. The obtained results allowed us to determine the separation efficiency for the tested bags at a level of over 99.99% and to state that this separation efficiency increased with the working time of the bag. The structure of the outlet dust from filters in the wood composites processing factory constitutes an element of the working environment if the purified air is returned in a recirculation circuit to the interior of the working area. Thanks to this, it is possible to predict the separation efficiency in the long-term use of filter dust collectors for wood dust in furniture factories.

Keywords: separation efficiency; wood dust; long-term filtration; dust filtration; maturation of filter bags

Citation: Dembiński, C.; Potok, Z.; Kučerka, M.; Kminiak, R.; Očkajová, A.; Rogoziński, T. The Dust Separation Efficiency of Filter Bags Used in the Wood-Based Panels Furniture Factory. *Materials* **2022**, *15*, 3232. https://doi.org/10.3390/ma15093232

Academic Editors: Frank Lipnizki and Ana Paula Piedade

Received: 14 February 2022
Accepted: 28 April 2022
Published: 29 April 2022

Publisher's Note: MDPI stays neutral with regard to jurisdictional claims in published maps and institutional affiliations.

Copyright: © 2022 by the authors. Licensee MDPI, Basel, Switzerland. This article is an open access article distributed under the terms and conditions of the Creative Commons Attribution (CC BY) license (https://creativecommons.org/licenses/by/4.0/).

1. Introduction

Technological progress that is taking place in the furniture industry, in addition to comprehensive benefits, also brings about threats in the form of increasing interference with the natural environment. One of its expressions is the increasing demand for wood and its products [1–4]. The growing amount of wood consumption and wood products produced worldwide is directly related to the increase in dust emissions from wood processing and wood composites. These pollutants pose a very serious threat to human health. Wood dust is one of the most dangerous pathogens found inside factories processing wood materials [5–10]. The dimensions of the dust particles and their properties are conducive to long-term floating in the air, which is a very serious exposure for people staying in it as the exposure is a harmful factor to the human body. Therefore, the dust concentration in the air is an amount that should be systematically controlled, and it should be ensured that its level does not exceed the permissible concentrations [11–15]. In order to prevent the harmful effects of wood dust, the permissible amount in the air surrounding the woodworking stations has been determined. Until 2018, the permissible concentration of beech and oak wood dust in the inhaled air was 2 mg·m^{-3}, and for the dust of other species, 4 mg·m^{-3}. In 2018, the regulations were harmonized, and as a result of the compromise, the standard of wood dust concentration in the inhaled air for all wood species was 3 mg·m^{-3} [16]. From January 2022, the maximum permissible concentration of the inhalable fraction will be only 2 mg·m^{-3}. To reduce the amount of wood dust in the air, it is worth paying attention to the sources of its formation and the phenomena occurring during air purification from dust particles. The amount of dust in the working environment under air recirculation conditions depends on an efficiently conducted filtration process. The quantity describing the filtration process is, apart from the filtration resistance, also the dust separation efficiency. In these works, various aspects related to the pulse-jet filter were investigated. Unfortunately, most of the work is not concerned with wood dust, which makes it necessary to conduct research to explain the phenomena occurring during the filtration of wood dust. Since it is a very complex process, it must be carefully directed. Problems related to dust filtration have been the subject of previous papers [17–21]. Increasing the filtration separation efficiency or the influence of the filter fabrics used has already been the subject of research [22,23]. However, it is difficult to find studies whose results would show the variability of filtration efficiency depending on the length of time the filter bags are used. In order to assess the filtration process and the air quality at the outlet of the filter, it is, therefore, necessary to know the ability of the filters to clean it with the assumed filtration efficiency in the period of long-term use in industrial conditions. This issue was investigated in the past by Thorpe and Brown [24], who undertook to investigate the emission and filtration efficiency of dust from hand sanders used in the wood industry, but their research was based on a process supervised in laboratory conditions, assuming that during the operation of the tested sander, attempts were made to simulate the conditions of industrial use. The researchers only had new filter materials at their disposal. However, the efficiency of filtration depends not only on the design of the filters and the type of filter fabric used in it but also on the length and conditions of use of the filters in factories [25]. It is its ability to retain dust during long-term work in industrial conditions that should be the subject of current research on filtration processes in the wood and wood processing industry.

The aim of the paper is to an experimental study of the separation efficiency of the filter fabric operating in two different industrial filters, considering the different service lives of filter bags made of this fabric. Moreover, an attempt to determine the influence of the bag working time in industrial conditions on its separation efficiency was made. Understanding this issue will allow the characterization of the nature of the filtration phenomenon in industrial conditions and the impact on it by the properties of waste, which, despite the constantly improved technology and conducted research [22,23,26–33] on dust extraction, still pass through the filter and pose the most significant air pollution hazard in the wood industry. Thus far, no studies have been carried out to determine the filtration efficiency of filter bags operating for a long time in industrial conditions. The obtained information will become the basis for determining the requirements for the use of non-woven filter fabrics for wood dust separation.

2. Materials and Methods

2.1. General Assumptions

The IKEA Industry furniture factory in Lubawa (Poland) was selected as the research site. The main reason for this choice is the very large scale of highly automated production, which for each type of processing has different technological solutions and dust extraction installations, which are an excellent source of obtaining a lot of information on the phenomena occurring during the separation of various types of dust waste from various wood composites machining operations. The tests were carried out for two different operating lines: surface treatment line and drilling centers line. Both lines are equipped with a JKF filter (Berzyna, Poland), in which the Gutsche filter bags (Fulda, Germany) were installed. Despite the fact that the same bags are assembled, both filters differ in design.

2.2. Samples Used for Testing

According to the assumptions, filter bags used in the IKEA Industry furniture factory in Lubawa were collected for the tests. Several new bags were inserted into the selected filters of two technological lines, which were removed at two-month intervals to perform the necessary experimental procedures. The bags, after being shortened to a length of 1500 mm, were installed on pilot-scale testing stand in the laboratory, and the separation properties of materials previously operated for a certain period in industrial conditions were tested. The list of the obtained samples of bags is presented in Table 1.

Table 1. Test bags obtained from the furniture factory.

Dust Exhaust Installation	Bag Working Time [Days]	Bag Producer
Narrow surfaces treatment line Drilling centers line	0, 67, 133, 272	Gutshe

2.3. Narrow Surfaces Treatment Line of Furniture Panels

Two narrow surface processing sublines are connected to the dedusting installation. The main subline is a set of two machine tools that format and edge band furniture elements on both sides simultaneously. The machine tools have their own transport of elements, which are cut, formatted, and edge band at individual stages depending on the needs. Both machines on the line are separated by a turntable, thanks to which, after turning the elements by an angle of 90°, it is possible to process four sides of an element. The processing is carried out with circular saws and cutters. The line is adapted to producing elements made of particleboard, MDF, or solid wood. The auxiliary line processes one narrow surface of the bed joint and other narrow pieces of furniture. The same tools are used here as in the mainline. The total amount of waste generated on the line is 250 kg·h^{-1} with an average particle size of 140.88 µm. It is equipped with a two-part dust exhaust installation with one filter at the end. The production scheme on this line is described in the publication by Dembiński et al. 2021 [32].

2.4. Drilling Centers Line

At this point, a set of five numerically controlled drilling machines, used mainly for making structural holes in furniture elements, is connected to the dust extraction installation. Movable tilting spindles are also used for cutting curved elements with end mills. The machine tools are also equipped with a module for edge banding narrow surfaces of furniture elements. Dust with an average particle size of 168.64 µm in the amount of about 100 kg·h^{-1} is generated on the line of CNC drilling machines. The operation of the line was described in the publication by Dembinski et al. 2021 [34].

2.5. Characteristics of Dust Extraction Installations, Filters, and Filter Fabrics

Both the line for the narrow surface treatment and the line of drilling centers line are served by two installations connected to filters serving individual lines. Dust extraction installations for both narrow surface treatment lines and drilling center lines are equipped with JK-90 MT fans (Berzyna, Poland). Their chambers contain a different number of filter bags (140 pcs. In the filter in the extraction system on the narrow surface treatment line and 162 pcs. The air demand in dust extraction systems for each line is over 55,000. m^3·h^{-1}. The filtration speed value in the filter for the narrow surface treatment line is 4.918 cm·s^{-1} and 4.653 cm·s^{-1} in the filter for drilling center line. The filter bags were regenerated with an interval of 606 s in the filter for the narrow surfaces treatment line, while in the filter for the line of drilling centers line, this interval was 690 s. The dust load was respectively

4.509 g·m^{-3} for the narrow surface treatment line and 1.671 g·m^{-3} for the drilling centers line. The exact operating parameters of filters and dust extraction installations are described by Dembiński et al. 2021 [32]. Gutsche filter bags were installed in the described filters. The basic technical parameters of the material of filter bags are presented in Table 2.

Table 2. Basic technical parameters of the filter bags used according to the manufacturer.

Parameter	Unit	Parameter Value
Bag producer		Gutshe
Material type/symbol		Polyester with PP film
Material weight	g·m^{-1}	400
Material thickness	mm	1.5
Tensile strength—lengthwise	daN·5 cm^{-1}	40
Tensile strength—across	daN·5 cm^{-1}	50
Air permeability	dm^3·min^{-1}·dm^{-2}	250
Surface finishing		Thermal stabilization, calendering
High-temperature resistance	°C	90
Acid resistance		Good
Alkali resistance		Sufficient
Water-resistant		Weak
Declared filtration efficiency for particles > 2.5 μm	%	99.998
Declared filtration efficiency for particles < 2.5 μm	%	99.957

2.6. Laboratory Research Stand

The filter bags obtained at the factory were tested on the stand for testing the filtration process on the pilot scale. The stand is used to determine the basic filtration parameters under set conditions. During the tests, the filtration parameters were maintained, corresponding to the actual conditions found in the industrial filter. In achieving these conditions, measurement and control systems were used:

- Dust dosing system in the intended amounts and concentration;
- Control valve for adjusting the volumetric capacity of the main air circulation fan;
- System controlling the frequency of regeneration of the filter element.

During the filtration at the testing stand, measurements of the number of dust particles at the outflow pipe were made at 5 min intervals. This measurement was performed using a Hiac/Royco 5250 A model laser particle counter (Rockford, IL, USA). Sampling was performed for 30 s five times during each filtration cycle. This device can automatically measure the number of particles in the purified air at the rated airflow through the measuring system equal to 2.831685×10^{-2} m^3·min^{-1}. Using a counter, you can determine the number of particles broken down into individual fractions from 0.5 to 25 μm. The counter shows the number of particles with the following sizes: 0.5; 1.0; 2.0; 3.0; 5.0; 10.0; 15.0; 25.0 μm. This gives the dimensional structure of the dust content in the purified air and further allows to determine the separation efficiency of the filter materials.

The diagram of the laboratory stand operation is presented in Figure 1.

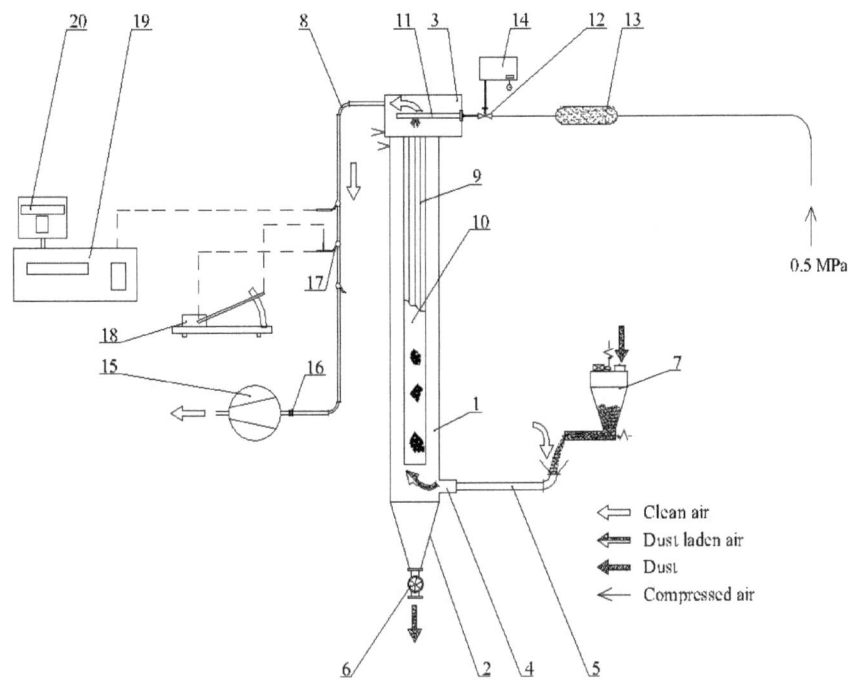

Figure 1. Test rig set-up: 1. filtering chamber, 2. hopper, 3. clean air chamber, 4. inlet, 5. dust inlet tube, 6. mucus feeder, 7. screw dust feeder DSK-I-04p (HYDRAPRESS, Białe Błota, Poland) 8. outflow pipe, 9. metal cage, 10. filtering bag, 11. cleaning nozzle, 12. electromagnetic valve, 13. compressed air tank, 14. The controlling device, 15. main fan, 16. gate valve, 17. Prandtl tube, 18. inclined-tube manometer type MPR-1 (ZAM Kety, Poland), 19. differential manometer type CMR-10 A (ZAM Kety, Poland), 20. printer.

All bags were tested under the same controlled filtration conditions as presented in Table 3. The filtration parameters were selected based on previous tests carried out for wood dust corresponding to the operating conditions of industrial filtration [19,27,28,33–37].

Table 3. Filtration parameters during testing of filtering non-wovens.

Parameter	Unit	Parameter Value
Maximum airflow velocity in the main fan duct (Figure 1 p. 15) w	$m \cdot s^{-1}$	4.290
Average velocity $\bar{w} = 0.85\, w$	$m \cdot s^{-1}$	3.646
Air volume flow V	$m^3 \cdot s^{-1}$	0.0286
	$m^3 \cdot h^{-1}$	103.0
Air to cloth ratio f	$m^3 \cdot (m^2 \cdot h)^{-1}$	145.8
Filtration velocity w_f	$m \cdot s^{-1}$	0.0405
Dust concentration	$G \cdot m^{-3}$	10

The observed results were then compared with each other with regard to the working time of the bag.

2.7. Dust Used for Laboratory Tests

The bags were tested using test beech dust with a bulk density of 177.8 kg·m^{-3} from a bent furniture factory. The particle-size distribution of dust was determined by the sieve method using a Retsch AS200 (Haan, Germany) sieve machine.

2.8. Separation Efficiency

In order to determine the separation efficiency, the mass of dust particles in the air on both sides of the filter bag was determined. The mass of dust particles in the inlet dust in fractions corresponding to the size ranges for the dust content in the purified air was determined based on the setting of the dust feeder to the filtration chamber (10 g·m^{-3}) and the determination of the particle size distribution of dust in sieve fraction < 0.032 mm carried out with the laser particle sizer Analysette 22 Micro-tec plus (Firtsch, Idar-Oberstein, Germany). The quantitative-dimensional structure of the dust in the incoming air to the bag in the test stand was determined concerning the assumed dimensional channels in accordance with the empirical particle distribution function, using the method described earlier in the study of Rogoziński (2016) [28] and presented in Table 4.

Table 4. Dust mass in the in 1 m^3 of inlet air.

Dimensional Range [µm]	Percentage [%]	Inlet Fraction Mass [g]	Inlet Fraction Mass [kg]
<0.5	0.173983	0.0173983	1.73983 × 10^{-5}
0.5–1	0.035953	0.003595325	3.59533 × 10^{-6}
1–2	0.008763	0.00087631	8.7631 × 10^{-7}
2–3	0.007876	0.000787628	7.87628 × 10^{-7}
3–5	0.100016	0.010001631	1.00016 × 10^{-5}
5–10	0.703310	0.070331044	7.0331 × 10^{-5}
10–15	0.980705	0.098070539	9.80705 × 10^{-5}
15–25	1.906999	0.190699982	0.0001907
Total Dimensional range from 0.5 µm to 25 µm		0.391760759	0.000391761
more than 25 µm	96.082392	9.608239241	0.009608239

The separation efficiency of the tested filter bags was calculated for particles in the range of 0–25 µm. The dust content and the shares of its individual fractions in the purified air were determined based on the results of measurements carried out using the Hiac laser particle counter. The results obtained from the measurement of the number of particles per 1 m^3 in the purified air were then converted into the mass of particles in individual fractions. The mass of dust particles in the filtered air in a given fraction per 1 m3 of air was determined according to Formula (1).

$$m_{1i} = V_{cz} \cdot m_{sd} \cdot a \ [kg] \quad (1)$$

where:

V_{cz}—dust particle volume [m^3]

m_{sd}—mass of wood substance (1500 kg·m^3)

a—the number of particles in each size range obtained from a particle number measurement performed.

The volume of the dust particle V_{cz} was calculated assuming that it takes the shape of a sphere with a diameter D equal to the arithmetic mean of the given size range in the measuring range of the particle counter according to Formula (2).

$$V_{cz} = \frac{\pi}{6} D^3 \quad (2)$$

After calculating the mass of particles in the purified air for individual fractions, their values were summed up to obtain the mass of dust of all 8 ranges in one cubic meter of purified air.

$$m_i = \sum_{i=1}^{8} m_{1i} \qquad (3)$$

The next step in the calculations was determining the separation efficiency for individual fractions in the filtered air. It was determined from the Equation (4).

$$\eta_i = \frac{m_{0i} - m_{1i}}{m_{0i}} \cdot 100\% \qquad (4)$$

where:

η_i—separation efficiency for individual fractions
m_{0i}—the mass of particles at the inlet to the filtration chamber of the given fraction
m_{1i}—the mass of particles leaving the chamber for a given fraction

The final stage was to determine the total efficiency. It was determined according to Formula (5).

$$\eta = \frac{\sum_{i=1}^{8}(m_{0i} - m_{1i})}{\sum_{i=1}^{8} m_{0i}} \qquad (5)$$

3. Results and Discussion

The particle-size distribution test was performed three times, and the result was averaged. The set of sieves used allowed for the separation of particles in size ranges from 0.00 to 0.032; from 0.032 to 0.063; from 0.063 to 0.125; from 0.125 to 0.500; and from 0.500 to 1.000 mm. The results are shown in Figure 2.

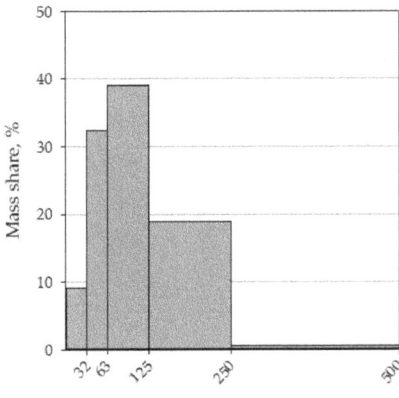

Figure 2. Dust particle-size distribution.

Based on the Equation (3), the mass of particles in 1 m³ of purified air was determined. The calculation of this value is necessary to determine the separation efficiency of tested materials. The results for both lines are presented in the graphs (Figures 3 and 4).

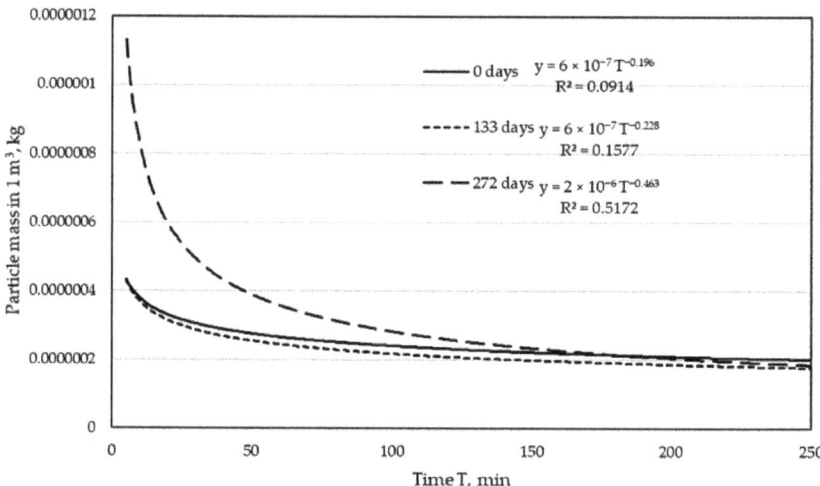

Figure 3. Particle mass in purified air for a drilling centers line.

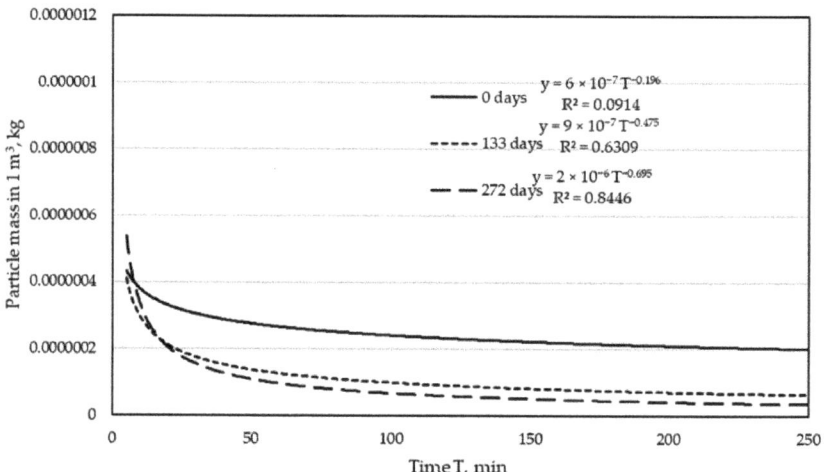

Figure 4. Particle mass in purified air for narrow surfaces treatment line.

When analyzing the graphs in Figures 3 and 4, it can be concluded that in the case of bags obtained from the filter in the installation connected to the drilling centers line (Figure 3), the total mass of particles in the purified air decreases with the extended period of industrial bag operation. The process is clearly different between a new bag and a bag that has been used for 133 and 272 days.

The analysis of the mass of particles in the purified air with the use of bags obtained from the filter in the dust extraction installation of the narrow surfaces treatment line (Figure 4) shows that the filtration process for a period of 133 days ran without significant changes in relation to the initial state in regards to the number of particles in the filtered air. The situation changed with the extension of the service life, and for 272 days, the mass of particles in the purified air decreased significantly in relation to the new bag. The decrease in the number of dust particles along with the bag operation time was found in their research by Dolny and Rogoziński in 2012 [27] and by Rogoziński [28]. The tests carried out by these authors confirmed that filtration time is one of the factors influencing filtration efficiency. These studies also showed a decrease in the filtration efficiency for particles that most penetrate the human respiratory tract, i.e., in the size (size) 2 and 3 μm.

The dependences presented in Figures 3 and 4 show the influence of the operation time of the bag in the industrial dust collector on the mass of the test dust in the outlet air from the test stand. In the case of the bags from the filter of the line of drilling centers line, it is clearly visible that with increasing the operation time of the bags in the industrial dust collector, the mass of test dust in the purified air during the test systematically decreases. The decrease in dust mass is faster in the initial filtering period and gradually slows down. In Figure 3, we see a large difference in dust mass between a new and a used bag for 133 days. This difference is much smaller if the total mass of dust is compared between the used bags of 133 and 272 days. Therefore, it can be concluded that the dust mass retained in the filter in the first phase of filtration is greater than in the later periods of operation. The dust mass in the filtered air looks slightly different for the bags from the filter in the narrow surfaces treatment line (Figure 4). While in the first phase of filtration (133 days), a decrease in the dust mass in the outlet air is noticeable, the further use of the bag (up to 272 days) does not change this mass significantly.

The next step in analyzing the results was to determine the separation efficiency for individual dust fractions according to Formula (3). By calculating the separation efficiency for each of the tested dust fractions, the dependence of this efficiency on the particle size was obtained. These dependencies are shown in Figures 5–8 as graphs of fractional separation efficiency.

Laboratory tests have shown that the tested filter bags have the lowest separation efficiency for particles of 2 μm. It is a characteristic effect of shaping the separation efficiency of filter wood dust collectors using textile filter materials. A similar observation was found during their research by Rogoziński and Trofimov [33] and Rogoziński [35]. However, they showed that one of the critical factors influencing the efficiency of wood dust separation is air humidity. It has been shown that with increasing air humidity, the filtration efficiency rises. It has also been confirmed that thermal modification of the filter fleece surface increases filtration efficiency. Nevertheless, non-woven filter materials remain the least effective for particles of this size.

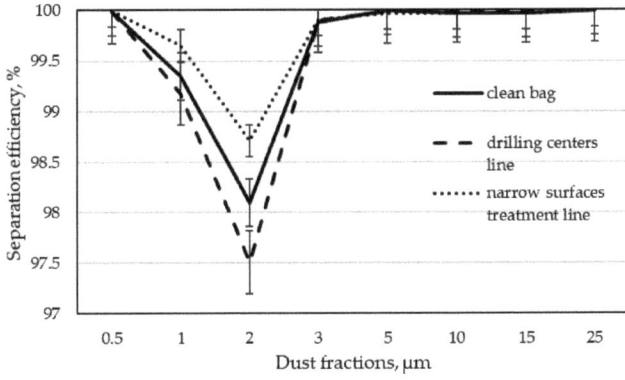

Figure 5. Fractional separation efficiency for 133 days of operation time of filter bags in individual dust extraction installations.

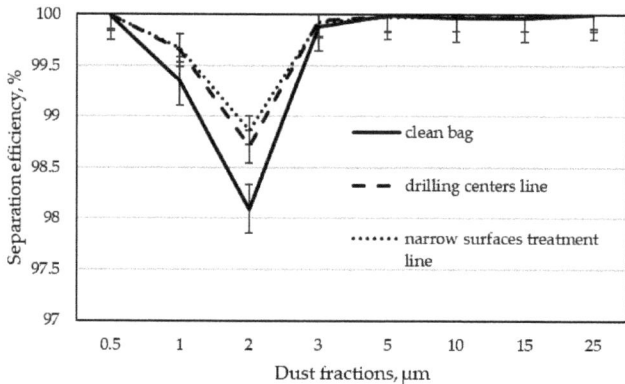

Figure 6. Fractional separation efficiency for 272 days of operation time of filter bags in individual dust extraction installations.

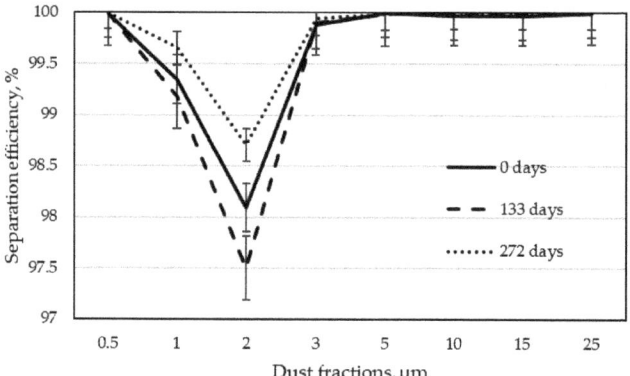

Figure 7. Fractional separation efficiency for filter bags from the dust extraction installations of drilling centers line in particular periods of use.

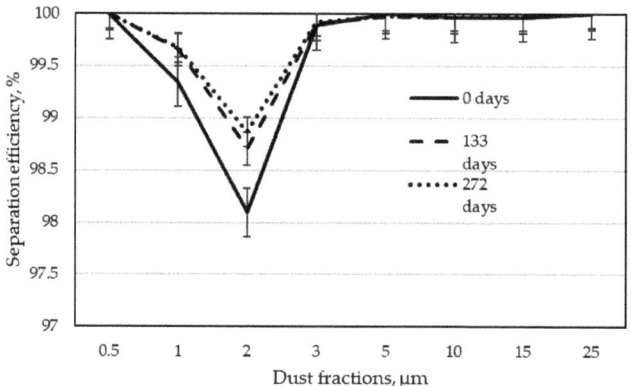

Figure 8. Fractional separation efficiency for filter bags from the dust extraction installation of a narrow surfaces treatment line in particular periods of use.

A decrease in the separation efficiency for particles in the range of 2 to 5 µm was also demonstrated in the studies by Jackiewicz and Gradoń [22]. These researchers focused on increasing dust removal efficiency by using various non-wovens, showing that the use of thinner fibers or the use of the electrostatic effect in separation significantly increases its efficiency for aerosol particles. Xiao et al. [38] came to similar conclusions. They found that in addition to the thickness of the fiber, the filtration efficiency is also influenced by the increase in the surface density of the non-woven layer while maintaining the same fiber diameter and the pore size in the non-woven fabric. These studies, however, did not concern the influence of the aging of non-wovens on filtration efficiency.

The characteristic "V" shape on the separation efficiency chart for the 2 µm fraction was also obtained during the tests of two unused different filter fabrics (non-woven polyester fabric with an anti-clogging thermo-bonded surface and acrylic polymer microporous coating over a polyester non-woven fabric) [23]. This corresponded to the lowest separation efficiency for the most penetrating particles, i.e., depending on the non-woven fabric used, from 2 to 4 µm. Unfortunately, also, in this case, we are not dealing with testing non-wovens used in a long-term manner.

A decrease in the separation efficiency for particles with dimensions of 0.1–0.2 µm was also observed by other researchers. Balgis et al. [32] showed a similar relationship for the filtration of cellulose triacetate using a non-woven fabric with porous structures.

The study of separation efficiency in plywood production was presented by Welling et al. [5]. They considered wide-belt sanders connected to a common suction system with a 100% polyester filter fabric with a weight of 420 $g \cdot m^{-2}$. The study compared the separation efficiency of an industrial filter with various filtering materials during work in laboratory conditions with MDF dust. In addition to factory filter fabric (100% polyester filter with a weight of 420 $g \cdot m^{-2}$), glass fiber filters, glass microfiber, and paper filters were also tested. In this case, the lowest separation efficiency was recorded for dust in the range of 2 to 4 µm for the filter used in the factory. The remaining materials were approximately 90% effective for this particle size. The separation efficiency increased with increasing particle size. Unfortunately, these authors do not provide the duration of the bags' operation in the factory, so their results cannot be directly compared to those presented in this work.

Graphs 7 and 8 show separation efficiency depending on the operation time of the non-woven fabric for individual wood dust fractions. It can be clearly seen that in the case of filter bags obtained from the narrow surfaces treatment line, the separation efficiency increases for all dust fractions with the increasing working time. The most significant increase in filtration efficiency over time was observed for the 2 µm particle size after 133 days. The separation efficiency of bags from the filter of the drilling centers line was slightly different. Here, in contrast to the filtration of dust from the narrow surfaces treatment line, the filtration efficiency for the most penetrating dust (2 µm) had the lowest values after 133 days. After 272 days of filtration, it increased, confirming the beneficial effect of filtration time on its efficiency. Such a result may indicate that the bag was mechanically damaged during operation in industrial conditions. It took a long time for the incoming dust to fill the damage and the separation efficiency for the most penetrating particles began to increase.

The final step was to determine the total filtration efficiency of the tested materials for both lines. The results were calculated on the basis of Formula (5) and presented in Figure 9. The results obtained for different working times of the filter materials were compared. In any case, the overall separation efficiency is very high. Similar results to those obtained in the research were obtained by Thorpe and Brown [24]. They examined the separation efficiency of beech dust generated when sanding wooden elements in simulated industrial conditions encountered in wooden furniture factories. These researchers showed that for hand sanders with external filter bags, the efficiency, depending on the granulation of the sanding paper used, ranged from 97.19 to 99.99%. Concerning the overall efficiency results, an analysis of variance (ANOVA) was performed. Thanks to it, it can be concluded that the use of the bag in the installation of the narrow surfaces treatment line for a period of 133 days does not show significant differences in separation efficiency compared to the control sample, i.e., a new, unused bag. Extending the operation time of the bags to

272 days in the same installation increased the separation efficiency, and similarity was found between its separation efficiency and the efficiency of other filters (narrow surfaces treatment line 133, drilling centers line 133, and drilling centers line 272). The bags from the dust extraction installations of CNC drills show a significantly higher filtration efficiency than the other tested ones.

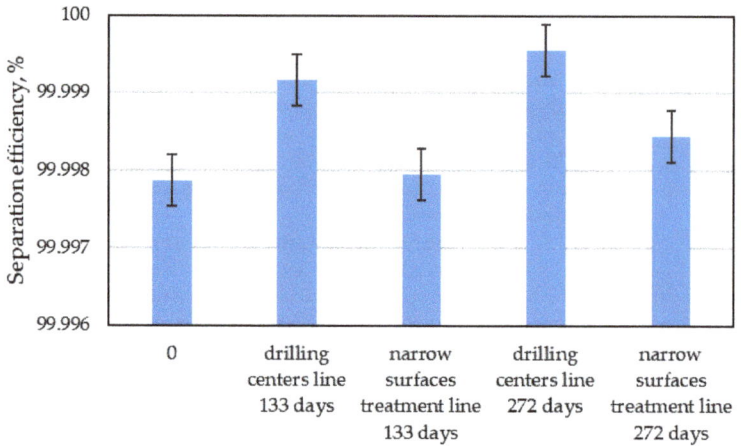

Figure 9. Total separation efficiency for all installations.

4. Conclusions

The conducted tests have shown that the filtration efficiency of bags operating in industrial conditions depends on the filtration conditions as well as the amount and properties of dust. The dust mass in the purified air decreases with the duration of use of the filter bags.

The analysis of the fractional dust separation efficiency clearly showed that for very small (0.5 and 1 μm) and large particles (over 3 μm), the efficiency reaches very high values. The least effective was for particles with a size of 2 μm.

It was also found that for the bags from both dust extraction installations, separation efficiency is higher with longer use of the bag, and so for the extraction installation in the line of CNC drilling machines, the efficiency increased earlier and to a higher level than for the filter of the installation in the narrow surfaces treatment line.

Author Contributions: Conceptualization, C.D. and T.R.; methodology, Z.P. and T.R.; software, M.K. and R.K.; validation, M.K., R.K., A.O. and T.R.; formal analysis, M.K. and T.R.; investigation C.D. and Z.P.; resources, C.D and M.K.; data curation, C.D. and Z.P.; writing—original draft preparation, C.D., Z.P. and T.R.; writing—review and editing, M.K., R.K., A.O. and T.R.; visualization, Z.P. and M.K.; supervision, A.O. and T.R.; project administration, M.K. and T.R.; funding acquisition, M.K. and T.R. All authors have read and agreed to the published version of the manuscript.

Funding: This research was supported by the Cultural and Educational Grant Agency of Ministry of Education, Science, Research and Sport of the Slovak Republic under contract no. KEGA 026UMB-4/2021 and by the grant agency VEGA under the project No. 1/0324/21 and project No. 1/0629/20. The study was also supported by the funding for statutory R&D activities as the research task No. 506.227.02.00 of the Faculty of Forestry and Wood Technology, Poznań University of Life Sciences.

Institutional Review Board Statement: Not applicable.

Informed Consent Statement: Not applicable.

Data Availability Statement: Not applicable.

Conflicts of Interest: The authors declare no conflict of interest.

References

1. Dukarska, D. Rośliny Alternatywne Jako Potencjalny Surowiec w Produkcji Płyt Wiórowych. *Biul. Inf. OBRPPD* **2013**, *54*, 5–17.
2. Górna, A.; Adamowicz, K. Predykcja cen surowca drzewnego na podstawie siedmioletniego modelu tendencji rozwojowej. *Sylwan* **2020**, *164*, 206–215. [CrossRef]
3. Antov, P.; Savov, V. Possibilities for Manufacturing Eco-Friendly Medium Density Fibreboards from Recycled Fibres—A Review. In Proceedings of the 30th International Conference on Wood Science and Technology-ICWST 2019 "Implementation of Wood Science in Woodworking Sector" and 70th Anniversary of Drvna industrija Journal, Zagreb, Croatia, 12–13 December 2019; pp. 18–24.
4. Sydor, M.; Wieloch, G. Construction Properties of Wood Taken into Consideration in Engineering Practice. *Drewno* **2009**, *52*, 63–73.
5. Welling, I.; Lehtimäki, M.; Rautio, S.; Lähde, T.; Enbom, S.; Hynynen, P.; Hämeri, K. Wood Dust Particle and Mass Concentrations and Filtration Efficiency in Sanding of Wood Materials. *J. Occup. Environ. Hyg.* **2008**, *6*, 90–98. [CrossRef] [PubMed]
6. Očkajová, A.; Kučerka, M.; Kminiak, R.; Krišťák, Ľ.; Igaz, R.; Réh, R. Occupational Exposure to Dust Produced when Milling Thermally Modified Wood. *Int. J. Environ. Res. Public Health* **2020**, *17*, 1478. [CrossRef]
7. Kminiak, R.; Kučerka, M.; Kristak, L.; Reh, R.; Antov, P.; Očkajová, A.; Rogoziński, T.; Pędzik, M. Granulometric Characterization of Wood Dust Emission from CNC Machining of Natural Wood and Medium Density Fiberboard. *Forests* **2021**, *12*, 1039. [CrossRef]
8. Mračková, E.; Krišťák, Ľ.; Kučerka, M.; Gaff, M.; Gajtanska, M. Creation of Wood Dust during Wood Processing: Size Analysis, Dust Separation, and Occupational Health. *BioResources* **2015**, *11*, 209–222. [CrossRef]
9. Pędzik, M.; Rogoziński, T.; Majka, J.; Stuper-Szablewska, K.; Antov, P.; Kristak, L.; Kminiak, R.; Kučerka, M. Fine Dust Creation during Hardwood Machine Sanding. *Appl. Sci.* **2021**, *11*, 6602. [CrossRef]
10. Majka, J.; Sydor, M.; Pędzik, M.; Antov, P.; Krišťák, Ľ.; Kminiak, R.; Kučerka, M.; Rogoziński, T. Quantifying the finest particles in dust fractions created during the sanding of untreated and thermally modified beech wood. *BioResources* **2021**, *17*, 7–20. [CrossRef]
11. Čavlović, A.; Beljo Lučić, R.; Ištvanić, J. Exposure to Wood Dust in Croatian Woodworking Industry. *Wood Res.* **2009**, *54*, 109–116.
12. Black, N.; Dilworth, M.; Summers, N. Occupational Exposure to Wood Dust in the British Woodworking Industry in 1999/2000. *Ann. Occup. Hyg.* **2007**, *51*, 249–260. [CrossRef] [PubMed]
13. Očkajová, A.; Kučerka, M.; Krišťák, Ľ.; Ružiak, I.; Gaff, M. Efficiency of Sanding Belts for Beech and Oak Sanding. *BioResources* **2016**, *11*, 5242–5254. [CrossRef]
14. Igaz, R.; Kminiak, R.; Krišťák, Ľ.; Němec, M.; Gergeľ, T. Methodology of Temperature Monitoring in the Process of CNC Machining of Solid Wood. *Sustainability* **2018**, *11*, 95. [CrossRef]
15. Očkajová, A.; Kučerka, M.; Kminiak, R.; Rogoziński, T. Granulometric Composition of Chips and Dust Produced from the Process of Working Thermally Modified Wood. *Acta Fac. Xylologiae* **2020**, *62*, 103–111.
16. Rozporządzenie Ministra Rodziny, Pracy i Polityki Społecznej z dnia 12 czerwca 2018 r. w sprawie najwyższych dopuszczalnych stężeń i natężeń czynników szkodliwych dla zdrowia w środowisku pracy. Journal of Laws of the Polish Government Legislation Center (Dz.U.2018 poz. 1286). Available online: https://isap.sejm.gov.pl/isap.nsf/DocDetails.xsp?id=WDU20180001286 (accessed on 13 February 2022).
17. Chen, X.; Mao, Y.; Fan, C.; Wu, Y.; Ge, S.; Ren, Y. Experimental investigation on filtration characteristic with different filter material of bag dust collector for dust removal. *Int. J. Coal Prep. Util.* **2021**, 1–16. [CrossRef]
18. Furumoto, K.; Narita, T.; Fukasawa, T.; Ishigami, T.; Kuo, H.-P.; Huang, A.-N.; Fukui, K. Influence of pulse-jet cleaning interval on performance of compact dust collector with pleated filter. *Sep. Purif. Technol.* **2021**, *279*, 119688. [CrossRef]
19. Potok, Z.; Rogoziński, T. Pilot-Scale Study on the Specific Resistance of Beech Wood Dust in a Pulse-Jet Filter. *Sustainability* **2020**, *12*, 4816. [CrossRef]
20. Lu, H.-C.; Tsai, C.-J. Influence of Different Cleaning Conditions on Cleaning Performance of Pilot-Scale Pulse-Jet Baghouse. *J. Environ. Eng.* **2003**, *129*, 811–818. [CrossRef]
21. Li, S.; Song, S.; Wang, F.; Jin, H.; Zhou, S.; Xie, B.; Hu, S.; Zhou, F.; Liu, C. Effects of cleaning mode on the performances of pulse-jet cartridge filter under varying particle sizes. *Adv. Powder Technol.* **2019**, *30*, 1835–1841. [CrossRef]
22. Jackiewicz, A.; Gradoń, L. Sposoby Zwiększania Sprawności Odpylania Filtrów Włókninowych. *Inżynieria I Apar. Chem.* **2011**, *50*, 42–43.
23. Simon, X.; Bémer, D.; Chazelet, S.; Thomas, D. Downstream particle puffs emitted during pulse-jet cleaning of a baghouse wood dust collector: Influence of operating conditions and filter surface treatment. *Powder Technol.* **2014**, *261*, 61–70. [CrossRef]
24. Thorpe, A.; Brown, R.C. Measurements of the effectiveness of dust extraction systems of hand sanders used on wood. *Ann. Occup. Hyg.* **1994**, *38*, 279–302. [CrossRef] [PubMed]
25. Mukhopadhyay, A. Theory, Selection and Design of Pulse-Jet Filter. In *Pulse-Jet Filtration: An Effective Way to Control Industrial Pollution/Arunangshu Mukhopadhyay*; Taylor & Francis: Abingdon, UK, 2009; ISBN 978-0-415-58103-5.
26. Bémer, D.; Regnier, R.; Calle, S. Separation efficiency of a wood dust collector-field measurement using a fluorescent aerosol. *Ann. Occup. Hyg.* **2000**, *44*, 173–183. [CrossRef]
27. Dolny, S.; Rogoziński, T. Efficiency of Beech Wood Dust Separation from Air with Increased Relative Humidity. *Acta Sci. Pol. Silvarum Colendarum Ratio Ind. Lignaria.* **2012**, *8*, 73–78.

28. Rogoziński, T. Wood dust collection efficiency in a pulse-jet fabric filter. *Drewno. Prace Naukowe. Doniesienia. Komun.* **2016**, *59*, 249–256.
29. Ward, M.; Siegel, J. Modeling Filter Bypass: Impact on Filter Efficiency. *ASHRAE Trans.* **2005**, *111*, 1091–1100.
30. Boskovic, L.; Agranovski, I.E.; Altman, I.S.; Braddock, R.D. Filter efficiency as a function of nanoparticle velocity and shape. *J. Aerosol Sci.* **2008**, *39*, 635–644. [CrossRef]
31. Azimi, P.; Zhao, D.; Stephens, B. Estimates of HVAC filtration efficiency for fine and ultrafine particles of outdoor origin. *Atmos. Environ.* **2014**, *98*, 337–346. [CrossRef]
32. Balgis, R.; Murata, H.; Ogi, T.; Kobayashi, M.; Bao, L. Enhanced Aerosol Particle Filtration Efficiency of Nonwoven Porous Cellulose Triacetate Nanofiber Mats. *ACS Omega* **2018**, *3*, 8271–8277. [CrossRef]
33. Rogoziński, T.; Trofimov, S. Principles of pulse-jet filters used in the woodworking industry. *Ann. WULS For. Wood Technol.* **2019**, *105*, 98–101. [CrossRef]
34. Dembiński, C.; Potok, Z.; Dolny, S.; Kminiak, R.; Rogoziński, T. Performance of Filter Bags Used in Industrial Pulse-Jet Baghouses in Wood-Based Panels Furniture Factory. *Appl. Sci.* **2021**, *11*, 8965. [CrossRef]
35. Rogoziński, T. An Approach to the Determination of Wood Dust Separation Efficiency in Pulse-Jet Filters. *Ann. WULS For. Wood Technol.* **2015**, *90*, 157–161.
36. Rogoziński, T. Pilot-scale study on the influence of wood dust type on pressure drop during filtration in a pulse-jet baghouse. *Process Saf. Environ. Prot.* **2018**, *119*, 58–64. [CrossRef]
37. Dolny, S.; Rogoziński, T. Air Pulse Pressure in Conditions of Air Cleaning Fromwood Dusts by Filtration. *Ann. WULS For. Wood Technol.* **2010**, *2010*, 134–141.
38. Xiao, Y.; Sakib, N.; Yue, Z.; Wang, Y.; Cheng, S.; You, J.; Militky, J.; Venkataraman, M.; Zhu, G. Study on the Relationship Between Structure Parameters and Filtration Performance of Polypropylene Meltblown Nonwovens. *Autex Res. J.* **2020**, *20*, 366–371. [CrossRef]

Article

Physical and Mechanical Properties of Particleboard Produced with Addition of Walnut (*Juglans regia* L.) Wood Residues

Marta Pędzik, Radosław Auriga, Lubos Kristak, Petar Antov and Tomasz Rogoziński

1. Wood Technology Centre, Łukasiewicz Research Network, Poznań Institute of Technology, 60-654 Poznań, Poland; marta.pedzik@pit.lukasiewicz.gov.pl
2. Faculty of Forestry and Wood Technology, Poznań University of Life Sciences, 60-627 Poznań, Poland
3. Institute of Wood Sciences and Furniture, Warsaw University of Life Sciences—SGGW, 02-787 Warsaw, Poland; radoslaw_auriga@sggw.edu.pl
4. Faculty of Wood Sciences and Technology, Technical University in Zvolen, 960 01 Zvolen, Slovakia; kristak@tuzvo.sk
5. Faculty of Forest Industry, University of Forestry, 1797 Sofia, Bulgaria; p.antov@ltu.bg
* Correspondence: tomasz.rogozinski@up.poznan.pl

Abstract: The depletion of natural resources and increased demand for wood and wood-based materials have directed researchers and the industry towards alternative raw materials for composite manufacturing, such as agricultural waste and wood residues as substitutes of traditional wood. The potential of reusing walnut (*Juglans regia* L.) wood residues as an alternative raw material in particleboard manufacturing is investigated in this work. Three-layer particleboard was manufactured in the laboratory with a thickness of 16 mm, target density of 650 kg·m^{-3} and three different levels (0%, 25% and 50%) of walnut wood particles, bonded with urea-formaldehyde (UF) resin. The physical properties (thickness swelling after 24 h) and mechanical properties (bending strength, modulus of elasticity and internal bond strength) were evaluated in accordance with the European standards. The effect of UF resin content and nominal applied pressure on the properties of the particleboard was also investigated. Markedly, the laboratory panels, manufactured with 50% walnut wood residues, exhibited flexural properties and internal bond strength, fulfilling the European standard requirements to particleboards used in load-bearing applications. However, none of the boards met the technical standard requirements for thickness swelling (24 h). Conclusively, walnut wood residues as a waste or by-product of the wood-processing industry can be efficiently utilized in the production of particleboard in terms of enhancing its mechanical properties.

Keywords: alternative raw material; walnut; applied pressure; wood residues; resin content; particleboard; physical and mechanical properties

Citation: Pędzik, M.; Auriga, R.; Kristak, L.; Antov, P.; Rogoziński, T. Physical and Mechanical Properties of Particleboard Produced with Addition of Walnut (*Juglans regia* L.) Wood Residues. *Materials* **2022**, *15*, 1280. https://doi.org/10.3390/ma15041280

Academic Editor: Dimitris S. Argyropoulos

Received: 22 December 2021
Accepted: 6 February 2022
Published: 9 February 2022

Publisher's Note: MDPI stays neutral with regard to jurisdictional claims in published maps and institutional affiliations.

Copyright: © 2022 by the authors. Licensee MDPI, Basel, Switzerland. This article is an open access article distributed under the terms and conditions of the Creative Commons Attribution (CC BY) license (https://creativecommons.org/licenses/by/4.0/).

1. Introduction

The global demand for wood and wood-based materials is constantly increasing. Using wood more efficiently to meet projected demands for the production of wood-based panels is a key circular economy principle [1,2]. The growing environmental concerns and recent legislative regulations, related to promoting the cascading use of natural resources, have posed new challenges to both the wood-based panel industry, related to the optimization of the available wood and other lignocellulosic raw materials, recycling, reusing wood and wood-based composites, and the search for alternative resources [3–5].

A problem for many companies and producers of wood and wood-based products is the insufficient amount of wood on the local market, which results in significant competition between wood-based industries. This competition will become more and more intense due to the expanding production capacities resulting in greater supply as a response to the growing demand [6,7]. The factors affecting the timber market and the increase in the price of timber are random and can occur at any time. The increase in wood prices may be caused by the global economic crisis, i.e., market and economic conditions. In turn, the fall in wood prices is often caused by natural disasters and factors, such as storms, e.g., in Italy or Austria [8]. The occurrence of bark beetle in spruce in the Czech Republic contributed to the deterioration of the quality of the raw material and the need for unplanned logging, and almost 100 million m^3 of wood was obtained, which resulted in losses of approximately EUR 1.12 billion in the forestry sector [9]. In 2020, the prices of industrial timber logs delivered in the United States increased by about 2.5% for all classes and species combined compared to the previous year [10]. The average selling price of sawmill softwood in September 2021 in the United Kingdom was around GBP 79 per cubic meter of bark in nominal terms, and a year earlier the price was around GBP 50 [11]. There is also an increase in the average price of round wood obtained, e.g., by the Polish forest inspectorates by 7.8% in 2020 compared to the previous year, and by 11.1% compared to 2016 [12–14]. The export of significant amounts of unprocessed wood is another reason for the limited availability of wood raw material. The import of wood from other countries is associated with additional transport costs and emissions of harmful compounds. In Europe, the emphasis is on pro-ecological activities and reducing CO_2 emissions. However, since 2015, the amount of exported industrial roundwood from European countries has increased from approx. 66 million m^3 to almost 78.5 million m^3 in 2020 [15].

The wood-based panel industry has certain flexibility about the use of raw materials, caused by the continuously changing wood raw material situation or regional variations of wood supplies. Moreover, the increased demand from other wood-based industries and the energy sector for wood previously used mainly for wood-based panel manufacturing has significantly increased worldwide [16]. These challenges have forced the wood-based sector to shift towards alternative raw materials, including recovered wood and by-products from other forest-based industries, as well as to optimize the technological production processes in order to maintain a consistent quality level.

Particleboards are one of the most important value-added panel products in the wood-based industry with a wide variety of applications [17,18]. Compared to the pulp and paper industry or construction, the production of particleboards can utilize low-quality raw materials. Proper waste management, including wood and wood-based by-products, is of great importance for the environment [19]. Many authors have investigated the potential applications of a particular material or the selection of appropriate manufacturing conditions, such as the type and amount of adhesive used or the temperature and pressure applied [20–27]. In the case of the expected deterioration of the technological properties of the boards with the addition of various alternative lignocellulosic raw materials, one possible way to counteract these undesirable effects is to increase the amount of binder [28]. The selection of the resin type and content is made on the basis of assumptions regarding the selected properties and projected applications. If the board is to have high water resistance, choose a resin other than UF or modify it with a different resin, e.g., PF (phenol-formaldehyde) or pMDI (polymeric 4,4′-methylenediphenyl isocyanate) [29–35]. This is a common procedure when using particles, e.g., annual plants, which will ensure good bonding and strength parameters [20,36]. The use of alternative raw materials contributes to the sustainable management of unused forest biomass, including bark, harvested and production residues, such as unprocessed sawmill by-products, i.e., less valuable wood waste from processing wood, such as sawdust, wood chips, shavings and wood pulp, and by-products of the food and agricultural industries, but also results in decreased panel production costs [3,37–40]. The increased shortage of wood raw materials, i.e., full-value wood and roundwood, justifies the wider industrial utilization of wood residues and by-products for wood-based panel manufacturing.

Particleboards can be fabricated from crushed lignocellulosic particles of one or more substitute raw materials, including post-consumer wood and wood from fruit trees and urban greenery [3,41–44]. Recycling of waste from construction and demolition in the form of residual medium density fiberboard (MDF), particleboard, cardboard and plywood is a viable option for producing boards suitable for furniture and interior applications [5,45]. Recently, many studies have been focused on the production of particleboard from alternative lignocellulosic raw materials, such as vine stalks [46], cotton stalks [26], bamboo and banana chips [47,48], poppy husks [49], wheat and straw [25,50], seaweed [51] and even chicken feathers [52,53]. Boards made with the addition of these materials should have comparable properties to industrially produced boards from softwood particles and comply with the requirements of the technical standard EN 312 about the specification of particleboards [54].

One of the possibilities of using alternative wood resources as a feedstock for particleboard production is the raw material created during care or liquidation of plantations as well as the cutting of old walnut trees. The total global area harvested of walnut plantations in 2019 amounted to approx. 9.3 million ha, including approx. 143,507 ha in Europe and 2270 ha in Poland [55]. These trees are cultivated worldwide not only for obtaining their edible nuts, but also for the production of decorative veneers, which is a very exacting and expensive process [56]. In turn, the price of wood residues suitable for particleboard is much lower. Due to the high price and low availability of walnut wood on the market, its use in particleboard production is feasible only in the case of wood residues and by-products from the wood-processing industry as a sustainable solution to the increased global demand for raw material.

The aim of this research work was to investigate the effect of the content of walnut (*Juglans regia* L.) wood residues on the physical and mechanical properties of three-layer particleboard as a way to alleviate the shortage of raw materials in the wood-based panel industry. The effect of reduced urea-formaldehyde (UF) resin content and applied pressure on the exploitation properties of the particleboard was also evaluated. This paper is a continuation of the research carried out by the authors on the possibility of using alternative raw materials for the production of particleboard.

2. Materials and Methods

2.1. Materials

Industrially produced softwood particles were obtained from the local particleboard factory. The walnut wood shavings were made from a medium-sized walnut tree stem obtained from a backyard cut in Dreglin, Poland. The harvested wood was debarked and then shredded into chips with a knife chipper. The produced chips were ground on a Pallmann laboratory cutter (Pallmann GmbH, Zweibrücken, Germany) in the form of particles and dried to a moisture content of approx. 8%, which was determined by the drying-weighing method according to the EN 322 standard [57]. The material was then sorted using screens with a mesh diameter of 4 mm and 2 mm, in order to select the material for the surface layers and the core layer. The desired fraction for the core layer consisted of particles retained on a sieve with a mesh size of 2 mm. Particles larger than 4 mm were reground and sorted, and particles smaller than 2 mm were used for the surface layers. The reason for the selection of the indicated sizes of vortices used in individual layers was the use of chip mixtures dimensionally similar to the dimensions of industrial chips used in the production of particleboards.

2.2. Adhesives

Commercial urea-formaldehyde (UF) resin with a molar ratio of 1.2, supplied by the factory Silekol Sp, z o.o. (Kędzierzyn-Koźle, Poland), was used for the production of the particleboards. The selected properties of the resin are presented in Table 1. Ammonium sulfate ((NH_4)$_2SO_4$) was used as a hardener at a 10% water solution, and mixed with the resin before spraying into the wood particles. The formulation of the adhesive was 50:15:1.5 parts by weight of the resin, water and hardener, respectively. The proportions were selected to obtain the appropriate gel time of the adhesive mass. In addition, 0.8 wt.% paraffin emulsion, based on the dry particle weight, was added to the resin in order to protect the produced boards against exposure to water and to maintain the dimensional stability.

Table 1. Selected properties of UF resin used in this work.

Characteristic	Value
Dry solids	67%
Relative density	1.30 g·cm^{-3}
pH	8.0
Gel time	50 s
Dynamic viscosity	0.5 Pa·s

2.3. Production of Panels

The particle mixes were glued with the UF resin at a 12% resin content of the surface layers (SL) and a 10% resin content of the core layer (CL), based on the mass of the oven dry wood particles, using pneumatic spraying. The differences in the resin content were due to the different sizes and surface area of the particles used for the individual layers. The mixture of wood particles and resin was manually formed into a mat in a frame with 320 mm × 320 mm dimensions and pressed using aluminum plates and spacer bars. Three-layer particleboard with a thickness of 16 mm, target density of 650 kg·m^{-3} and three different addition levels (0%, 25% and 50%) of walnut particles, bonded with UF resin, were produced under laboratory conditions. The share of surface layers in the panel was 35 wt.%. The hot-pressing process was carried out in a ZUP-NYSA PH-1LP25 single opening hydraulic laboratory press using standard particleboard manufacturing conditions, i.e., a pressing temperature of 180 °C, unit pressure of 2.5 N·mm^{-2}, and pressing time of 20 s·mm^{-1} of the board thickness.

The manufacturing parameters of the laboratory-fabricated particleboard are given in Table 2.

Table 2. Manufacturing parameters of particleboard fabricated from industrial wood particles and residual walnut wood particles bonded with UF resin.

Share of Walnut Wood Particles [%]	UF Resin Content of the Surface Layers and Core Layer (SL_CL) [%]	Unit Pressure [N·mm^{-2}]
0		
25	10_8	
50		
0		1.5
25	12_10	
50		
0		
25	10_8	
50		
0		2.5
25	12_10	
50		

The manufactured boards were conditioned at a temperature of 20 ± 2 °C and a relative air humidity of 65 ± 5% for 7 days. The manufactured boards were cut into the required test size in accordance with the relevant standards and subjected to tests to evaluate their physical and mechanical properties. The bending strength (MOR) and modulus of elasticity (MOE) were determined according to the EN 310 standard; the internal bond (IB) was determined according to EN 319. The thickness swelling (TS) after 24 h of soaking in water was determined according to EN 317. Each property was determined on 10 replicates for a given variant of the boards. The significance of the differences between the values of the individual panel parameters was calculated using the Tukey's post hoc HSD test.

The experimental data was statistically analyzed using STATISTICA 13.3 software (TIBCO Software Inc., Palo Alto, CA, USA).

3. Results

Graphical representation of the results obtained for the mechanical properties (MOR, MOE and IB) of the laboratory-produced three-layered particleboards, fabricated with a different share of residual walnut wood particles, is presented in Figures 1–3. Table 3 presents the minimum requirements that the boards must meet in terms of mechanical and swelling properties in order to qualify them to the particular types, P5, P6 or P7. The results obtained for the TS (24 h) are presented in Table 4. All assumed variants of boards, both in terms of the level of substitution and technological factors, were successfully produced in accordance with the experimental design. Samples for testing were obtained from them and the determination of each property was made for at least 10 replicates for a given variant of the board.

Table 3. Requirements in terms of mechanical properties and swelling for particleboards.

* Property	Unit	Requirements for a Thickness Range > 13 to 20 mm		
		Type P5	Type P6	Type P7
Bending strength	N/mm^2	16	18	20
Modulus of elasticity in bending	N/mm^2	2400	3000	3100
Internal bond	N/mm^2	0.45	0.50	0.70
Swelling in thickness, 24 h	%	10	15	10

* Own study based on the EN 312 standard.

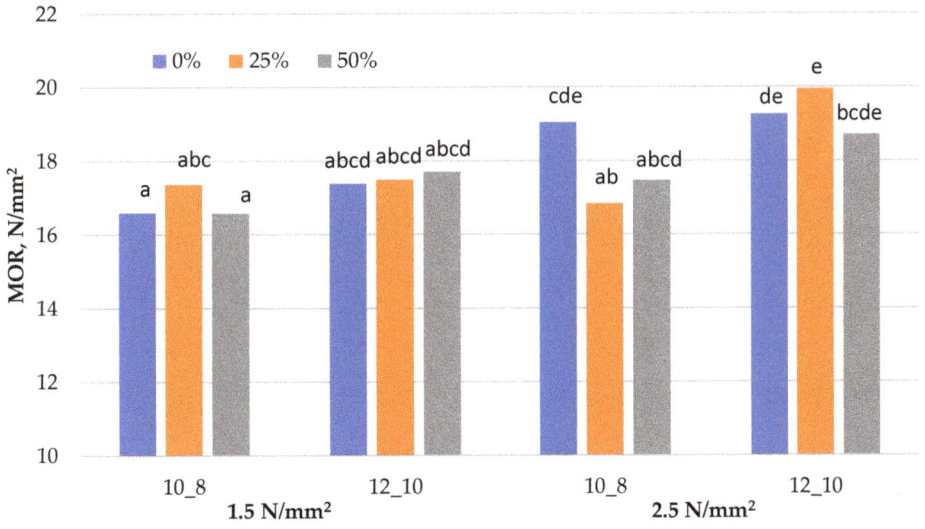

Figure 1. Bending strength (MOR) of particleboards produced.

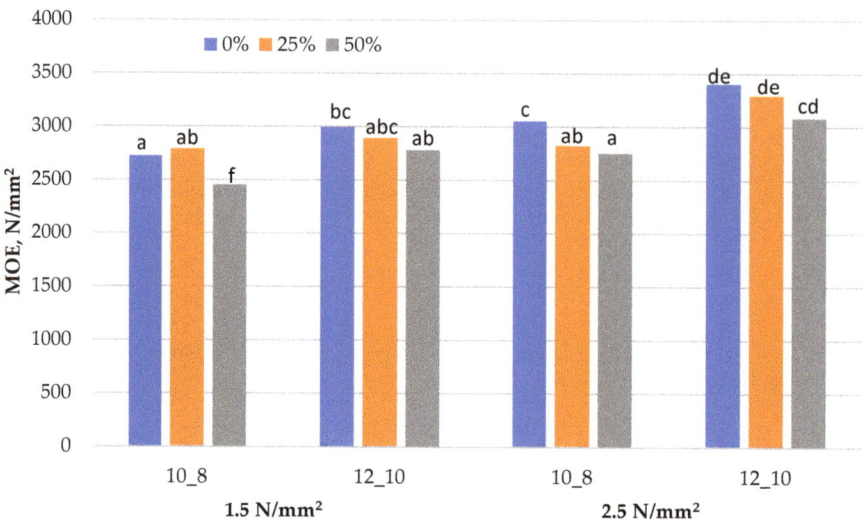

Figure 2. Modulus of elasticity (MOE) of particleboards produced.

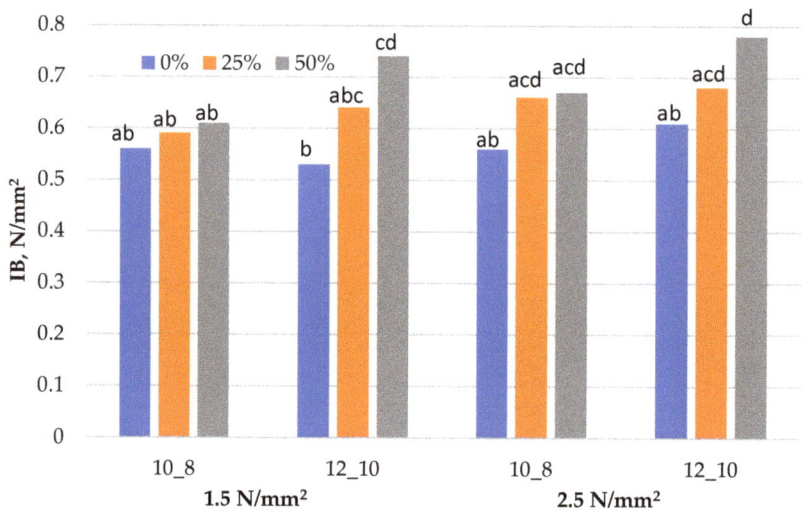

Figure 3. Internal bond (IB) of particleboards produced.

Table 4. Results of the TS (24 h) of the particleboards produced.

Unit Pressure, N·mm^{-2}	UF Resin Content of the Surface Layers and Core Layer (SL_CL), %	TS, %		
		The Share of Residue Walnut Wood Particles, %		
		0	25	50
1.5	10_8	25.9 (1.32) bc	24.7 (1.08) b	27.2 (1.25) c
2.5		24.3 (0.96) b	24.8 (0.91) b	26.7 (1.21) c
1.5	12_10	21.6 (1.44) a	21.3 (1.42) a	21.4 (1.25) a
2.5		21.7 (1.18) a	20.7 (1.59) a	21.5 (1.27) a

Different letters denote a significant difference. Means followed by the same letter do not statistically differ from each other ($p \leq 0.05$) according to Tukey's post hoc test.

3.1. Bending Strength (MOR)

The analysis of the obtained results demonstrates that the addition of walnut wood did not significantly affect the MOR value of the boards, even with the 50% addition of walnut particles (Figure 1). At low applied pressure, neither the walnut content nor the degree of sizing changed the MOR value. On the other hand, increasing the pressing pressure significantly improved the MOR for softwood boards. The increased pressure and resin content resulted in increased MOR values of the boards, fabricated with the addition of walnut wood particles. For these technological parameters, no statistical differences were found for boards with a 25% and 50% share of walnut wood.

The boards produced at a pressure of 1.5 $N \cdot mm^{-2}$ in the entire range of substitution demonstrated similar MOR values, which were also confirmed by the statistical analysis. These values were assigned to the homogeneous group marked with the letter a in Figure 1. Significant differences to the boards produced with a low applied pressure were observed for the MOR values of the boards produced with low resin content and higher applied pressure. The MOR value of particleboard fabricated of softwood particles was only approx. 19 $N \cdot mm^{-2}$ for both resin contents used at a pressure of 2.5 $N \cdot mm^{-2}$. Similar values were obtained for boards manufactured with walnut wood particles at the highest values of production parameters. The 25% and 50% share of walnut wood particles in the composition of laboratory-fabricated panels resulted in decreased MOR value of the boards bonded with resin content of 10% and 8%, and of the softwood particleboard by 11.6% and 8%, respectively, resulting in the achievement of parameters characteristic for the boards from the group with an applied pressure of 1.5 $N \cdot mm^{-2}$. These differences, although slight, were confirmed by the statistical analysis. As a result, eight variants of panels were classified into the homogeneous group marked with the letter a, and the remaining 4 were classified as group e.

Boards manufactured according to the standard conditions in the entire range of walnut wood particles substitution, and the board manufactured with softwood particles under standard pressure conditions, with a lower resin content, met the high requirements of technical standards for construction boards, which is type P6 (heavy-duty load-bearing boards for use in dry conditions). The remaining boards complied with the standard requirements for P5 type boards, i.e., load-bearing boards for use in humid conditions.

Different letters denote a significant difference. Means followed by the same letter do not statistically differ from each other ($p \leq 0.05$) according to Tukey's post hoc test.

3.2. Modulus of Elasticity (MOE)

In terms of MOE, all of the manufactured boards met the minimum standard requirements for P5, i.e., load-bearing boards for use in humid conditions—2400 $N \cdot mm^{-2}$. Boards made of softwood particles only demonstrated the highest MOE values for most variants in the range of 2721–3403 $N \cdot mm^{-2}$, as shown in Figure 2. The addition of residual walnut wood particles resulted in decreased MOE values in all variants produced. However, there was a slight decrease of up to 10% between the board without walnut wood and the board with its 50% substitution. However, in many cases, these changes were statistically insignificant. With the increase of both resin content and the applied pressure, the MOE value of the produced boards increased. More significant effect on the stiffness of the boards was observed for the applied pressure change. For the board fabricated with a 12% resin content of the SL and 10% of the CL, along with increasing the applied pressure, an increase in MOE values by 13.6%, 13.8% and 10.9%, about the respective share of residual walnut wood, was observed, respectively.

Different letters denote a significant difference. Means followed by the same letter do not statistically differ from each other ($p \leq 0.05$), according to Tukey's post hoc test.

3.3. Internal Bond (IB) Strength

Due to the relatively lower number of particles in the boards with increased walnut wood content, there is a better coating of each particle with the UF resin. This resulted in higher IB values, determined for the particleboard fabricated with the 50% share of walnut particles and higher, in particular, with an additionally increased degree of sizing. The analysis of the IB results obtained demonstrated that the substitution of softwood particles with walnut wood particles increased the tensile strength for each variant of the boards. The highest IB values in each variant of applied pressure and resin content were obtained for the particleboard manufactured with a 50% share of walnut wood, i.e., minimum 0.61 N·mm^{-2} and maximum 0.78 N·mm^{-2}, and 0.59 N·mm^{-2} and 0.68 N·mm^{-2} for the boards fabricated with a 25% walnut wood share, respectively. The improvement of IB was particularly noticeable for boards bonded with the higher resin content, i.e., 12% and 10%. At a 50% share of walnut particles, the IB value was about 28% higher than the reference board produced at a pressure of 1.5 N·mm^{-2} and almost 40% higher than the same variants produced at the higher pressure.

All laboratory-produced particleboards fulfilled the requirements of technical standards for P6 boards, i.e., heavy-duty load-bearing boards for use in dry conditions, and the boards with maximum walnut particles substitution produced at a pressure of 2.5 N·mm^{-2} also met the most stringent standard minimum requirements for boards of type P7, i.e., heavy-duty load-bearing boards for use in humid conditions, exceeding it significantly.

Different letters denote a significant difference. Means followed by the same letter do not statistically differ from each other ($p \leq 0.05$) according to Tukey's post hoc test.

3.4. Thickness Swelling (TS)

The results of the TS (24 h) of the laboratory-produced three-layer particleboards are presented in Table 4.

At a lower degree of sizing of 50%, the addition of walnut wood deteriorated the TS properties due to the swelling of the particles, which have a higher density. By increasing the amount of resin, the absorption of water by the particles is made difficult, even if the resin used does not have high water resistance. With the standard resin content of 12% of the SL and 10% of the CL, this parameter remained at the level of 20.7–21.7% for the entire range of wood particle substitution and both variants of applied pressure.

These slight differences were statistically insignificant for all panels manufactured with this resin content. The reduction in the amount of the UF resin resulted in deteriorated dimensional stability of the boards. The TS value of boards produced with the resin content of 10% of the SL and 8% of the CL was significantly higher than the boards with the standard resin content. In addition, the 50% share with the residue walnut wood particles contributed the most to increasing the TS value of boards produced with the lower applied pressure. TS values of the boards produced with 50% share of walnut wood and bonded with a reduced resin content were 5% higher at 1.5 N·mm^{-2} and 10% higher at 1.5 N·mm^{-2}, compared to the board fabricated from only pine particles. With regards to the board manufactured under standard conditions, the difference was 23% and 26%, respectively. None of the boards met the minimum requirements of technical standards in terms of TS (24 h) values, amounting to 14% for the P3 type board, i.e., non-load-bearing boards for use in humid conditions, and 10% for the P5 type board, i.e., load-bearing boards for use in humid conditions.

4. Discussion

Taking into account the variable content of walnut wood, two resin contents were selected in the study. This was to check whether this treatment would compensate for the decrease in mechanical properties associated with the use of wood other than pine. It turned out that the addition of walnut wood residues did not decrease the mechanical properties. In addition to the improved mechanical properties, the increased amount of UF resin improved the dimensional stability of the boards by reducing the water absorption of the lignocellulosic particles.

Urea-formaldehyde (UF) adhesives are the most widely used thermosetting resins for the production of various types of wood-based composites [58]. The wide industrial use of these resins is due to their good adhesion performance, high reactivity, water solubility, short press times and a relatively low price [17,59]. Due to the very variable and complex nature of the wood raw material, the properties of wood-based panels are largely determined by the characteristics of the resin [58].

The mechanical properties, e.g., MOE, MOR and IB values of boards manufactured with willow (*Salix viminalis* L.) wood for the core layer, improved with an increased resin content from 8% to 9.5% by 10% and 20%, respectively [27]. It was also found that increasing the resin content by up to 16% and the pressing pressure had a significant impact on the mechanical and physical properties of the particleboard fabricated from non-wood raw materials. This results in a better filling of the voids between the particles, improving the compaction of the panels, thus leading to the improvement of the mechanical and physical properties [60]. In addition, increasing the amount of resin from 8–10% to 10–12% may also lower the surface roughness values in boards with particles and dust wood [61].

However, a significant disadvantage of this commonly used thermosetting amino resin is the low water-resistance, which has been confirmed by tests. Low water resistance values were observed for all variants of the produced experimental boards, which mean that the manufactured products are intended to be used mainly in internal conditions. Increasing the gluing degree of the boards resulted in a decrease in TS value from 25.9% to 21.6% for a pine board and from 24.7% to 21.3% for a board with a 25% share of walnut. Additionally, increasing the pressure from 1.5 $N \cdot mm^{-2}$ to 2.5 $N \cdot mm^{-2}$ lowered the TS value to 21.7%, 20.7% and 21.5% for the board with 0%, 25% and 50% walnut content, respectively. The lowest TS value of 20.7% was determined for the boards manufactured with 25% share of walnut wood particles. The properties of particleboard can vary depending on the resin content. Perhaps, if an adhesive with greater resistance to water, e.g., phenol-formaldehyde resin, was used, the results would be much better.

For spruce, sunflower and topinambour (*Helianthus tuberosus*) particles, a slight improvement in TS was also obtained, by a maximum of 8% and approx. 2% for the remaining ones [62]. However, for pine and white mustard straw, increasing the amount of glue from 10% to 14% resulted in a reduction of TS from 51% to 22% and from 62% to 39%, respectively [28]. Nevertheless, better results for panels made of lignocellulosic particles, including annual plants and cereals, can be achieved by using adhesives more dedicated to such raw materials, i.e., pMDI [62,63].

5. Conclusions

The conducted research proved that it is possible to add walnut wood residues to commonly produced softwood particleboard. However, it should not be expected that there will be a lot of this wood, because it is a remnant, not a wholesome quantitatively and qualitatively assortment, similarly to recycled wood particles, residues from fruit orchards, urban wastes or residues from the production of wood products or food production waste.

The boards produced with higher resin content and higher applied pressure achieved the highest MOR values. For boards produced with a reduced applied pressure of 1.5 $N \cdot mm^{-2}$, no significant differences were found between the MOR values in terms of the share of residue walnut wood, even up to 50%, and a change in the resin content. The greater the addition of walnut wood particles, the more visible was the deterioration of MOE, and the greater the improvement of IB values. A higher proportion of the UF resin had a positive effect on the dimensional stability of the boards, while the effect of the addition of the walnut wood particles and the change in pressure were not statistically confirmed. Perhaps, if the resin used were modified to add water-resistant resins, the boards would meet the requirements of the standards throughout, also in terms of thickness swelling.

In terms of mechanical properties, all laboratory-produced boards exhibited very high MOR, MOE and IB values, and even the panels, fabricated with 50% walnut wood content, met the requirements of the technical standards for P5 boards, i.e., load-bearing boards for use in humid conditions, and in some cases, even in P6 and P7 variants, i.e., for load-bearing applications. It can be concluded that residual walnut wood particles can successfully replace softwood particles in the production of three-layer particleboard, which can be used in structural applications. In order to improve the resource efficiency and achieve enhanced valorization of waste biomass, future research should be aimed at the rational use of the available wood and lignocellulosic raw materials, the search for alternative resources and the optimization of production parameters.

Author Contributions: Conceptualization, M.P.; methodology, R.A.; investigation, M.P. and R.A.; data curation, R.A.; writing—original draft preparation, review and editing, T.R., M.P., P.A. and L.K.; visualization, M.P. and T.R.; supervision, T.R.; funding acquisition, P.A., L.K. and T.R. All authors have read and agreed to the published version of the manuscript.

Funding: T The study was supported by the funding for statutory R&D activities as the research task No. 506.227.02.00 of the Faculty of Forestry and Wood Technology, Poznań University of Life Sciences. This research was also supported by the Slovak Research and Development Agency under contracts No. APVV-18-0378 and APVV-19-0269 and APVV-20-0159. This research was also supported by the project No. НИС-Б-1145/04.2021, "Development, Properties and Application of Eco-Friendly Wood-Based Composites", carried out at the University of Forestry, Sofia, Bulgaria.

Institutional Review Board Statement: Not applicable.

Informed Consent Statement: Not applicable.

Data Availability Statement: Not applicable.

Acknowledgments: Some of the raw data used in the article were collected with help of Daniel Piasecki, and some of these data could have been used in Piasecki's diploma thesis.

Conflicts of Interest: The authors declare no conflict of interest.

References

1. Antov, P.; Krišťák, L.; Réh, R.; Savov, V.; Papadopoulos, A.N. Eco-Friendly Fiberboard Panels from Recycled Fibers Bonded with Calcium Lignosulfonate. *Polymers* **2021**, *13*, 639. [CrossRef] [PubMed]
2. Janiszewska, D.; Frackowiak, I.; Mytko, K. Exploitation of Liquefied Wood Waste for Binding Recycled Wood Particleboards. *Holzforschung* **2016**, *70*, 1135–1138. [CrossRef]
3. Pędzik, M.; Janiszewska, D.; Rogoziński, T. Alternative Lignocellulosic Raw Materials in Particleboard Production: A Review. *Ind. Crops Prod.* **2021**, *174*, 114162. [CrossRef]
4. Mirski, R.; Dukarska, D.; Walkiewicz, J.; Derkowski, A. Waste Wood Particles from Primary Wood Processing as a Filler of Insulation Pur Foams. *Materials* **2021**, *14*, 4781. [CrossRef] [PubMed]
5. Jivkov, V.; Simeonova, R.; Antov, P.; Marinova, A.; Petrova, B.; Kristak, L. Structural Application of Lightweight Panels Made of Waste Cardboard and Beech Veneer. *Materials* **2021**, *14*, 5064. [CrossRef]
6. Pedzik, M.; Bednarz, J.; Kwidzinski, Z.; Rogozinski, T.; Smardzewski, J. The Idea of Mass Customization in the Door Industry Using the Example of the Company Porta KMI Poland. *Sustainability* **2020**, *12*, 3788. [CrossRef]
7. Kwidziński, Z.; Bednarz, J.; Pędzik, M.; Sankiewicz, Ł.; Szarowski, P.; Knitowski, B.; Rogoziński, T. Innovative Line for Door Production Technoporta—Technological and Economic Aspects of Application of Wood-Based Materials. *Appl. Sci.* **2021**, *11*, 4502. [CrossRef]
8. Gejdoš, M.; Lieskovský, M.; Giertliová, B.; Němec, M.; Danihelová, Z. Prices of raw-wood assortments in selected markets of central Europe and their development in the future. *BioResources* **2019**, *14*, 2995–3011. [CrossRef]
9. Toth, D.; Maitah, M.; Maitah, K.; Jarolínová, V. The Impacts of Calamity Logging on the Development of Spruce Wood Prices in Czech Forestry. *Forests* **2020**, *11*, 283. [CrossRef]
10. University in Kentucky. Delivered Timber Prices. Available online: https://forestry.ca.uky.edu/delivered-timber-prices?fbclid=IwAR2sJYiNPt3xiGr0cggpCULPspu0r2Wb5Yj7MNMO7Vg-kF5t2tURCX_HbYM (accessed on 22 January 2022).
11. *Timber Price Indices Data to September 2021*; Forest Research; Silvan House: Edinburgh, UK, 2021.
12. *Average Price of Wood Sale in the First Three Quarters 2016*; Statistics Poland: Warsaw, Poland, 2016.
13. *Average Price of Wood Sale in the First Three Quarters 2020*; Statistics Poland: Warsaw, Poland, 2020.
14. *Average Price of Wood Sale in the First Three Quarters 2021*; Statistics Poland: Warsaw, Poland, 2021.
15. Food and Agriculture Organisation of the United Nations. Forestry Production and Trade. Available online: https://www.fao.org/faostat/en/#data/FO (accessed on 22 January 2022).
16. Hildebrandt, J.; Hagemann, N.; Thrän, D. The Contribution of Wood-Based Construction Materials for Leveraging a Low Carbon Building Sector in Europe. *Sustain. Cities Soc.* **2017**, *34*, 405–418. [CrossRef]

17. Bekhta, P.; Noshchenko, G.; Réh, R.; Kristak, L.; Sedliačik, J.; Antov, P.; Mirski, R.; Savov, V. Properties of Eco-Friendly Particleboards Bonded with Lignosulfonate-Urea-Formaldehyde Adhesives and PMDI as a Crosslinker. *Materials* **2021**, *14*, 4875. [CrossRef] [PubMed]
18. Bekhta, P.; Korkut, S.; Hiziroglu, S. Effect of Pretreatment of Raw Material on Properties of Particleboard Panels Made from Wheat Straw. *BioResources* **2013**, *8*, 4766–4774. [CrossRef]
19. Faraca, G.; Boldrin, A.; Astrup, T. Resource Quality of Wood Waste: The Importance of Physical and Chemical Impurities in Wood Waste for Recycling. *Waste Manag.* **2019**, *87*, 135–147. [CrossRef] [PubMed]
20. Dukarska, D.; Pędzik, M.; Rogozińska, W.; Rogoziński, T.; Czarnecki, R. Characteristics of Straw Particles of Selected Grain Species Purposed for the Production of Lignocellulose Particleboards. *Part. Sci. Technol.* **2021**, *39*, 213–222. [CrossRef]
21. Papadopoulos, A.N. Property Comparisons and Bonding Efficiency of UF and PMDI Bonded Particleboards as Affected by Key Process Variables. *BioResources* **2006**, *1*, 201–208. [CrossRef]
22. Sandak, A.; Sandak, J.; Janiszewska, D.; Hiziroglu, S.; Petrillo, M.; Grossi, P. Prototype of the Near-Infrared Spectroscopy Expert System for Particleboard Identification. *J. Spectrosc.* **2018**, *2018*, 6025163. [CrossRef]
23. Gumowska, A.; Wronka, A.; Borysiuk, P.; Robles, E.; Sala, C.; Kowaluk, G. Production of Layered Wood Composites with a Time-Saving Layer-By-Layer Addition. *BioResources* **2018**, *13*, 8089–8099. [CrossRef]
24. Borysiuk, P.; Tetelewska, A.; Auriga, R.; Jenczyk-Tołłoczko, I. The Influence of Temperature on Selected Strength Properties of Furniture Particleboard. *Ann. WULS For. Wood Technol.* **2019**, *108*, 128–134. [CrossRef]
25. Li, X.; Cai, Z.; Winandy, J.E.; Basta, A.H. Selected Properties of Particleboard Panels Manufactured from Rice Straws of Different Geometries Hammer-Milled Rice Straw Particles of Six Different Categories and Two Types of Resins. The Results Show. *Bioresour. Technol.* **2010**, *101*, 4662–4666. [CrossRef]
26. Nazerian, M.; Beyki, Z.; Gargarii, R.M.; Kool, F. The Effect of Some Technological Production Variables on Mechanical and Physical Properties of Particleboard Manufactured from Cotton (*Gossypium hirsutum*) Stalks. *Maderas. Cienc. Tecnol.* **2016**, *18*, 167–178. [CrossRef]
27. Warmbier, K.; Wilczyński, A.; Danecki, L. Effects of Density and Resin Content on Mechanical Properties of Particleboards with the Core Layer Made from Willow *Salix viminalis*. *For. Wood Technol.* **2013**, *84*, 284–287.
28. Dukarska, D.; Bartkowiak, M.; Stachowiak-Wencek, A. White Mustard Straw as an Alternative Raw Material in the Manufacture of Particleboards Resinated with Different Amounts of Urea-Formaldehyde Resin. *Drewno* **2015**, *58*, 49–63. [CrossRef]
29. Dziurka, D.; Mirski, R. Properties of Liquid and Polycondensed UF Resin Modified with PMDI. *Drv. Ind.* **2014**, *65*, 115–119. [CrossRef]
30. Mansouri, H.R.; Pizzi, A.; Leban, J.-M. Improved Water Resistance of UF Adhesives for Plywood by Small PMDI Additions. *Holz Roh Werkst.* **2006**, *64*, 218–220. [CrossRef]
31. Simon, C.; George, B.; Pizzi, A. Copolymerization in UF/PMDI Adhesives Networks. *J. Appl. Polym. Sci.* **2002**, *86*, 3681–3688. [CrossRef]
32. Lee, S.H.; Lum, W.C.; Zaidon, A.; Maminski, M. Microstructural, Mechanical and Physical Properties of Post Heat-Treated Melamine-Fortified Urea Formaldehyde-Bonded Particleboard. *Eur. J. Wood Wood Prod.* **2015**, *73*, 607–616. [CrossRef]
33. Choudhary, C.L.; Negi, A.; Yadav, S.M.; Sihag, K. Role of Resin Content in the Manufacture of Particleboard from Mixed Plantation Species. *Int. J. Biol. Sci.* **2015**, *6*, 132–135.
34. Lubis, M.A.R.; Park, B.-D.; Lee, S.-M. Performance of Hybrid Adhesives of Blocked-PMDI/Melamine-Urea-Formaldehyde Resins for the Surface Lamination on Plywood. *J. Korean Wood Sci. Technol.* **2019**, *47*, 200–209. [CrossRef]
35. Sutiawan, J.; Hadi, Y.S.; Nawawi, D.S.; Abdillah, I.B.; Zulfiana, D.; Lubis, M.A.R.; Nugroho, S.; Astuti, D.; Zhao, Z.; Handayani, M.; et al. The Properties of Particleboard Composites Made from Three Sorghum (*Sorghum bicolor*) Accessions Using Maleic Acid Adhesive. *Chemosphere* **2021**, *290*, 133–163. [CrossRef]
36. Dukarska, D.; Czarnecki, R.; Dziurka, D.; Mirski, R. Construction Particleboards Made from Rapeseed Straw Glued with Hybrid PMDI/PF Resin. *Eur. J. Wood Wood Prod.* **2017**, *75*, 175–184. [CrossRef]
37. Mirski, R.; Derkowski, A.; Dziurka, D.; Wieruszewski, M.; Dukarska, D. Effects of Chip Type on the Properties of Chip–Sawdust Boards Glued with Polymeric Diphenyl Methane Diisocyanate. *Materials* **2020**, *13*, 1329. [CrossRef] [PubMed]
38. Martins, R.S.F.; Gonçalves, F.G.; de Segundinho, P.G.A.; Lelis, R.C.C.; Paes, J.B.; Lopez, Y.M.; Chaves, I.L.S.; de Oliveira, R.G.E. Investigation of Agro-Industrial Lignocellulosic Wastes in Fabrication of Particleboard for Construction Use. *J. Build. Eng.* **2021**, *43*, 102903. [CrossRef]
39. Grigorov, R.; Mihajlova, J.; Savov, V. Physical and mechanical properties of combined wood-based panels with participation of particles from vine sticks in core layer. *Innov. Woodwork. Ind. Eng. Des.* **2020**, *17*, 42–52.
40. Iždinský, J.; Vidholdová, Z.; Reinprecht, L. Particleboards from Recycled Wood. *Forests* **2020**, *11*, 1166. [CrossRef]
41. Auriga, R.; Borysiuk, P.; Gumowska, A.; Smulski, P. Influence of Apple Wood Waste from the Annual Care Cut on the Mechanical Properties of Particleboards. *Ann. WULS For. Wood Technol.* **2019**, *105*, 47–53. [CrossRef]
42. Auriga, R.; Borysiuk, P.; Smulski, P. Apple Wood from an Annual Care Cut as a Raw Material Additive for Particleboard Production. *Biul. Inf. OB-RPPD* **2019**, *1–2*, 17–24.
43. Auriga, R.; Borysiuk, P.; Misiura, Z. Evaluation of the Physical and Mechanical Properties of Particle Boards Manufactured Containing Plum Pruning Waste. *Biul. Inf. OB-RPPD* **2021**, *1–2*, 5–11.
44. Iždinský, J.; Reinprecht, L.; Vidholdová, Z. Particleboards from Recycled Pallets. *Forests* **2021**, *12*, 1597. [CrossRef]

45. da Azambuja, R.R.; de Castro, V.G.; Trianoski, R.; Iwakiri, S. Recycling Wood Waste from Construction and Demolition to Produce Particleboards. *Maderas. Cienc. Tecnol.* **2018**, *20*, 681–690. [CrossRef]
46. Yeniocak, M.; Göktas, O.; Erdil, Y.Z.; Özen, E. Investigating the Use of Vine Pruning Stalks (*Vitis vinifera* L. CV. *Sultani*) as Raw Material for Particleboard Manufacturing. *Wood Res.* **2014**, *59*, 167–176.
47. de Araújo, P.C.; Arruda, L.M.; del Menezzi, C.H.S.; Teixeira, D.E.; de Souza, M.R. Lignocellulosic Composites from Brazilian Giant Bamboo (*Guadua magna*): Part 2: Properties of Cement and Gypsum Bonded Particleboards. *Maderas. Cienc. Tecnol.* **2011**, *13*, 297–306. [CrossRef]
48. Papadopoulos, A. Banana Chips (*Musa acuminata*) as an Alternative Lignocellulosic Raw Material for Particleboard Manufacture. *Maderas. Cienc. Tecnol* **2018**, *20*, 395–402. [CrossRef]
49. Kucuktuvek, M.; Kasal, A.; Kuskun, T.; Ziya Erdil, Y. Utilizing Poppy Husk-Based Particleboards as an Alternative Material in Case Furniture Construction. *BioResources* **2017**, *12*, 839–852.
50. Mo, X.; Cheng, E.; Wang, D.; Sun, X.S. Physical Properties of Medium-Density Wheat Straw Particleboard Using Different Adhesives. *Ind. Crops Prod.* **2003**, *18*, 47–53. [CrossRef]
51. Rammou, E.; Mitani, A.; Ntalos, G.; Koutsianitis, D.; Taghiyari, H.R.; Papadopoulos, A.N. The Potential Use of Seaweed (Posidonia Oceanica) as an Alternative Lignocellulosic Raw Material for Wood Composites Manufacture. *Coatings* **2021**, *11*, 69. [CrossRef]
52. Taghiyari, H.R.; Majidi, R.; Esmailpour, A.; Samadi, Y.S.; Jahangiri, A.; Papadopoulos, A.N. Engineering Composites Made from Wood and Chicken Feather Bonded with UF Resin Fortified with Wollastonite: A Novel Approach. *Polymers* **2020**, *12*, 857. [CrossRef]
53. Taghiyari, H.R.; Militz, H.; Antov, P.; Papadopoulos, A.N. Effects of Wollastonite on Fire Properties of Particleboard Made from Wood and Chicken Feather Fibers. *Coatings* **2021**, *11*, 518. [CrossRef]
54. EN 312:2010; Particleboards—Specifications. European Committee for Standardization: Brussels, Belgium, 2010.
55. FAO. 2021 Food and Agriculture Organisation of the United Nations. Crops and Livestock Products. Available online: https://Fao.Org/Faostat/En/#data/FO/Visualize (accessed on 9 December 2021).
56. Conservation Commission of Missouri. Timber Price Trends April–June 2021. Available online: https://research.mdc.mo.gov/project/forest-economics-missouri/timber-price-trends-apr-june-2021?fbclid=IwAR0uXGofkgt7KXxUV2KdbcwtYlhP_b9eajfULC4ggZFg4ITrJGF35oYv9OI (accessed on 22 January 2022).
57. EN 322:1993; Wood-Based Panels—Determination of Moisture Content. European Committee for Standardization: Brussels, Belgium, 1993.
58. Dunky, M. Adhesives in the Wood Industry. In *Handbook of Adhesive Technology*, 2nd ed.; Revised and Expanded; Pizzi, A., Mittal, K.L., Eds.; Marcel Dekker, Inc.: New York, NY, USA; Basel, Switzerland, 2003; 71p.
59. Pizzi, A.; Papadopoulos, A.N.; Policardi, F. Wood Composites and Their Polymer Binders. *Polymers* **2020**, *12*, 1115. [CrossRef]
60. Taha, I.; Elkafafy, M.S.; el Mously, H. Potential of Utilizing Tomato Stalk as Raw Material for Particleboards. *Ain Shams Eng. J.* **2018**, *9*, 1457–1464. [CrossRef]
61. Nemli, G.; Aydın, I.; Zekoviç, E. Evaluation of Some of the Properties of Particleboard as Function of Manufacturing Parameters. *Mater. Des.* **2007**, *28*, 1169–1176. [CrossRef]
62. Klímek, P.; Meinlschmidt, P.; Wimmer, R.; Plinke, B.; Schirp, A. Using Sunflower (*Helianthus annuus* L.), Topinambour (*Helianthus tuberosus* L.) and Cup-Plant (*Silphium perfoliatum* L.) Stalks as Alternative Raw Materials for Particleboards. *Ind. Crops Prod.* **2016**, *92*, 157–164. [CrossRef]
63. Klímek, P.; Wimmer, R.; Meinlschmidt, P.; Kúdela, J. Utilizing Miscanthus Stalks as Raw Material for Particleboards. *Ind. Crops Prod.* **2018**, *111*, 270–276. [CrossRef]

Article

Characterisation of Wood Particles Used in the Particleboard Production as a Function of Their Moisture Content

Dorota Dukarska [1,*], Tomasz Rogoziński [2], Petar Antov [3], Lubos Kristak [4] and Jakub Kmieciak [1]

1. Department of Mechanical Wood Technology, Poznań University of Life Sciences, 60-627 Poznań, Poland; j.kmieciak24@gmail.com
2. Department of Furniture Design, Poznań University of Life Sciences, 60-627 Poznań, Poland; tomasz.rogozinski@up.poznan.pl
3. Faculty of Forest Industry, University of Forestry, 1797 Sofia, Bulgaria; p.antov@ltu.bg
4. Faculty of Wood Sciences and Technology, Technical University in Zvolen, 960 01 Zvolen, Slovakia; kristak@tuzvo.sk
* Correspondence: dorota.dukarska@up.poznan.pl

Citation: Dukarska, D.; Rogoziński, T.; Antov, P.; Kristak, L.; Kmieciak, J. Characterisation of Wood Particles Used in the Particleboard Production as a Function of Their Moisture Content. *Materials* **2022**, *15*, 48. https://doi.org/10.3390/ma15010048

Academic Editor: Tomasz Sadowski

Received: 2 December 2021
Accepted: 20 December 2021
Published: 22 December 2021

Publisher's Note: MDPI stays neutral with regard to jurisdictional claims in published maps and institutional affiliations.

Copyright: © 2021 by the authors. Licensee MDPI, Basel, Switzerland. This article is an open access article distributed under the terms and conditions of the Creative Commons Attribution (CC BY) license (https://creativecommons.org/licenses/by/4.0/).

Abstract: The properties of particleboards and the course of their manufacturing process depend on the characteristics of wood particles, their degree of fineness, geometry, and moisture content. This research work aims to investigate the physical properties of wood particles used in the particleboard production in dependence on their moisture content. Two types of particles currently used in the production of three-layer particleboards, i.e., microparticles (MP) for the outer layers of particleboards and particles for the core layers (PCL), were used in the study. The particles with a moisture content of 0.55%, 3.5%, 7%, 10%, 15%, and 20% were tested for their poured bulk density (ρ_p), tapped bulk density (ρ_t), compression ratio (k), angle of repose (α_R), and slippery angle of repose (α_s). It was found that irrespective of the fineness of the particles, an increase in their moisture content caused an increase in the angle of repose and slippery angle of repose and an increase in poured and tapped bulk density, while for PCL, the biggest changes in bulk density occurred in the range up to 15% of moisture content, and for MP in the range above 7% of moisture content, respectively. An increase in the moisture content of PCL in the range studied results in a significant increase in the compression ratio from 47.1% to 66.7%. The compression ratio of MP increases only up to 15% of their moisture content—a change of value from 47.1% to 58.7%.

Keywords: wood particles; moisture content; angle of repose; slippery angle of repose; poured bulk density; tapped bulk density

1. Introduction

The physicomechanical properties of particleboards, except their technological parameters, depend significantly on the characteristics of the raw materials used in their production [1]. These are the type and amount of the adhesive system used for their bonding, the type of the raw material, its degree of fragmentation and geometry (size and shape), and moisture content [2–12]. The right choice of the moisture content of the particles, independently of their fineness degree, is also necessary for assuring the correct industrial process of particleboards. The excessive moisture content of particles may cause delamination of the particleboards during their pressing [13]. In turn, overly dry particles increase the risk of fires during drying and also contribute to the formation of wood dust, which disturbs the process of particles bonding or mat densification during hot pressing. One of the stages of wood particles preparation for particleboards production, independently of the material they originate from and bonding agents used, is the energy-consuming drying operation. Therefore, the possibility of producing boards from particles with higher moisture content is an opportunity to optimise energy consumption in production plants, e.g., by reducing the work of dryers or selecting appropriate drying conditions and thereby reducing the production costs [14,15]. Furthermore, the aspect of wood particles' moisture content also seems to be of interest due to the trend of using isocyanate adhesives in the production process. Conventional thermosetting aminoplastic resins show deteriorated water resistance, which results from the hydrolysis of the methylene bridges [16,17]. However, polymeric 4,4'-diphenylmethane diisocyanate (pMDI) adhesives show high reactivity and the ability to a chemical reaction with wood components and the water they contain. It profitably influences the bonding quality of the particles and thereby the physicomechanical properties of particleboards [4,18–20]. These properties were studied by Jiang and Lu [21] for producing boards meeting the requirements of the EN 312 standard [22] prescribed for P2 type boards from particles with a moisture content of 25%, bonded with melamine–urea–formaldehyde resin (MUF) modified with different additional proportions of polyurethane prepolymer. As studies showed, the use of pure MUF resin for bonding particles with moisture content above 20% leads to the blowout of particleboards [13].

While preparing particles for the production of particleboards, it should also be considered that the moisture content of the mat increases together with the moisture content of the particles, which influences the parameters of the bonding process and the quality of particles bonding and, as a result, the properties of the finished boards. The moisture content of the mat is one of the most important factors influencing the heat transfer in the mat [23]. The rate of heat penetration in the mat determines the pressing time, which is critical for the efficiency of the production process [24]. It influences the increase in the temperature in the mat, which determines the cure rate of the adhesive resin. The combination of high temperature, moisture content, and time may cause an excessive increase in the vapour pressure in the mat and trigger an explosion when the press is opened [23,25]. However, the technological problems connected with the excessive moisture content of wood particles and vapour pressure during pressing can be effectively solved. Murayama et al. [26] investigated the temperature variability and vapour pressure during pressing of the particleboards and concluded that this problem can be solved by choosing an optimal moisture content of board layers. The increased moisture content of the face-layer and the lower face-layer thickness was expected to reduce the time of reaching the required temperature in the hot-pressing process. The usage of air-injection during the pressing of boards can also be a solution [19]. In the developed method, the air-injection press, which has holes punched in the heating plates, injects high-pressure air into the board through the holes of one plate and releases the air through the holes of the other plate. The advantage of this way of board pressing is allowing for reducing the pressing time required for manufacturing boards from high-moisture-content particles. Unfortunately, the air-injection press could not improve the properties of the particleboards [13].

As it results from the above considerations, the issue of the influence of particle moisture on the properties of finished particleboards and the course of their production process is practically well known. However, an often overlooked issue is the effect of particle moisture content on their physical properties such as poured bulk density, tapped bulk density, angle of repose, and slippery angle of repose. These properties are relevant because they influence the method and conditions of their storage and transport. They also influence the course of the production process, including the operation of transport and dosing devices and the choice of proper technological parameters during the bonding and pressing of particles, and consequently the properties of finished particleboards. Moisture content affects characteristics of bulk solids including wood and lignocellulosic particles such as particle-size distribution and bulk density [10,27,28]. The range of changes in these properties depending on the moisture content of the particles should be known in order to be able to assess its possible influence on the course of the production process and the properties of particleboards. Considering the above aspects, this work aimed to evaluate the physical properties of wood particles commonly used in the production process of three-layer particleboards depending on their moisture content.

2. Materials and Methods

2.1. Materials

Two types of wood particles used in the production of three-layer particleboards were used in the study, i.e., so-called microparticles intended for outer layers of particleboards and particles of the core layer (Figure 1).

Figure 1. Images of wood particles: (**a**) microparticles (MP), (**b**) particles of the core layer particleboards (PCL).

These particles have been produced under industrial conditions mainly from middle- and small-sized softwoods and selected low-density hardwood species. The raw material was also residues from the sawmilling industry in the form of shavings, sawdust, and chips. The initial moisture content of wood particles, determined by the drying–weighing method, was 7 ± 0.5%. To achieve the intended goal, the particles were submitted to drying or moisturising to the moisture content of 0.55%, 3.5%, 7%; 10%, 15%, and 20% ± 0.5%. In effect, the moisturising of different types of particles can be conducted directly or indirectly by increasing the moisture content of the environment [29]. In the present study, the first method was used i.e., wetting of the particles by spraying them with an appropriate amount of water and in order to homogenise the moisture content in the whole mass by seasoning for a period of 72 h. The moisture analyser MA R. 50 (Radom, Radwag, Poland) was used for the control of the moisture content of the particles. It determines the moisture content of the material on the basis of weight losses of the tested sample during its heating at a determined drying temperature (105 °C was used in the study). For preliminary characterisation of the raw materials used in this study, their fractional composition was determined for particles with 0.55% moisture content and additionally for particles with 20% moisture content. In the case of PCL particles, the fractional composition was determined based on sieve analysis, with the use of flat sieves made of mesh with square perforations of 6.3, 5.0, 4.0, 2.5, 1.6, 1.0, and 0.5 mm. In turn, for MP particles, sieves with mesh sizes of 3.15, 1.25, 1.0, 0.63, 0.4, and 0.315 mm were used. For PCL particles, due to the greater differences in the shape and size of individual particles, the additional dimensional analysis was carried out, for which 250 particles of the predominant fraction (from a 2.5 mm mesh sieve) with the moisture content of 0.55% were drawn. Their length, width, and thickness were determined. This allowed the estimation of the basic shape factors of this type of particles, i.e., the degree of slenderness (λ_s), flatness (ψ), width coefficient (m), and specific surface area (F_w) estimated according to the equations shown below [10]:

$$\lambda_s = \frac{l}{h} \tag{1}$$

$$\psi = \frac{b}{h} \qquad (2)$$

$$m = \frac{l}{b} \qquad (3)$$

$$F_w = \frac{0.002n}{w}(lh + lb + bh) \qquad (4)$$

where l—mean length of wood particles (mm), h—mean thickness of wood particles (mm), b—mean width of wood particles (mm), w—mean weight of dry wood particles (g), n—number of wood particles selected for analysis, 0.002—coefficient taking into consideration the fact that wood particles have two surfaces and a unit converter from mm to m.

Such analysis in the case of particles with a diverse geometry (Figure 1) is justified because of the fact that the biomass particles are mostly inhomogeneous in terms of size and shape [30]. As a result, two particles going through the same sieve with the same mesh size may differ in shape. Therefore, the information obtained from the sieving process may not fully reflect the geometry of the biomass particles with such an irregular shape [31]. The parameters characterising the geometry of the PCL particles are presented in Table 1.

Table 1. Characteristics of PCL particles of the predominant fraction.

Parameter	Value
Average dimensions (mm):	
length (l)	19.1 * ± 6.2 **
width (w)	3.2 ± 1.0
thickness (h)	1.38 ± 0.4
Shape factors:	
degree of slenderness (λ_s)	13.84
flatness (ψ)	2.32
width coefficient (m)	6.23

* mean value, ** standard deviation.

Subsequently, the prepared material was tested for the influence of the moisture content of the particles on their poured bulk density (ρ_p), tapped bulk density (ρ_t), compression ratio (k), angle of repose (α_R), and slippery angle of repose (α_s).

2.2. Poured Bulk Density and Tapped Bulk Density of Wood Particles

The first parameter, the poured bulk density, was expressed as the ratio of the weight of loosely poured wood particles to their volume. To determine the effect of the moisture content of the tested particles on their tapped bulk density, particles loosely poured into a pot equipped with a volume scale were densified on a lab electromagnetic vibratory sieve shaker AS200 (Retsch GmbH, Haan, Germany) in the time of 10 min and with the vibration amplitude of 2 mm. The tapped bulk density was expressed as the ratio of the weight of poured wood particles to their volume recorded after the tapping. Based on the obtained results of the poured bulk density and tapped bulk density, the compression ratio (k) of the tested particles was determined, depending on their moisture content according to the equation [10]:

$$k = \left(\frac{\rho_t}{\rho_p} \cdot 100\right) - 100 \qquad (5)$$

where k—compression ratio (%), ρ_p—poured bulk density (kg/m^3), and ρ_t—tapped bulk density (kg/m^3).

The average values of poured bulk density and tapped bulk density for each tested variant were determined based on five unitary measurements.

2.3. Angle of Repose and Slippery Angle of Repose of Wood Particles

In general, the angle of repose is defined as the angle between the slant height and the base of a cone created during the loose falling of bulk material at right angles to the ground. However, due to the differences in the fineness and therefore the size of the tested particles, other procedures were used to determine the influence of the moisture content on the angle of repose of the particles. In the case of microparticles, the analysis of the angle of repose (according to PN-74 Z-04002.07 standard [32]) was based on pouring them in a steel discharge hopper with a calibrated hole with the diameter of 22 mm, which was attached to the base with a gear train (Figure 2a). Next, the hopper was being lifted with a linear movement to the moment of pouring of the microparticles on a plate with the diameter of 120 mm. After piling up a stable cone, its height was measured. In contrast, in the case of PCL-type particles, the angle of repose was determined by pouring them into a cylinder with the diameter of 120 mm and the height of 100 mm, and then by lifting it, a cone from the particles was formed whose height was decoded from the millimetre scale attached to the base of the device (Figure 2b) [33].

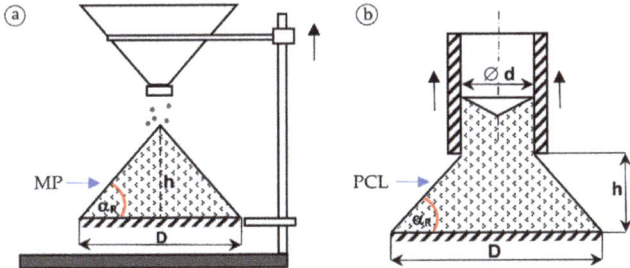

Figure 2. Scheme of the device measuring the angle of repose of: (**a**) microparticles, (**b**) particles of the core layer of particleboards.

The angle of repose for both types of wood particles tested was estimated based on the equation:

$$tg\ \alpha_R = \frac{2h}{D - d} \quad (6)$$

where α_R—angle of repose [°], h—cone height [mm], and D—cone base diameter [mm]. In the case of MP (Figure 2a), the value of $d = 0$.

The analysis of the slippery angle of repose was based on evenly pouring particles with the weight of about 200 g (MP and PCL) on a levelled and flat surface of the measuring device and then by lifting its side edge, finding the minimal rake angle that causes pouring of the layer of material. The slippery angle of repose value was directly decoded from a protractor pitch attached to the base of the device (Figure 3).

Figure 3. Scheme of the device measuring the slippery angle of repose of tested particles.

The mean values of angle of repose and slippery angle of repose were determined based on five unitary measurements.

2.4. Statistical Analysis

The obtained results of the tests selected for wood particles were statistically analysed using the STATISTICA v.13.1 software (StatSoft Inc., Tulsa, OK, USA). The mean values of the determined parameters were compared in the one-way analysis of variance—Tukey's post hoc test, in which homogeneous groups of mean values for each parameter were identified for $p = 0.05$. In the case of the fractional composition, a two-factor analysis of variance was used, assuming the size of the fraction and wood particle moisture as a qualitative factor.

3. Results

3.1. Fractional Composition of Wood Particles

The wood raw materials used in the study, independently of the degree of their fineness, posed a mixture of particles with diverse shapes and sizes. It is proclaimed by images of particles (Figure 1), the estimated coefficient of PCL shape (Table 1), and their fractional compositions (Figure 4), which were determined for dry particles (0.55%) and additionally for particles with the moisture content of 20%. Based on these diagrams, it can be concluded that in the case of MP independently of their moisture content, the particles of fractions 0.63 mm and 0.4 mm had the largest weight share. At a moisture content of 0.55%, the share of 0.63 mm fraction was 41.9%, and at a moisture content of 20%, it was slightly more, i.e., about 46.8%. In contrast, for the MP of fraction 0.4 mm, the share is greater by approximately 25%, which was observed for the moisture content of 0.55% in relation to the moisture content of 20% (respectively 30.4% and 24.2%). A similar dependence can be observed in the case of MP of a finer fraction, i.e., <0.4 mm. In contrast, the tests of the PCL particles showed that the particles of fractions 2.5 and 1.4 mm had the biggest share in the whole mixture. The content of dry particles (i.e., with the moisture content of 0.55%) of fraction 2.5 mm remained on the level of about 30%, whereas dry particles of fraction 1.4 mm were on the level of about 25.6%. Analogically to MP, for the particles of this type, it was also observed that the increase in their moisture content caused a significant differentiation of their fractional composition. In general, it can be stated that the increase in the moisture content of particles resulted in a decreased share of the finer fractions (smaller than the predominant fraction) and increased share of the larger fractions. The observed differences were statistically significant, which was confirmed by the two-factor post hoc analysis, which in the case of MP of fractions 0.315–1.0 mm allowed for distinguishing six homogenous groups. In the case of PCL, statistically the greatest differences were observed for the fractions 1.4, 2.5, and 6.3 mm. The observed changes in the fractional composition of the tested particles can be explained by the fact that at higher moisture content of the particles, an insufficient separation of MP occurs due to their adhesion to larger particles. Finer particles show the ability to agglomerate with the larger ones, which escalated with the increase in their moisture content [34].

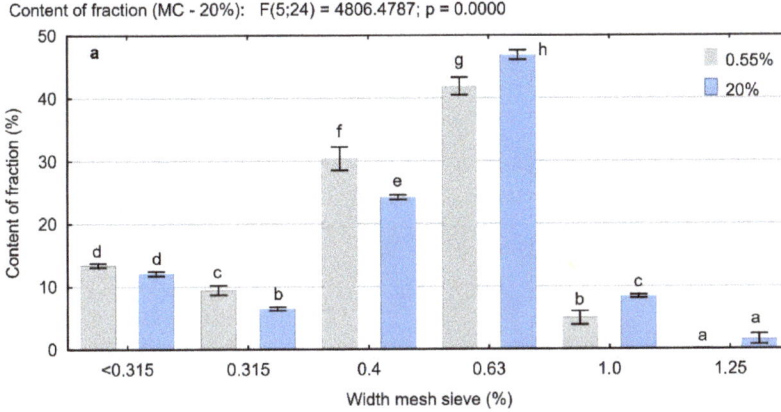

Figure 4. Fractional composition of microparticles (**a**) and particles of the core layer of particleboards (**b**) depending on their moisture content (a, b, c … —homogeneous groups as determined by the Tukey test).

3.2. Poured Bulk Density and Tapped Bulk Density of Wood Particles

The poured bulk density of particles of different biomass types are dependent on their shape, size, the way of their forming in the mass, and the friction between the particles [35,36]. In the case of material such as wood particles, wood chips, or wood dust, the poured bulk density also depends on the absolute density of the wood itself [37]. Figure 5 presents the results of tests on the influence of particle moisture content on the formation of their poured bulk density and tapped bulk density. It was determined that in the case of MP particles, the increase in moisture content from 0.55% to 7% did not cause statistically significant differences in the values of the poured bulk density. The observed mean values ρ_p in Tukey's test were classified in the same homogenous group (a). The further increase in the moisture content resulted in a gradual increase in the poured bulk density.

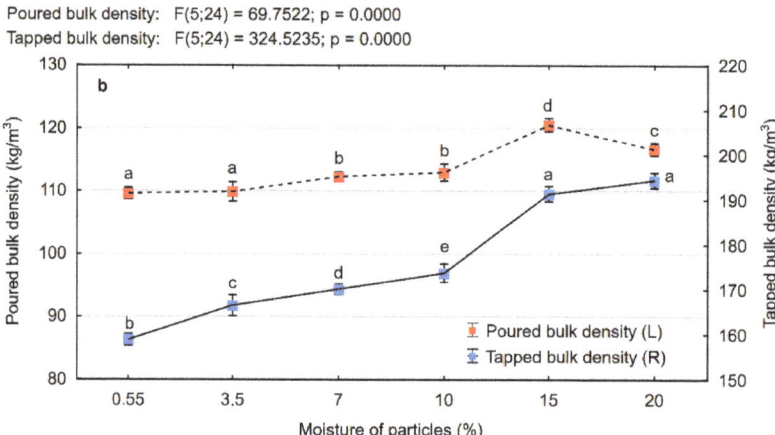

Figure 5. Poured bulk density and the tapped bulk density of microparticles (**a**) and wood particles of the core layer of particleboard (**b**) depending on their moisture content (a, b, c ... —homogeneous groups as determined by the Tukey test).

The MP-type particles with a moisture content of 20% showed the ρ_p value higher by about 14% than the ones with the lowest moisture content, whereby the highest increase in density occurred by changing the moisture content from 15% to 20%. In the case of PCL, the increase in poured bulk density was observed just by 9% but to the moisture content of 15%. Regarding the PCL particles with a moisture content of 20%, a small but statistically significant decrease ρ_p was observed. Previous studies on the influence of particle moisture content of different types of biomass (almond nut, sunflower seed, flaxseed, straw) on their physical properties showed that this relationship can develop in different ways [2,10,38–40]. For example, Dukarska et al. [10] and Aviara et al. [40] proved that the bulk density of seeds (*Moringa oleifera*) or straw particles of selected grain species increased with respect to an increase in moisture content. In contrast, other researchers such as Littelield et al. [41] showed that the bulk density of pecan shells decreased with an increase in moisture content. According to the authors, the decrease in the poured bulk density of the particles is caused by the increase in their size under the influence of the moisture content and thereby their volume, which was following on faster than the weight increase as a result of increasing the moisture content. As expected, along with the increase in the moisture content of particles, their tapped bulk density also increased. Considering the statistical analysis, the influence of the moisture content of particles on their tapped bulk density was greater than in the case of poured bulk density. It can be observed especially with regard to the PCL particles. Analysing the results obtained for MP, it can be stated that this density increased more with the increase in the moisture content from 0.55 to 3.5% and from 10% to 15%. However, the increase in the moisture content of MP from 0.55% to 20% caused the increase in their tapped bulk density by about 12% and in the case of PCL by about 22%. It results from the fact that the higher moisture content of MP (similar to wood dust) contributes to an increase in their volume, which causes the decrease in free space around them and the increase in consistency of the whole mass of particles [10,27,28]. In practice, the raw material with higher moisture content, independently of their degree of fineness, requires an increase in the volume essential to their storage or transport [41]. Moreover, it can be concluded that at higher moisture content, the PCL particles are more susceptible than MP to compaction by vibration, which can cause some difficulties in the technological process of producing particleboards and during their transport.

Comparing the results from Figure 5, it was also established that in the tested range of the moisture content, MP were marked by higher poured and tapped bulk density than the particles used for the core layers of particleboards. The literature shows that fine particles (such as MP characterised by their small size) are better at filling empty spaces during their pouring than larger particles. As the sizes of the particles increase, larger particles cannot sufficiently fill the empty spaces during tapping, which causes the decrease in their tapped bulk density [41]. This corresponds with the works of other authors who, studying the effect of particle size of different types of biomass, also observed that the value ρ_t increases with decreasing particle size of cereal straw, corn straw, switchgrass, and nutshells [35,41,42]. As demonstrated by Dukarska et al. [10], in the case of cereal straw, stem morphology is also an important factor that is related to the size and geometry of the particles, in particular, their thickness and degree of slenderness.

3.3. Compression Ratio

The changes in poured and tapped bulk density caused by the changes in the moisture content of the tested wood particles are reflected also in their compression ratio. A graphical representation of the compression ratio of wood particles depending on their moisture content is presented in Figure 6. It can be stated that in the case of PCL, along with the increase in their moisture content from 0.55% to 20%, a gradual and significant increase in the value of this parameter by about 48% was observed. The increase in the degree of fineness of the wood raw material to MP caused the decrease in vulnerability to its compressibility. It was observed that the increase in their moisture content resulted in the increase in the compression ratio from 44.9% to 58.7%, so by about 25% in comparison to the dry particles, however just in the moisture content range to 15%. By 20% moisture content of the MP, the decrease in its moisture content was set down to the level, which had been set down for dry particles. This phenomenon probably results from the small sizes of MP and their significant compaction in the whole mass. Moreover, increasing their moisture content resulted in reduced free spaces between particles and increased their adhesion to each other. When a particulate system becomes damp, the cohesion increases due to the creation of the liquid bridge bonds between the particles. The system remains stable until the moisture content is too high for strong bridges. When the particulate system becomes more and more wet, the material reaches the state of a slurry. During drying out the dump particulate system, the solid bonds and bridges between the particles can be formed. As a result of this, the material will become cohesive again. Changes in moisture content and resulting changes in other properties of a particulate system can cause serious problems in handling particles in an industrial installation [43]. In addition, biological materials (which may also include wood particles) become softer; thus, deformation is greater with an increase in moisture content [41]. In contrast to MP, increasing the moisture content of PCL chips from 15% to 20% results in a further increase in their compression ratio up to 66.7%. This might be attributed to the changes in the dimensions of the particles on account of their swell influenced by the increasing moisture content. Along with increasing dimensions of the particles, the volume of the space between them in a layer also increased and can be filled with the particles during tapping.

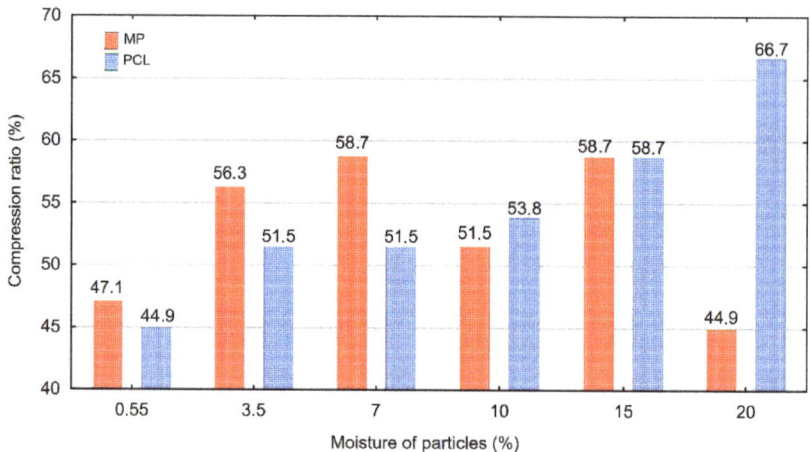

Figure 6. The compression ratio of wood particles depending on their moisture content.

3.4. Angle of Repose and Slippery Angle of Repose of Wood Particles

The angle of repose and slippery angle of repose are important physical parameters characterising wood particles. Graphical representation of the angle of repose and the slippery angle of repose of MP and PCL depending on their moisture content is presented analytically in Figure 7. Independently of the fineness of the particles, along with increasing their moisture content, the angle of repose and slippery angle of repose also increased. However, it can be observed that the biggest changes in values were recorded for the angle of repose of PCL particles. In this case, the increased moisture content from 0.55% to 20% resulted in increased α_R from 36.6° to 47.6°, i.e., by about 30%, whereby, as the variance analysis showed, statistically significant differences were set down for the moisture content higher than 3.5%. This is evidenced by the results of the analysis of variance, which identified five different homogeneous groups for each PCL moisture content above this value. Increasing the moisture content had much less effect on the angle of repose of the MP, for which the maximum value of α_R totals just 7%. Markedly, the greatest differences between MP and PCL particles were observed in the range of lower moisture content (up to 7%). This is an important observation in view of the fact that in the industrial practice, depending on the technology, the moisture content of the particles for the outer layers of the three-layer particleboards varies from 2% to 8% and in the case of the core layers, it varies from 1% to 6%. The results of the multivariate significance tests ANOVA presented in Table 2 are a confirmation that the lack of interactions between the tested effects (the type and moisture content of particles) were in the range of the higher moisture content, over 10% ($p > 0.05$). The results obtained for the angle of repose and the slippery angle of repose can be explained by the differences in their sizes and shape as well as by their consistency and looseness. The physical properties of wood particles, e.g., the size, shape, or roughness of the surface significantly affect the looseness of the particles of different types of biomass [30,31,36]. According to the subject literature, the increased particle size is connected with the decrease in their tenacity and therefore the value of the angle of repose [30]. In turn, fine particles, which can include wood MP, have a higher specific surface area, which increases the contact and cohesiveness among the particles, which may cause difficulties in their flow in technological conditions [31].

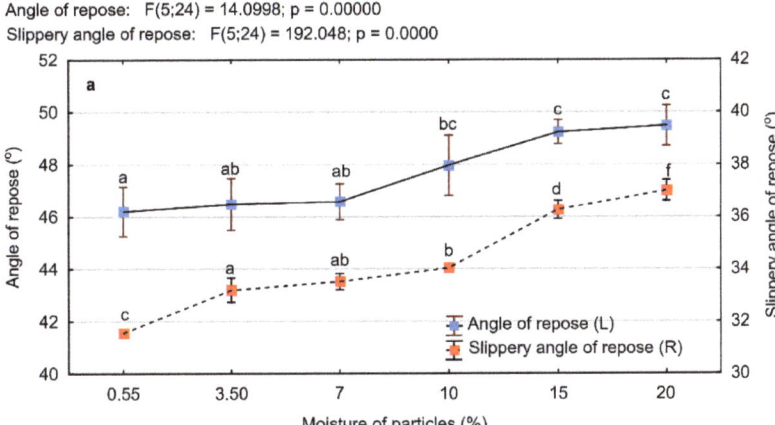

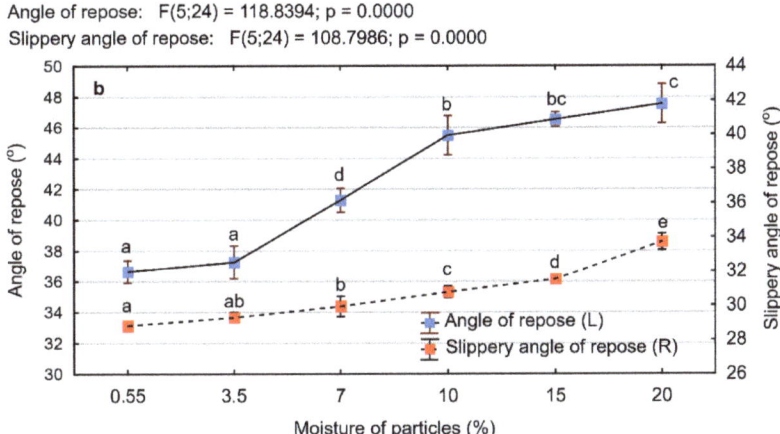

Figure 7. The angle of repose and the slippery angle of repose of microparticles (**a**) and particles of the core layer of particleboard (**b**) depending on their moisture content (a, b, c ... —homogeneous groups as determined by the Tukey test).

Table 2. Multivariate significance tests of the MP and PCL particle angle of repose.

Effect	F	P
Moisture content range: 0.55–7%		
Type of particles	F (1; 482.9) = 623.94	0.000
Particle moisture content	F (2; 17.79) = 22.99	0.000
Type of particles × particle moisture content	F (2; 14.01) = 18.11	0.000
Moisture content range: 10–20%		
Type of particles	F (1; 41.89) = 44.85	0.001
Particle moisture content	F (2; 8.13) = 8.71	0.00
Type of particles × particle moisture content	F (2; 0.4) = 0.43	0.65

With regard to the slippery angle of repose, no significant differences were determined. Independently from the degree of fineness of the particles the increase in the value α_s was on the same level, i.e., 17% on average. The increase in the slippery angles of repose along with the increased moisture content of particles may indicate an increase in their traction and thereby restriction of their mobility. Comparing the data obtained for both tested types of wood particles, it can be concluded that the values of MP's slippery angles of repose were greater than the PCL particles of particleboards.

4. Conclusions

Based on the conducted research, the influence of the moisture content variation on the physical properties of PCL and MP was determined. It was demonstrated that the increase in the moisture content of the particles resulted in increased dimensions of wood particles, independently of their degree of fineness, and increased values of the slippery angle of repose and the angle of repose. Furthermore, the increased moisture content of wood particles led to enhanced tapped bulk density for both types of studied wood particles for the production of three-layer particleboards. Regarding the poured bulk density, increased values were determined at 15% moisture content and over 7% moisture content for the PCL particles and MP, respectively. The PCL particle compression ratio was also increased. With regard to the MP particles, this increase was determined only at moisture content values up to 15%. The results of the present study can be utilised in the industrial practice of the wood-based panel industry to optimise the technological parameters and production costs related to particleboard manufacturing.

Author Contributions: Conceptualisation, D.D. and T.R.; methodology, D.D. and T.R.; validation, D.D., T.R., P.A. and L.K.; formal analysis, P.A. and L.K.; investigation, D.D., T.R. and J.K; resources, D.D. and T.R.; writing—original draft preparation, D.D. and T.R.; writing—review and editing P.A. and L.K.; visualisation, D.D., T.R. and J.K. All authors have read and agreed to the published version of the manuscript.

Funding: The study was supported by the funding for statutory R&D activities as the research task No. 506.224.02.00 and 506.227.02.00 of the Faculty of Forestry and Wood Technology, Poznań University of Life Sciences. This research was also supported by the project No. НИС-Б-1145/04.2021 "Development, Properties and Application of Eco-Friendly Wood-Based Composites" carried out at the University of Forestry, Sofia, Bulgaria. This research was also supported by the Slovak Research and Development Agency under the contracts No. APVV-18-0378, APVV-19-0269.

Institutional Review Board Statement: Not applicable.

Informed Consent Statement: Not applicable.

Data Availability Statement: Not applicable.

Conflicts of Interest: The authors declare no conflict of interest.

References

1. Pędzik, M.; Janiszewska, D.; Rogoziński, T. Alternative Lignocellulosic Raw Materials in Particleboard Production: A Review. *Ind. Crops Prod.* **2021**, *174*, 114162. [CrossRef]
2. Wang, B.; Li, D.; Wang, L.; Huang, Z.; Zhang, L.; Chen, X.D.; Mao, Z. Effect of Moisture Content on the Physical Properties of Fibered Flaxseed. *Int. J. Food Eng.* **2007**, *5*. [CrossRef]
3. Cosereanu, C.N.; Brenci, L.-M.N.G.; Zeleniuc, O.I.; Fotin, A.N. Effect of Particle Size and Geometry on the Performance of Single-Layer and Three-Layer Particleboard Made from Sunflower Seed Husks. *BioResources* **2015**, *10*, 1127–1136. [CrossRef]
4. Dukarska, D.; Czarnecki, R.; Dziurka, D.; Mirski, R. Construction Particleboards Made from Rapeseed Straw Glued with Hybrid PMDI/PF Resin. *Eur. J. Wood Prod.* **2017**, *75*, 175–184. [CrossRef]
5. Benthien, J.T.; Ohlmeyer, M.; Schneider, M.; Stehle, T. Experimental Determination of the Compression Resistance of Differently Shaped Wood Particles as Influencing Parameter on Wood-Reduced Particleboard Manufacturing. *Eur. J. Wood Prod.* **2018**, *76*, 937–945. [CrossRef]
6. Istek, A.; Aydin, U.; Özlüsoylu, I. The Effect of Chip Size on the Particleboard Properties. In Proceedings of the International Congress on Engineering and Life Science (ICELIS), Kastamouno, Turkey, 26–29 April 2018; pp. 26–29.
7. Istek, A.; Aydin, U.; Ozlusoylu, I. The Effect of Mat Layers Moisture Content on Some Properties of Particleboard. *Drv. Ind.* **2019**, *70*, 221–228. [CrossRef]
8. Benthien, J.T.; Ohlmeyer, M. Effects of Flat-Shaped Face Layer Particles and Core Layer Particles of Intentionally Greater Thickness on the Properties of Wood-Reduced Particleboard. *Fibers* **2020**, *8*, 46. [CrossRef]

9. Barbu, M.C.; Montecuccoli, Z.; Förg, J.; Barbeck, U.; Klímek, P.; Petutschnigg, A.; Tudor, E.M. Potential of Brewer's Spent Grain as a Potential Replacement of Wood in PMDI, UF or MUF Bonded Particleboard. *Polymers* **2021**, *13*, 319. [CrossRef] [PubMed]
10. Dukarska, D.; Pędzik, M.; Rogozińska, W.; Rogoziński, T.; Czarnecki, R. Characteristics of Straw Particles of Selected Grain Species Purposed for the Production of Lignocellulose Particleboards. *Part. Sci. Technol.* **2021**, *39*, 213–222. [CrossRef]
11. Kristak, L.; Ruziak, I.; Tudor, E.M.; Barbu, M.C.; Kain, G.; Reh, R. Thermophysical Properties of Larch Bark Composite Panels. *Polymers* **2021**, *13*, 2287. [CrossRef]
12. Tudor, E.M.; Kristak, L.; Barbu, M.C.; Gergeľ, T.; Němec, M.; Kain, G.; Réh, R. Acoustic Properties of Larch Bark Panels. *Forests* **2021**, *12*, 887. [CrossRef]
13. Korai, H.; Ling, N.; Osada, T.; Yasuda, O.; Sumida, A. Development of an Air-Injection Press for Preventing Blowout of Particleboard I: Effects of an Air-Injection Press on Board Properties. *J. Wood Sci.* **2011**, *57*, 401–407. [CrossRef]
14. Sarı, B.; Nemli, G.; Ayrilmis, N.; Baharoğlu, M.; Bardak, S. The Influences of Drying Temperature of Wood Particles on the Quality Properties of Particleboard Composite. *Dry. Technol.* **2013**, *31*, 17–23. [CrossRef]
15. Park, Y.; Chang, Y.-S.; Park, J.-H.; Yang, S.-Y.; Chung, H.; Jang, S.-K.; Choi, I.-G.; Yeo, H. Energy Efficiency of Fluidized Bed Drying for Wood Particles. *J. Korean Wood Sci. Technol.* **2016**, *44*, 821–827. [CrossRef]
16. Pizzi, A.; Mittal, K.L. Urea-Formaldehyde Adhesives. In *Handbook of Adhesive Technology*; Marcel Dekker: New York, NY, USA, 2003.
17. Dorieh, A.; Mahmoodi, N.O.; Mamaghani, M.; Pizzi, A.; Zeydi, M.M. New Insight into the Use of Latent Catalysts for the Synthesis of Urea Formaldehyde Adhesives and the Mechanical Properties of Medium Density Fiberboards Bonded with Them. *Eur. Polym. J.* **2019**, *112*, 195–205. [CrossRef]
18. Li, X.; Cai, Z.; Winandy, J.E.; Basta, A.H. Selected Properties of Particleboard Panels Manufactured from Rice Straws of Different Geometries. *Bioresour. Technol.* **2010**, *101*, 4662–4666. [CrossRef] [PubMed]
19. Korai, H.; Saotome, H. Blowout Conditions and Properties of Isocyanate Resin Bonded Particleboard Manufactured from High-Moisture Particles Using an Air-Injection Press. *J. Wood Sci.* **2013**, *59*, 42–49. [CrossRef]
20. Bekhta, P.; Noshchenko, G.; Réh, R.; Kristak, L.; Sedliačik, J.; Antov, P.; Mirski, R.; Savov, V. Properties of Eco-Friendly Particleboards Bonded with Lignosulfonate-Urea-Formaldehyde Adhesives and PMDI as a Crosslinker. *Materials* **2021**, *14*, 4875. [CrossRef]
21. Jiang, J.; Lu, X. Improving Characteristics of Melamine–Urea–Formaldehyde Resin by Addition of Blocked Polyurethane Prepolymer. *Eur. J. Wood Wood Prod.* **2017**, *75*, 185–191. [CrossRef]
22. European Committee for Standardization. *EN 312. Particleboards. Specifications*; European Committee for Standardization: Brussels, Belgium, 2003.
23. Rofii, M.N.; Yamamoto, N.; Ueda, S.; Kojima, Y.; Suzuki, S. The Temperature Behaviour inside the Mat of Wood-Based Panel during Hot Pressing under Various Manufacturing Conditions. *J. Wood Sci.* **2014**, *60*, 414–420. [CrossRef]
24. Cai, Z.; Muehl, J.H.; Winandy, J.E. Effects of Panel Density and Mat Moisture Content on Processing Medium Density Fiberboard. *For. Prod. J.* **2006**, *56*, 20–25.
25. Dai, C.; Wang, S. Press Control for Optimized Wood Composite Processing and Properties. In *Fundamentals of Composite Processing*; Gen. Tech. Rept. FPL-GTR-149; Winandy, J., Kamke, F.A., Eds.; USDA Forest Serv., Forest Products Lab.: Madison, WI, USA, 2004; pp. 54–64.
26. Murayama, K.; Kukita, K.; Kobori, H.; Kojima, Y.; Suzuki, S.; Miyamoto, K. Effect of Face-Layer Moisture Content and Face–Core–Face Ratio of Mats on the Temperature and Vapor Pressure Behavior during Hot-Pressing of Wood-Based Panel Manufacturing. *J. Wood Sci.* **2021**, *67*, 42. [CrossRef]
27. Rogozinski, T.; Dolny, S. Influence of Moisture Content on the Apparent Densities of Dust from Sanding of Alder Wood. *Trieskové Beztrieskové Obrábanie Dreva* **2004**, *4*, 205–208.
28. Dolny, S.; Rogoziński, T. Influence of Moisture Content on the Physical and Aerodynamic Properties of Dusts from Working of Particleboards. *Ann. Wars. Univ. Life Sci.–SGGW For. Wood Technol.* **2010**, *71*, 138–141.
29. Kalman, H.; Portnikov, D. Analyzing Bulk Density and Void Fraction: B. Effect of Moisture Content and Compression Pressure. *Powder Technol.* **2021**, *381*, 285–297. [CrossRef]
30. Tannous, K.; Lam, P.S.; Sokhansanj, S.; Grace, J.R. Physical Properties for Flow Characterization of Ground Biomass from Douglas Fir Wood. *Part. Sci. Technol.* **2013**, *31*, 291–300. [CrossRef]
31. Rezaei, H.; Lim, C.J.; Lau, A.; Sokhansanj, S. Size, Shape and Flow Characterization of Ground Wood Chip and Ground Wood Pellet Particles. *Powder Technol.* **2016**, *301*, 737–746. [CrossRef]
32. Polski Komitet Normalizacji i Miar. *PN-74 Z-04002.07. Ochrona Czystości Powietrza. Badania Fizycznych Właściwości Pyłów. Oznaczanie Kąta Nasypu Pyłu*; Polski Komitet Normalizacji i Miar: Warszawa, Polska, 1974.
33. Трофимов, С.П. Цеховые системыаспирации и пневмотранспорта измельченных древесныхотходов; Белорусский государственный технологический университет: Минск, Беларусь, 2010; ISBN 978-985-434-929-9.
34. Grubecki, I. Airflow versus Pressure Drop for a Mixture of Bulk Wood Chips and Bark at Different Moisture Contents. *Biosyst. Eng.* **2015**, *139*, 100–110. [CrossRef]
35. Lam, P.S.; Sokhansanj, S.; Bi, X.; Lim, C.J.; Naimi, L.J.; Hoque, M.; Mani, S.; Womac, A.R.; Narayan, S.; Ye, X.P. Bulk Density of Wet and Dry Wheat Straw and Switchgrass Particles. *Appl. Eng. Agric.* **2008**, *24*, 351–358. [CrossRef]
36. Wu, M.R.; Schott, D.; Lodewijks, G. Physical Properties of Solid Biomass. *Biomass Bioenergy* **2011**, *35*, 2093–2105. [CrossRef]

37. Eisenbies, M.H.; Volk, T.A.; Therasme, O.; Hallen, K. Three Bulk Density Measurement Methods Provide Different Results for Commercial Scale Harvests of Willow Biomass Chips. *Biomass Bioenergy* **2019**, *124*, 64–73. [CrossRef]
38. Santalla, E.; Mascheroni, R. Note: Physical Properties of High Oleic Sunflower Seeds. *Food Sci. Technol. Int.* **2003**, *9*, 435–442. [CrossRef]
39. Aviara, N.A.; Oluwole, F.A.; Haque, M.A. Effect of Moisture Content on Some Physical Properties of Sheanut [Butyrospernum Paradoxum]. *Int. Agrophysics* **2005**, *19*, 193–198.
40. Aviara, N.A.; Power, P.P.; Abbas, T. Moisture-Dependent Physical Properties of Moringa Oleifera Seed Relevant in Bulk Handling and Mechanical Processing. *Ind. Crops Prod.* **2013**, *42*, 96–104. [CrossRef]
41. Littlefield, B.; Fasina, O.O.; Shaw, J.; Adhikari, S.; Via, B. Physical and Flow Properties of Pecan Shells—Particle Size and Moisture Effects. *Powder Technol.* **2011**, *212*, 173–180. [CrossRef]
42. Mani, S.; Tabil, L.G.; Sokhansanj, S. Grinding Performance and Physical Properties of Wheat and Barley Straws, Corn Stover and Switchgrass. *Biomass Bioenergy* **2004**, *27*, 339–352. [CrossRef]
43. McGlinchey, D. (Ed.) *Characterisation of Bulk Solids*; Blackwell Publishing Ltd.: Oxford, UK, 2005; ISBN 978-1-4443-0545-6.

Article

Upcycling and Recycling Potential of Selected Lignocellulosic Waste Biomass

Anita Wronka [1], Eduardo Robles [2] and Grzegorz Kowaluk [1,*]

[1] Institute of Wood Sciences and Furniture, Warsaw University of Life Sciences—SGGW, Nowoursynowska St. 159, 02-776 Warsaw, Poland; anita_wronka@sggw.edu.pl
[2] University of Pau and the Adour Region, E2S UPPA, CNRS, Institute of Analytical and Physicochemical Sciences for the Environment and Materials (IPREM-UMR 5254), 40004 Mont de Marsan, France; eduardo.robles@univ-pau.fr
* Correspondence: grzegorz_kowaluk@sggw.edu.pl; Tel.: +48-22-59-38-546

Citation: Wronka, A.; Robles, E.; Kowaluk, G. Upcycling and Recycling Potential of Selected Lignocellulosic Waste Biomass. *Materials* **2021**, *14*, 7772. https://doi.org/10.3390/ma14247772

Academic Editors: Ľuboš Krišťák, Petar Antov and Réh Roman

Received: 23 November 2021
Accepted: 12 December 2021
Published: 16 December 2021

Publisher's Note: MDPI stays neutral with regard to jurisdictional claims in published maps and institutional affiliations.

Copyright: © 2021 by the authors. Licensee MDPI, Basel, Switzerland. This article is an open access article distributed under the terms and conditions of the Creative Commons Attribution (CC BY) license (https://creativecommons.org/licenses/by/4.0/).

Abstract: This research aimed to confirm the ability to reduce carbon dioxide emissions by novel composite production using plantation waste on the example of lignocellulosic particles of black chokeberry (*Aronia melanocarpa* (Michx.) Elliott) and raspberry (*Rubus idaeus* L.). Furthermore, to characterize the particles produced by re-milled particleboards made of the above-mentioned alternative raw materials in the light of further recycling. As part of the research, particleboards from wooden black chokeberry and raspberry were produced in laboratory conditions, and select mechanical and physical properties were examined. In addition, the characterization of raw materials (particles) on the different processing stages was determined, and the fraction share and shape of particles after re-milling of the produced panels was provided. The tests confirmed the possibility of producing particleboards from the raw materials used; however, in the case of boards with raspberry lignocellulose particles, their share cannot exceed 50% so as to comply with the European standards regarding bending strength criterion. In addition, the further utilization of chips made of re-milled panels can be limited due to the significantly different shape and fraction share of achieved particles.

Keywords: biopolymer; wood; upcycling; composite; recycling; mechanical properties; physical properties; carbon storage; raspberry; black chokeberry; bio waste

1. Introduction

Every year, society's awareness of caring for the Earth is growing. The growing amount of waste is a problem, with its storage and greater carbon dioxide emissions. In the case of fruit bushes grown in Poland, which are pruned each year, their branches are often left in the field or are burned. It can be used as a biofuel to avoid wasting energy, but it is not yet a common practice in Poland. Another way to use orchard waste is to produce three-layer particleboards for the furniture industry. Even though the tree species used for wood products are renewable, it should not be limited only to it because renewable does not mean that it is infinite. Because of this, it is necessary to explore using other lignocellulosic materials that will fully or partially replace the wood raw material. This attempt to move into the broad utilization of renewable biopolymers was also suggested by Bari and collaborators [1]. Some attempts have already been made that have been proven to be more or less effective, for this purpose, materials such as pepper stalks [2], sugarcane [3,4], almond shell [5], apple and plum branches [6], bamboo chips [7], straw [8], wheat straw and corn pith [9], kiwi prunings [10], coffee husk [11], flax shiv [12], acai (*Euterpe oleracea* Mart.) fruit [13], oil palm empty fruit bunch [14], and kenaf [15] were used. The use of wooden lignocellulosic parts of fruit plant waste allows for the binding of carbon dioxide in the form of particleboards, without emitting it into the atmosphere. In this field, good examples are raspberry (*Rubus idaeus* L.) and chokeberry (*Aronia melanocarpa* (Michx) Elliott) plantation waste. These represent substantial waste in Polish fields, as

Polish production accounts for 60–70% of the world's production potential. The cultivation area of chokeberry is about 40 km^2 per year and the annual harvest of fruits is from 40 to 60 thousand tonnes. The main recipients of chokeberry fruits are China, Japan, and South Korea [16]. Whereas the area of raspberry cultivation in Poland is over 290 km^2, placing it at fifth place in the world's raspberry producers and third in Europe (after Russia and Serbia) [17].

The fruits of these shrubs are cultivated for their taste and health benefits. Raspberry fruits are rich in anthocyanins and have anti-inflammatory and anticancer properties, so it is often recommended to drink raspberry juice during colds [18]. Medicinal values also characterize black chokeberry fruits; just like raspberries, they have an antioxidant effect, and their consumption is recommended to prevent chronic diseases [19]. This added value for the fruits allows for assuming that the potential availability of lignocellulosic resources of those above-mentioned alternative raw materials will grow shortly. Therefore, it seems worth researching the development of long-term storage regarding the carbon fixed in these raw materials, such as producing particleboards and attempting to upcycle these wooden wastes and recycle the produced composites.

This investigation aimed to determine the ability to utilize raspberry and chokeberry lignocellulosic particles to produce particleboards for furniture purposes and to characterize the wooden particles produced by the re-milling particleboards mentioned regarding further recycling. As a result, the following hypothesis has been investigated: the lignocellulosic particles of raspberry and black chokeberry are valuable raw materials to produce the particleboards and obtain particles from re-milled panels, which can potentially be re-used in particleboard production.

2. Materials and Methods

2.1. Materials

Raspberry (*Rubus idaeus* L.) (Figure 2) and black chokeberry (*Aronia melanocarpa* (Michx.) Elliott) (Figure 3) wooden stalks were used for the current work. Two year old raspberry stalks, as well as four year old chokeberry rods, were collected from Polish fields. The raw materials were dried in a chamber drier under 70 °C to air-dry the moisture content (about 10–12%), and the bark content (*w/w*) was measured by manual debarking about 2 kg of each tested material. The wooden branches of the chokeberry and raspberries were shredded on saw blade in separate batches (50 mm long chips) and then grounded into a fine fraction using a laboratory three-knife drum mill (laboratory prototype) with an outlet equipped with 6 × 12 mm^2 mesh to form particles. The bulk density of the particles was calculated as the weight of a selected fraction, divided by the measuring cylinder's capacity (in volume). The measurement was repeated five times for every fraction. The produced particles were sorted on mesh of size 0.5 and 1 mm (face layers), and 8 mm and 2 mm (core layer) to exclude the oversized and undersized particles. The pictures of the cross-cuts of the investigated raw materials were taken with a NIKON SMZ 1500 (Kabushiki-gaisha Nikon, Minato, Tokyo, Japan) optical microscope.

2.2. Elaboration of Composites

Three-layer composites were produced as particleboards (PB) with different black chokeberry and raspberry contents. The lignocellulosic particles were dried to a moisture content of 5%. As a result, particleboards with a nominal density of 600 kg m^{-3}, 32% (*w/w*) of face layer content, and a total thickness of 16 mm were produced. The following content (*w/w*) of alternative raw materials was applied: 0% (reference panels, 100% of industrial (coniferous) particles), 10%, 25%, 50%, and 100%. The industrial urea-formaldehyde resin Silekol S-123 (Silekol Sp. z o.o., Kędzierzyn—Koźle, Poland) was used to resinate the particles, where the resination of particles for the face layer was 12% and the core layer was 10%. No hydrophobic agent (like paraffin emulsion) was added. The curing was done for 82 s inside an oven at 100 °C. Panels were pressed on a hydraulic press (ZUP-NYSA PH-1P125) at a maximum pressure of 2.5 MPa, with a temperature of 200 °C, and a time factor of 20 s mm^{-1}. The produced boards were conditioned before the tests in a climatic chamber (producer: Research and Development Centre for Wood-Based Panels Sp. z o. o. in Czarna Woda, Poland) at 20 °C and 65% air humidity, until a constant mass was obtained. The main steps of the material flow and samples preparation are presented in Figure 1.

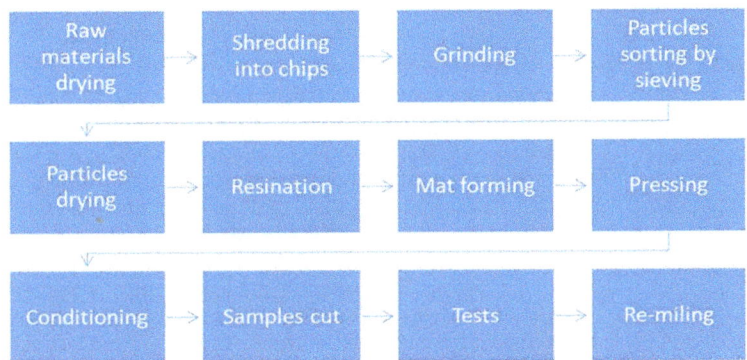

Figure 1. The process of material flow and samples preparation chart.

2.3. Characterization of the Elaborated Panels

All of the elaborated PB were conditioned at 20 °C, and the test specimens were cut on a saw blade, as required by European standards EN-326-2 [20] and EN-326-1 [21]. The modulus of rupture (MOR) and elasticity (MOE) were determined according to EN 310 [22], and the internal bond (IB) was determined according to EN 319 [23]. All the mechanical properties were examined with an INSTRON 3369 (Instron, Norwood, MA, USA) laboratory-testing machine, and, whenever applicable, the results were referred to standards [24]. Board density was determined according to EN 323 [25], thickness swelling (TS) to EN 317 [26], and surface water absorption was done following EN 381-1 [27]. The density profiles of the tested PB (three types: reference, 100% of raspberry, and 100% of chokeberry) were measured on a GreCon DAX 5000 device (Fagus-GreCon Greten GmbH and Co. KG, Alfeld/Hannover, Germany).

2.4. Raw Material Recycling and Characterization

The composites were re-milled on a laboratory knife mill (laboratory prototype delivered by Research and Development Centre for Wood-Based Panels Sp. z o. o. in Czarna Woda, Poland) equipped with three knives, two contra-knives, and a 6 × 12 mm^2 mesh. The fraction of chips taken from the re-milled particleboards was tested with an IMAL (Imal s.r.l., San Damaso (MO), Italy) vibrating laboratory sorter with seven sieves. The selected sieve sizes were 8, 4, 2, 1, 0.5, 0.25, and <0.25 mm. The amount of tested material for each fraction was about 100 g, and the set time of continuous vibrating was 5 min. As many as five repetitions were done for every tested material.

2.5. Statistical Analysis

Analysis of variance (ANOVA) and t-tests calculations were used to test (α = 0.05) for significant differences between factors and levels, where appropriate, using IBM SPSS statistic base (IBM, SPSS 20, Armonk, NY, USA). A comparison of the means was performed when the ANOVA indicated a significant difference by employing the Duncan test. The statistically significant differences for the achieved results are given in the Results and Discussion paragraph whenever the data were evaluated.

3. Results and Discussion

3.1. Materials Characterization

The bark content (w/w) was 7.4% for raspberry and 18.0% for chokeberry. According to [28], the average bark content of pine (*Pinus sylvestris* L.), which is the main raw material for particleboard production in Poland, is about 6.7% (w/w). Significant differences in bark content of the investigated materials were found. Such a high content of bark in the case of chokeberry could influence the mechanical properties of the produced PB [29]. It was found by Kowaluk et al. [6] that the bark density of orchard trees can be remarkably lower than the density of the wood. As confirmed in the case of single-layer particleboards produced from *Quercus cerris* bark [30], these panels had remarkably low mechanical properties when compared to the commercial particleboards. What was also confirmed by the mentioned researchers, is that the panels produced from *Quercus cerris* bark had low TS. The bark particles, being highly brittle, could also raise the fine particles/dust production when milled, which could negatively influence the mechanical properties of the panels. On the other hand, a fine bark particle can be upcycled and utilized, as was proven by Mirski and collaborators [31].

Concerning the anatomy of the investigated raw materials, raspberry (Figure 2) has a large amount (volumetric) of foamy parenchyma pith. However, this part of the material can be easily disintegrated mechanically, and it is not easy to separate the particles produced from the remaining particles. Furthermore, as a brittle and soft tissue, it produces a large amount of fine particles, characterized by a large specific surface. This feature is not desirable in PB production, since this fraction requires a high amount of resin to be added. If the resination is not tuned regarding these fine fractions, the mechanical parameter of produced PB drops down.

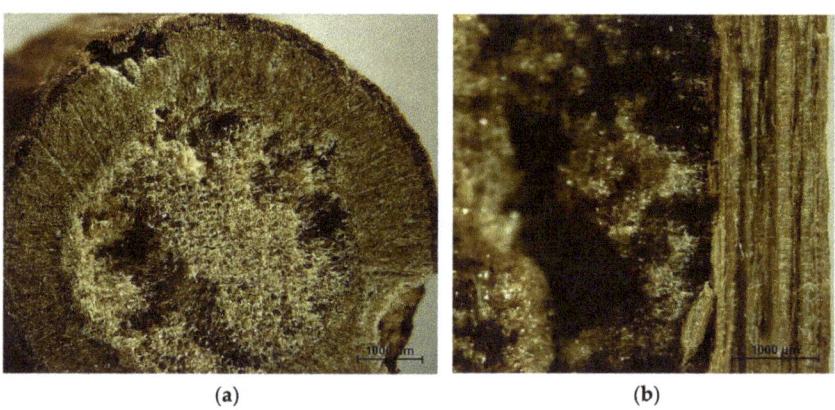

Figure 2. Cross-cut: (**a**) across and (**b**) along the fibers of a raspberry stalk.

The cross-cut of chokeberry (Figure 3) can be referred to as broadleaf plants. The year rings (Figure 3a) are clearly visible, and wood rays are going horizontally between bark and pith (Figure 3b; bark on the left, pith on the right). It is worth pointing out that the pith is also in foam form, which was found for raspberry stalks, but here the amount of foam pith was significantly lower than for raspberry.

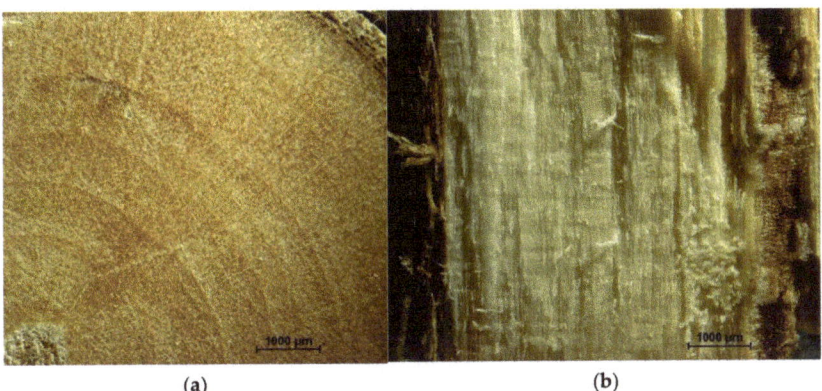

Figure 3. Cross-cut: (**a**) across and (**b**) along the fiber of the chokeberry stalk.

In Figure 4, the results of the measurement of the bulk density of particles used to produce the tested composites and those produced by re-milling of the tested composites are presented. In the case of the face layers' intended particles, the highest bulk density was found for chokeberry particles (164 kg m^{-3}). A 2.4% lower bulk density (when referred to highest value) was found for industrial face layer particles (160 kg m^{-3}). The lowest bulk density value among the tested particles was registered for raspberry particles (83 kg m^{-3}), which means an almost 50% lower density for chokeberry. When analyzing the core layer purpose particles, the results were as follows (descending order): industrial (157 kg m^{-3}), chokeberry (121 kg m^{-3}), and raspberry (89 kg m^{-3}). The results of the measurement of the bulk density of re-milled particles show that the bulk density of these particles was higher than for the primary particles, and, what should be pointed out, is that the differences between the tested materials were less than 2% when considering the lowest value. The achieved average bulk density values were statistically significantly different when compared within the same group (face, core, and re-milled).

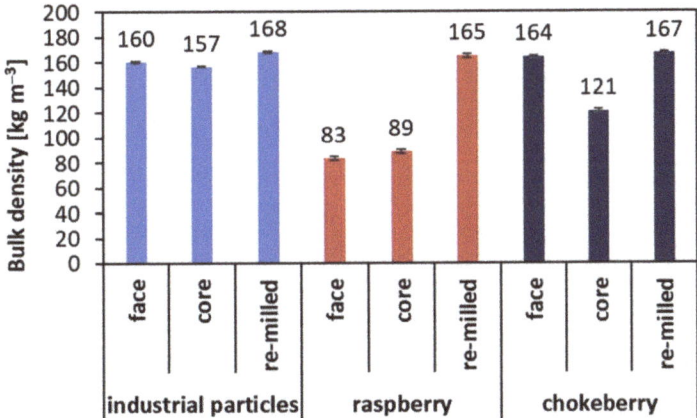

Figure 4. Bulk density of particles used to produce composites, considering the particles produced by the re-milling of composites.

The achieved density values of raspberry and chokeberry particles were low compared to other alternative lignocellulosic raw materials [6]. This is promising information, as, in the case of compressed lignocellulosic composites, a low bulk density leads to better densification, creating more spots where separate particles are connected. Thus, the produced composite structure is more even, less porous, and has higher mechanical properties. This can also lead to lower water absorption. However, it was confirmed by Papadopoulos et al. [7] that a lower bulk density can reduce the mat permeability due to densification during hot pressing, and the heat transfer through such a mat can be significantly slower.

3.2. Modulus of Rupture and Modulus of Elasticity

As shown in Figure 5, the modulus of rupture values decreased when the content of alternative raw materials increased. The MOR decrease was higher for composites produced of raspberry particles (from over 15 N mm^{-2} when 0% of raspberry particles to less than 10 N mm^{-2} for 100% raspberry composite). In the same conditions, MOR decreased for chokeberry composites that had reached over 12.1 N mm^{-2}. When compared within the same raw materials, the only statistically significant differences between average MOR values were found for the highest and lowest content of raw materials. When referring to the EN standard [24], it was found that in the case of raspberry, the content of alternative raw material should not exceed 50% for meeting the standard requirements.

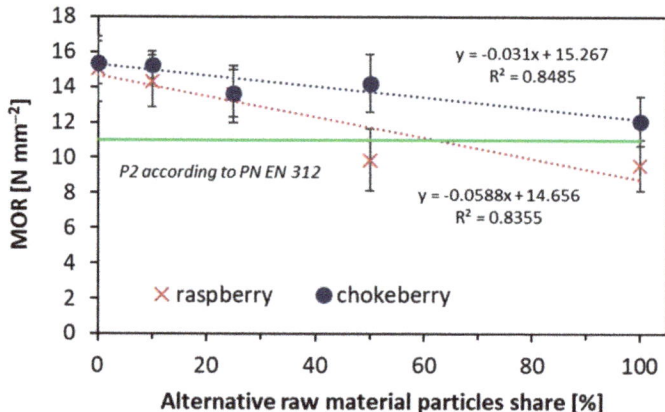

Figure 5. Modulus of rupture of the tested composites.

Similar tendencies for a reduction of MOE when the alternative raw material contend grows are presented in Figure 6. The reduction of MOE from the value of reference composite, 2805 N mm^{-2}, was to 1800 N mm^{-2} for the fully chokeberry composite and to 1617 N mm^{-2} for the 100% raspberry composite. It is worth adding that when referring to the EN standard [24], in the case of raspberry, the content of alternative raw material should not exceed about 80% for meeting the standard requirements. Furthermore, statistically significant differences for the average MOR values for chokeberry were found for all composite types excluding 25 and 50% alternative raw material particles share, when, in the case of raspberry, there were no statistically significant differences between the composites of 10 and 25%, as well as between 50 and 100%.

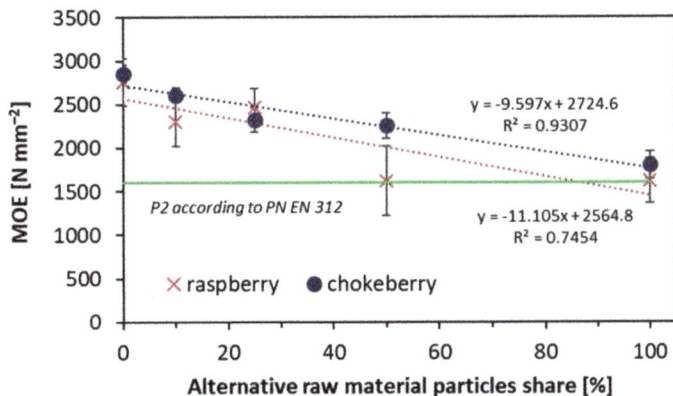

Figure 6. Modulus of elasticity of the tested composites.

Raspberry panels were expected to have the best mechanical properties because of their lower bulk density; however, they presented low mechanical features for MOR and MOE. The reason for this can be the content of the core, as can be appreciated in Figure 2. This part of the raw material can influence the production of particles with a low bulk density, but these particles do not allow for carrying a high mechanical load when the samples are bent, the face layers are strained/compressed, and core layers are under shear stress. Moreover, the geometry of the particles used to produce the composites can play a role here. As Wronka and Kowaluk demonstrated [32], the raspberry particles are shorter and have blunt (wide) ends when compared to industrially used softwood particles. Furthermore, because of the structure of the raspberry stalk, where the region of higher mechanical properties is on the external zone of the rod, the particles produced from this raw material are of a lower length to thickness ratio (slenderness), which is not desirable for particle composites. It has been confirmed [33] that the best mechanical properties for composites are achievable with a high length to thickness ratio.

3.3. Internal Bond

The positive effect of a low bulk density of raspberry particles has been found when analyzing the IB values of the tested composites. As shown in Figure 7, the IB was significantly raised when the content raspberry particles rose. The reference composite IB value was 0.72 N mm^{-2}, and, for 100% raspberry composite, the IB was 1.04 N mm^{-2}, while for 100% chokeberry composite, it was 0.53 N mm^{-2}. It should be pointed out that when comparing the achieved results of IB, all of the tested composites met the requirements of a specified European standard [24]. Furthermore, the statistical analyses within the alternative raw materials mentioned show no statistically significant differences between IB average values of 10% and 25%, 25%, and 50% for raspberry, as well as between 10% and 50%, and between 25% and 100% for chokeberry.

3.4. Thickness Swelling and Water Absorption

The results of the measurement of thickness swelling of the tested composites after 2 h and 24 h of soaking in water are presented in Figure 8. As can be seen, in the case of raspberry composites, the swelling in thickness significantly grew with the alternative raw material content increase. After 2 h of soaking, the lowest TS for the reference composite (0% of raspberry particles) was 18%, while for the 100% raspberry panel, the TS was 33%, which is an increase of more than 83%. After 24 h of soaking, the TS of the reference composite was below 20%, and for the 100% raspberry composite, the TS was over 36% (89% growth). In the case of chokeberry, the increase in alternative raw material particles content caused a decrease in thickness swelling. After 24 h of soaking of the chokeberry composites, the TS was 16%, which was an almost 16% reduction of TS. The only statistically significant differences for the average values of TS after 2 h of raspberry composites were found between the 0%, 50%, and 100% panels and the same composites after 24 h of soaking. Regarding chokeberry, statistically significant differences after 2 h of soaking were found for composites of 0% and 100%, and the same after 24 h of soaking. It should be highlighted that when referring to the achieved results of TS, none of the tested composites met the requirements of the European standard [24].

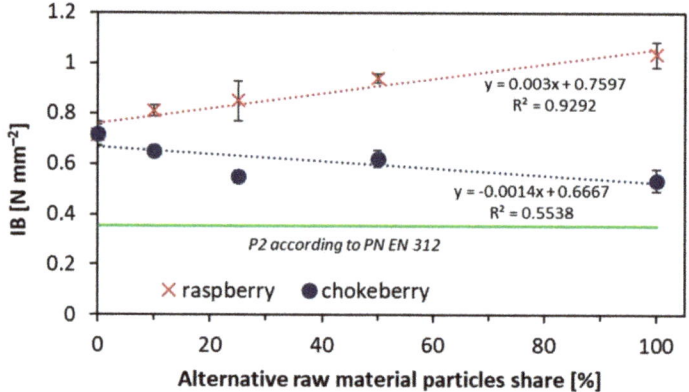

Figure 7. Internal bond of the tested composites.

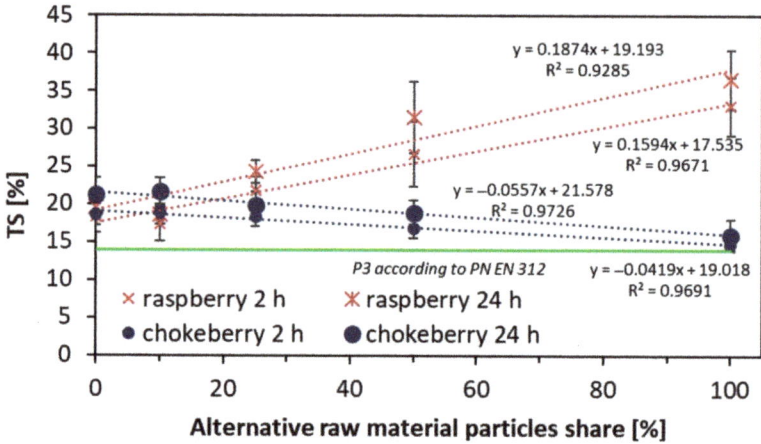

Figure 8. Thickness swelling of the tested composites.

Such a significant rise of TS of composites made of raspberry particles can be explained by the low bulk density of raspberry particles, as presented in Figure 4. Although a low bulk density helps with better densification of the pressed mat, this highly compressed mat can be easily decompressed by water penetrating the composite in light of the swelling in thickness. Thus, the material, which was more densified during hot pressing (composite preparation), has a potential of higher TS. On the other hand, the opposite situation was found in the case of chokeberry composites, where the mat densification was lower due to the higher bulk density of the chokeberry particles.

The WA values of the tested composites of different contents of alternative raw materials are presented in Figure 9. The high water absorption values after 24 h of soaking for the raspberry samples, from over 77% for the reference composite to over 108% for 100% raspberry composite, can be explained by the presence of low-density core particles, which can react with water like a sponge. The higher increase of WA for the samples with a higher content of raspberry particles after 24 h compared to WA after 2 h of soaking means that the structure of the samples is less penetrative (tighter) against water, and more time is needed to reach the deeper zones of the samples. This can be explained by the higher densification of the mat built by low bulk density particles. When evaluating the WA of chokeberry composites, it can be found that with the rising content of chokeberry particles, the WA slightly rose after 2 h of soaking, whereas, after 24 h of soaking, the WA decreased with the increase in chokeberry particles content. This means that chokeberry particles cause lower water absorption. A specific type of composite here can be the 100% chokeberry panel, where the maximum WA was reached after 2 h of soaking and did not raise even after 24 h of total soaking. One of the reasons could be the high bulk density of chokeberry particles, which lead to lower compression of particles during pressing, and leave more unfilled (empty) zones in the composite structure. These zones can be filled with water in a short time. Another reason is that the deciduous wood has a five times higher potential to transfer the water due to the larger dimensions of the vessels [28]. Statistically significant differences of average WA for raspberry composites after 2 h of soaking were between 0% and 100%, and between 0%, 10%, and 25% against 50% and 100% composites after 24 h. For chokeberry, these differences were found between samples of 0%, 10%, and 25% against 100% after 2 h, and 0%, 10%, and 25% against 50% and 100% composites after 24 h.

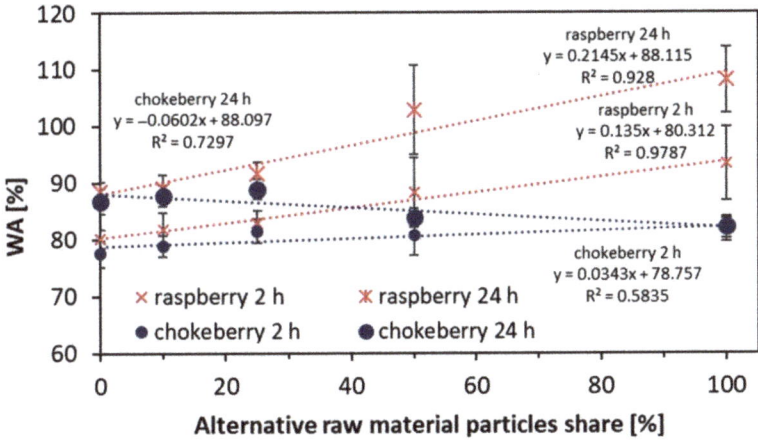

Figure 9. Water absorption of the tested composites.

3.5. Density and Density Profiles

The results of the density profile measurement of the tested composites are presented in Figure 10. Since the tested composites are symmetrical, half of the density profile is shown to improve the readability of the plots. As can be seen, the highest values of density in the face zone, over 950 kg m^{-3}, located about 0.7 mm in deep from the surface, were found for the industrial particles composite. On the other hand, the highest density zone of the raspberry composite, about 815 kg m^{-3}, was found about 1.8 mm under the composite surface. A similar zone for the chokeberry composite, but with a lower density, about 780 kg m^{-3}, was found at 0.4 mm under the surface. In the case of the lowest density in the core layers (middle of the thickness), the lowest value, about 550 kg m^{-3}, was registered for the raspberry composite, when the remaining composites had a similar core layer density, which was about 590 kg m^{-3}. It should be mentioned here that all of the tested samples were of the same average density of about 600 kg m^{-3}.

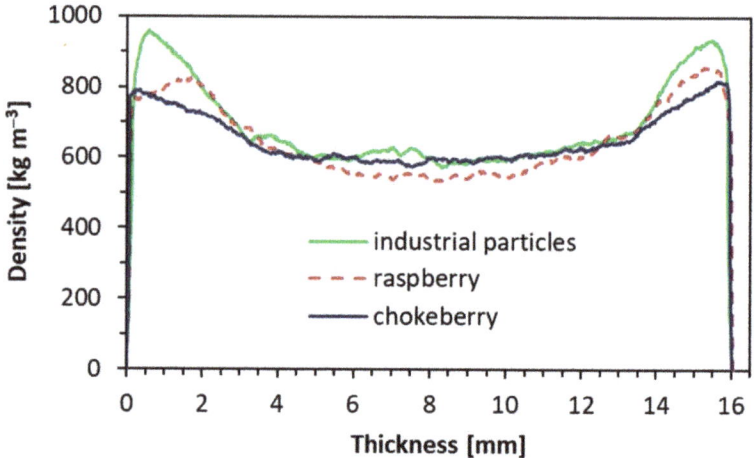

Figure 10. Density profiles of the tested composites.

A high density of face layers, which was found for the reference composites (industrial particles), can significantly influence the bending properties of composites as these face layers are generally responsible for tension and compression stresses when the material is bent. This remark can be confirmed on figures presenting MOR (Figure 5) and MOE (Figure 6) values. It was also confirmed for fibrous composites of different density profiles [34]. However, as the differences between the density values of core layers of the tested panels are low, it can be hard to refer to the remaining features of the tested composites.

3.6. Recycled Material Characterization

The pictures of different particles produced by re-milling the tested composites and industrial (not re-milled) particles are shown in Figure 11. It can be found that in the case of industrial particles, for all fractions excluding dust (< 0.25 mm), the particles had a high length to width ratio, which can be estimated on the level of 4:1 for 8 mm fraction and even higher, about 20:1 for 2 mm fraction. On the other hand, the pictures of re-milled particles of fractions 8 and 4 mm show that the particles were not elongated anymore, and these were more rounded or square, with a length-to-width ratio of about 1:1. For a fraction of 2 mm, a significant difference was found for chokeberry particles, which are more similar to industrial particles. In addition, the smaller chokeberry particles (1 mm and below) are closer to industrial (not re-milled) particles. When re-milled, industrial and raspberry particles are short and of higher width. These remarks can be valuable in the light of further use of re-milled particles, since, as confirmed, the shape of the particles can significantly influence the properties of the produced particle composites [35,36].

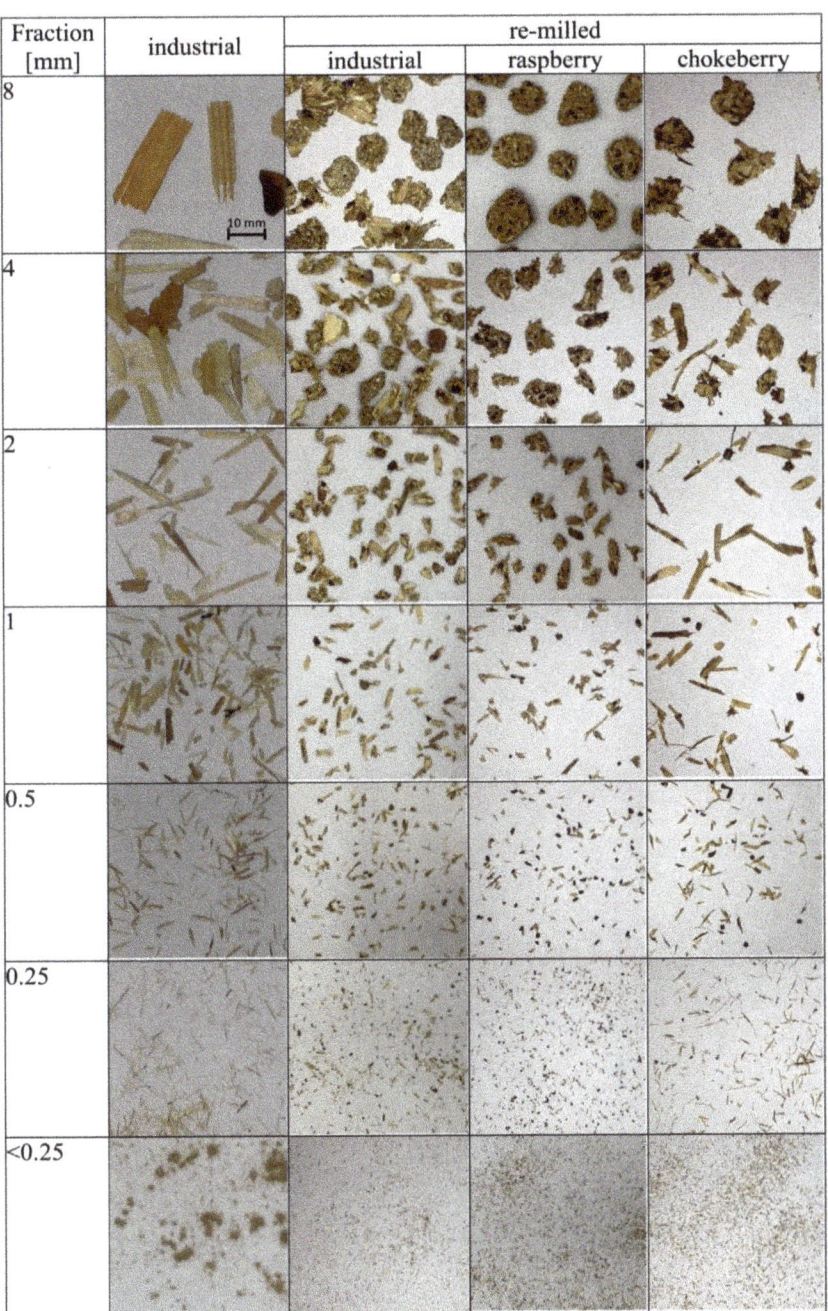

Figure 11. Pictures of the morphology of particles produced by re-milling the tested composites compared to industrial particles (each picture dimension is 50 mm × 50 mm).

The mass fraction share of particles produced by the re-milling of investigated composites and the fraction share of primary industrial particles is shown in Figure 12. As can be seen, in the case of raw industrial particles, the largest share is for particles of size 1 and 2 mm (over 74%), and 17% of size 4 mm. The remaining fractions are less than 9%. When analyzing the fraction share of re-milled particles, it can be stated that the fraction share of industrial and chokeberry particles is similar. The difference is between the distribution of fractions smaller and larger than 1 mm: for industrial re-milled particles, a more significant amount of fractions smaller than 1 mm were found, and a smaller amount of fractions were larger than 1 mm. The opposite distribution was found for chokeberry particles. Significant differences in fraction share were found for raspberry re-milled particles. These particles had many fractions of bigger dimensions, where the content of fractions of 4 mm + 8 mm was about 47%. This type of material also provides a large amount of smaller fractions: the sum of fractions 0.5 mm, 0.25 mm, and < 0.25 mm was about 30%, whereas, in the case of remaining re-milled materials, it was about 23% for industrial re-milled, 18% for re-milled chokeberry, and 8% for primary industrial particles. It should be pointed that a high amount of small fractions is not profitable when considering the achieved particles to be used as a raw material to produce similar particle composites. Since the specific surface of the particles grows with the particle size decrease, and, thus, a larger amount of binder is needed to cover the particle surface adequately, such small fractions should be separated and subjected to alternative processing/utilization.

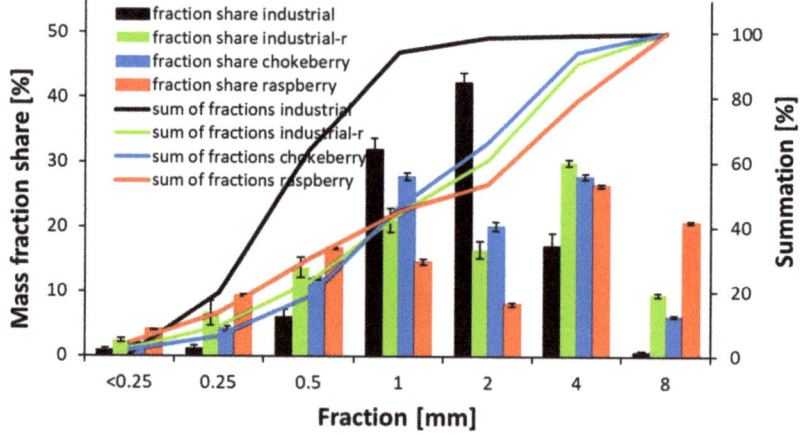

Figure 12. Mass fraction share of particles produced by re-milling of composites considering industrial particles.

4. Conclusions

According to the conducted research and the analysis of the achieved results, the following conclusions and observations can be drawn:

1. It has been confirmed that lignocellulosic particles of black chokeberry (*Aronia melanocarpa* (Michx.) Elliott) and raspberry (*Rubus idaeus* L.), being an orchard waste, can be successfully upcycled and used to produce lignocellulosic composites, thus having a positive contribution to carbon storage.
2. The bulk density of chokeberry particles on the outer layers is slightly higher than that of industrial particles. The inverse relationship occurs in the case of particles on the core layers. The particles in both layers are characterized by a lower density than the reference (industrial) particles for raspberry.
3. With an increase in the proportion of black chokeberry or raspberry particles in the particleboard, the bending strength and modulus of elasticity decreases.

4. A significant influence of the content of black chokeberry and raspberry particles was found on the perpendicular tensile strength (IB) of the tested composites: a significant increase with raspberry particles increasing and decrease with chokeberry particles increasing.
5. The thickness swelling of raspberry-containing composites increases after 2 h and 24 h of soaking in water. In the same conditions, the increase of chokeberry particle contents causes a lower thickness swelling.
6. The water absorption test showed increasing dynamics of water absorption for boards with a higher proportion of chokeberry and raspberry particles, but in the long run, boards made of chokeberry particles absorb less water than the reference and raspberry composites.
7. The highest density of face layers has been found for reference composites made of industrial particles, which influence the bending features of the tested composites.
8. Further use of particles produced from re-milled composites can be limited due to the shape of the re-milled particles, which, in the case of industrial and raspberry particles, is significantly different from unprocessed particles.

Author Contributions: A.W. took part in designing the experiments and performed the experiments, analyzed the data, and wrote the first draft of the paper; E.R. analyzed data statistically and wrote the final version of the paper; and G.K. designed the experiments, and analyzed the data. All the authors assisted in writing and improving the paper. All authors have read and agreed to the published version of the manuscript.

Funding: The research was financed by the own funds of the Institute of Wood Sciences and Furniture at the Warsaw University of Life Sciences—SGGW. E.R. wants to acknowledge the tenure track position "Bois: Biobased materials", part of E2S UPPA supported by the "Investissements d'Avenir" French program managed by the ANR (ANR-16-IDEX-0002).

Institutional Review Board Statement: Not apllicable.

Informed Consent Statement: Not applicable.

Data Availability Statement: The data presented in this study are available on request from the corresponding author.

Acknowledgments: Some of the mentioned tests have been completed within the Student Furniture Scientific Group (Koło Naukowe Meblarstwa), Faculty of Wood Technology, Warsaw University of Life Sciences, SGGW.

Conflicts of Interest: The authors declare no conflict of interest.

References

1. Bari, E.; Sistani, A.; Morrell, J.J.; Pizzi, A.; Akbari, M.R.; Ribera, J. Current strategies for the production of sustainable biopolymer composites. *Polymers* **2021**, *13*, 2878. [CrossRef]
2. Oh, Y.S.; Yoo, J.Y. Properties of particleboard made from chili pepper stalks. *J. Trop. For. Sci.* **2011**, *23*, 473–477.
3. Nadhari, W.N.A.W.; Karim, N.A.; Boon, J.G.; Salleh, K.M.; Mustapha, A.; Hashim, R.; Sulaiman, O.; Azni, M.E. Sugarcane (Saccharum officinarium L.) bagasse binderless particleboard: Effect of hot pressing time study. *Mater. Today Proc.* **2020**, *31*, 313–317. [CrossRef]
4. Robledo-Ortíz, J.R.; Martín Del Campo, A.S.; Blackaller, J.A.; González-López, M.E.; Pérez Fonseca, A.A. Valorization of sugarcane straw for the development of sustainable biopolymer-based composites. *Polymers* **2021**, *13*, 3335. [CrossRef] [PubMed]
5. Gürü, M.; Tekeli, S.; Bilici, I. Manufacturing of urea-formaldehyde-based composite particleboard from almond shell. *Mater. Des.* **2006**, *27*, 1148–1151. [CrossRef]
6. Kowaluk, G.; Szymanowski, K.; Kozlowski, P.; Kukula, W.; Sala, C.; Robles, E.; Czarniak, P. Functional Assessment of Particleboards Made of Apple and Plum Orchard Pruning. *Waste Biomass Valorization* **2019**, *11*, 2877–2886. [CrossRef]
7. Papadopoulos, A.N.; Hill, C.A.S.; Gkaraveli, A.; Ntalos, G.A.; Karastergiou, S.P. Bamboo chips (Bambusa vulgaris) as an alternative lignocellulosic raw material for particleboard manufacture. *Holz. Roh. Werkst.* **2004**, *62*, 36–39. [CrossRef]
8. Grigoriou, A.H. Straw as alternative raw material for the surface layers of particleboards. *Holzforsch. Holzverwert.* **1998**, *50*, 32–34.
9. Wang, D.; Sun, X.S. Low density particleboard from wheat straw and corn pith. *Ind. Crops Prod.* **2002**, *15*, 43–50. [CrossRef]
10. Nemli, G.; Kirci, H.; Serdar, B.; Ay, N. Suitability of kiwi (Actinidia sinensis Planch.) prunings for particleboard manufacturing. *Ind. Crops Prod.* **2003**, *17*, 39–46. [CrossRef]

11. Bekalo, S.A.; Reinhardt, H.W. Fibers of coffee husk and hulls for the production of particleboard. *Mater. Struct. Constr.* **2010**, *43*, 1049–1060. [CrossRef]
12. Papadopoulos, A. The potential for using flax (Linum usitatissimum L.) shiv as a lignocellulosic raw material for particleboard. *Ind. Crops Prod.* **2003**, *17*, 143–147. [CrossRef]
13. de Lima Mesquita, A.; Barrero, N.G.; Fiorelli, J.; Christoforo, A.L.; De Faria, L.J.G.; Lahr, F.A.R. Eco-particleboard manufactured from chemically treated fibrous vascular tissue of acai (Euterpe oleracea Mart.) Fruit: A new alternative for the particleboard industry with its potential application in civil construction and furniture. *Ind. Crops Prod.* **2018**, *112*, 644–651. [CrossRef]
14. Nadhari, W.N.A.W.; Ishak, N.S.; Danish, M.; Atan, S.; Mustapha, A.; Karim, N.A.; Hashim, R.; Sulaiman, O.; Yahaya, A.N.A. Mechanical and physical properties of binderless particleboard made from oil palm empty fruit bunch (OPEFB) with addition of natural binder. *Mater. Today Proc.* **2020**, *31*, 287–291. [CrossRef]
15. Juliana, A.H.; Paridah, M.T.; Rahim, S.; Nor Azowa, I.; Anwar, U.M.K. Properties of particleboard made from kenaf (Hibiscus cannabinus L.) as function of particle geometry. *Mater. Des.* **2012**, *34*, 406–411. [CrossRef]
16. Anonym Czas na aronię! Available online: http://polskiesuperowoce.pl/67632-czas-na-aronie (accessed on 1 November 2021).
17. Anonym Sady Ogrody. Rynek Owoców i Warzyw. Available online: www.sadyogrody.pl (accessed on 1 November 2021).
18. Bowen-Forbes, C.S.; Zhang, Y.; Nair, M.G. Anthocyanin content, antioxidant, anti-inflammatory and anticancer properties of blackberry and raspberry fruits. *J. Food Compos. Anal.* **2010**, *23*, 554–560. [CrossRef]
19. Jurikova, T.; Mlcek, J.; Skrovankova, S.; Sumczynski, D.; Sochor, J.; Hlavacova, I.; Snopek, L.; Orsavova, J. Fruits of black chokeberry aronia melanocarpa in the prevention of chronic diseases. *Molecules* **2017**, *22*, 944. [CrossRef]
20. EN 326-2:2010+A1 Wood-based panels. Sampling, cutting and inspection. In *Initial Type Testing and Factory Production Control*; European Committee for Standardization: Brussels, Belgium, 2014.
21. EN 326-1 Wood-based panels. Sampling, cutting and inspection. In *Sampling and Cutting of Test Pieces and Expression of Test Results*; European Committee for Standardization: Brussels, Belgium, 1993.
22. *EN 310 Wood-Based Panels—Determination of Modulus of Elasticity in Bending and of Bending Strength*; European Committee for Standardization: Brussels, Belgium, 1993.
23. *EN 319 Particleboards and Fibreboards—Determination of Tensile Strength Perpendicular to the Plane of the Board*; European Committee for Standardization: Brussels, Belgium, 1993.
24. *EN 312 Particleboards—Specifications*; European Committee for Standardization: Brussels, Belgium, 2010.
25. *EN 323 Wood-Based Panels—Determination of Density*; European Committee for Standardization: Brussels, Belgium, 1993.
26. *EN 317 Particleboards and Fibreboards—Determination of Swelling in Thickness after Immersion in Water*; European Committee for Standardization: Brussels, Belgium, 1993.
27. *EN 382-1 Fiberboards. Determination of Surface Absorption. Test Method for Dry Process Fiberboards*; European Committee for Standardization: Brussels, Belgium, 1993.
28. Krzysik, F. *Nauka o drewnie*; Państwowe Wydaw. Nauk.: Warsaw, Poland, 1975; pp. 97–103.
29. Yemele, M.C.N.; Blanchet, P.; Cloutier, A.; Koubaa, A. Effects of bark content and particle geometry on the physical and mechanical properties of particleboard made from black spruce and trembling aspen bark. *For. Prod. J.* **2008**, *58*, 48–56.
30. Lakreb, N.; As, N.; Gorgun, V.; Sen, U.; Gomes, M.G.; Pereira, H. Production and characterization of particleboards from cork-rich Quercus cerris bark. *Eur. J. Wood Wood Prod.* **2018**, *76*, 989–997. [CrossRef]
31. Mirski, R.; Kawalerczyk, J.; Dziurka, D.; Wieruszewski, M.; Trocinski, A. Effects of using bark particles with various dimensions as a filler for urea-formaldehyde resin in plywood. *BioResources* **2020**, *15*, 1692–1701.
32. Wronka, A.; Kowaluk, G. Influence of density on selected properties of furniture particleboards made of raspberry Rubus idaeus L. lignocellulosic particles. *Ann. WULS For. Wood Technol.* **2019**, *105*, 113–124. [CrossRef]
33. Lunguleasa, A.; Dumitrascu, A.E.; Spirchez, C.; Ciobanu, V.D. Influence of the strand characteristics on the properties of oriented strand boards obtained from resinous and broad-leaved fast-growing species. *Appl. Sci.* **2021**, *11*, 1784. [CrossRef]
34. Wong, E.D.; Zhang, M.; Wang, Q.; Han, G.; Kawai, S. Formation of the density profile and its effects on the properties of fiberboard. *J. Wood Sci.* **2000**, *46*, 202–209. [CrossRef]
35. Hashim, R.; Saari, N.; Sulaiman, O.; Sugimoto, T.; Hiziroglu, S.; Sato, M.; Tanaka, R. Effect of particle geometry on the properties of binderless particleboard manufactured from oil palm trunk. *Mater. Des.* **2010**, *31*, 4251–4257. [CrossRef]
36. Sackey, E.K.; Semple, K.E.; Oh, S.W.; Smith, G.D. Improving core bond strength of particleboard through particle size redistribution. *Wood Fiber Sci.* **2008**, *40*, 214–224.

Article

Enhancing Thermal and Mechanical Properties of Ramie Fiber via Impregnation by Lignin-Based Polyurethane Resin

Sucia Okta Handika [1], Muhammad Adly Rahandi Lubis [2,*], Rita Kartika Sari [1,*], Raden Permana Budi Laksana [2], Petar Antov [3], Viktor Savov [3], Milada Gajtanska [4,*] and Apri Heri Iswanto [5]

1. Department of Forest Products, Faculty of Forestry and Environment, IPB University, Bogor 16680, Indonesia; sucia_okta@apps.ipb.ac.id
2. Research Center for Biomaterials, National Research and Innovation Agency, Cibinong 16911, Indonesia; raden.budi.biomaterial@gmail.com
3. Faculty of Forest Industry, University of Forestry, 1797 Sofia, Bulgaria; p.antov@ltu.bg (P.A.); victor_savov@ltu.bg (V.S.)
4. Faculty of Wood Sciences and Technology, Technical University in Zvolen, 96001 Zvolen, Slovakia
5. Department of Forest Product, Faculty of Forestry, Universitas Sumatera Utara, Medan 20155, Indonesia; apri@usu.ac.id
* Correspondence: marl@biomaterial.lipi.go.id (M.A.R.L.); rita_kartikasari@apps.ipb.ac.id (R.K.S.); gajtanska@tuzvo.sk (M.G.)

Citation: Handika, S.O.; Lubis, M.A.R.; Sari, R.K.; Laksana, R.P.B.; Antov, P.; Savov, V.; Gajtanska, M.; Iswanto, A.H. Enhancing Thermal and Mechanical Properties of Ramie Fiber via Impregnation by Lignin-Based Polyurethane Resin. *Materials* **2021**, *14*, 6850. https://doi.org/10.3390/ma14226850

Academic Editor: Francisco Javier Espinach Orús

Received: 18 October 2021
Accepted: 10 November 2021
Published: 13 November 2021

Publisher's Note: MDPI stays neutral with regard to jurisdictional claims in published maps and institutional affiliations.

Copyright: © 2021 by the authors. Licensee MDPI, Basel, Switzerland. This article is an open access article distributed under the terms and conditions of the Creative Commons Attribution (CC BY) license (https://creativecommons.org/licenses/by/4.0/).

Abstract: In this study, lignin isolated and fractionated from black liquor was used as a pre-polymer to prepare bio-polyurethane (Bio-PU) resin, and the resin was impregnated into ramie fiber (*Boehmeria nivea* (L.) Gaudich) to improve its thermal and mechanical properties. The isolated lignin was fractionated by one-step fractionation using two different solvents, i.e., methanol (MeOH) and acetone (Ac). Each fractionated lignin was dissolved in NaOH and then reacted with a polymeric 4,4-methane diphenyl diisocyanate (pMDI) polymer at an NCO/OH mole ratio of 0.3. The resulting Bio-PU was then used in the impregnation of ramie fiber. The characterization of lignin, Bio-PU, and ramie fiber was carried out using several techniques, i.e., Fourier-transform infrared spectroscopy (FTIR), differential scanning calorimetry (DSC), thermogravimetric analysis (TGA), pyrolysis-gas-chromatography-mass-spectroscopy (Py-GCMS), Micro Confocal Raman spectroscopy, and an evaluation of fiber mechanical properties (modulus of elasticity and tensile strength). Impregnation of Bio-PU into ramie fiber resulted in weight gain ranging from 6% to 15%, and the values increased when extending the impregnation time. The reaction between the NCO group on Bio-PU and the OH group on ramie fiber forms a C=O group of urethane as confirmed by FTIR and Micro Confocal Raman spectroscopies at a wavenumber of 1600 cm^{-1}. Based on the TGA analysis, ramie fiber with lignin-based Bio-PU had better thermal properties than ramie fiber before impregnation with a greater weight residue of 21.7%. The mechanical properties of ramie fiber also increased after impregnation with lignin-based Bio-PU, resulting in a modulus of elasticity of 31 GPa for ramie-L-isolated and a tensile strength of 577 MPa for ramie-L-Ac. The enhanced thermal and mechanical properties of impregnated ramie fiber with lignin-based Bio-PU resins could increase the added value of ramie fiber and enhance its more comprehensive industrial application as a functional material.

Keywords: fractionated lignin; bio-polyurethane resin; ramie fiber; impregnation; thermal stability; mechanical properties

1. Introduction

Polyurethane (PU) resin is a polymer that contains urethane linkages (R−N−H−C=O−R) in the main polymer chain. PU can be functionalized with reactive groups such as isocyanate, amine, epoxy, acrylate, or carboxylic acid groups [1]. Modified PU resins are widely used in various applications, such as coatings, foams, medicinal products, thermal insulation, and adhesives [2–5]. Due to its superior properties compared to other polymers, such as high mechanical strength, flexibility at low temperatures, and the ability to be formed as a rigid, semi-rigid, or flexible foam with various densities, PU has become one of the most important type of resins used in various value-added applications, with an estimated global market volume of approximately 24 million tons in 2020 [6,7]. The synthesis of PU is carried out through a condensation reaction between isocyanates and polyols, where the main ingredients of the polymer to produce polyurethane are derived from refining crude oil and coal [8]. Hence, the main environmental problem in the production process is related to the formation of toxic substances such as polyisocyanate and its phosgene intermediate. The increased industrial PU demands and growing environmental concerns have posed new requirements related to the utilization of raw materials from renewable, bio-based sources to reduce declining fossil reserves. The environmental approaches in the development of sustainable PUs have resulted in the replacement of petrochemical polyols with bio-polyols, such as vegetable oil, oil waste, or lignocellulosic biomass [9]. Major techniques for converting vegetable oils to polyols are hydroformylation, transesterification/amidation, epoxidation/oxirane-ring opening, and ozonolysis [10–13]. Lignocellulosic biomass can be converted to liquid polyols through liquefaction processes or oxypropylation for polyurethane applications [14]. Bio-polyurethanes (Bio-PU) can also be obtained by using polyols from renewable materials such as lignin and tannin for various applications [15,16].

Lignin is the second most abundant polymer source in nature after cellulose, accounting for around 15–30% in wood biomass (bark typically contains large amounts of lignin ca. 40%) and 10–20% in grass biomass [17–19]. Nowadays, lignin is mainly produced as a waste or by-product of the pulp and paper industry with an estimated annual volume of about 50–70 million tons of which less than 2% is converted into value-added applications [20,21]. Lignin has a heterogeneous, amorphous structure, and consists of three-dimensional non-uniform polymeric tissue, constructed with repeating units [9]. Lignin is a complex aromatic polymer of three types of phenylpropane units (monolignols), namely p-hydroxyphenyl, syringyl, and guaiacyl. These units are randomly linked by ether and carbon bonds, forming a three-dimensional structural network with several reactive groups in structures, such as phenolic/aliphatic hydroxyl, methoxyl, carbonyl, and carboxyl [22]. The hydroxyl group and free position in the aromatic rings are the principal functions of lignin.

Lignin has emerged as a good alternative to polyol in resins, foams, and adhesives The NCO content of lignin Bio-PU decreases compared to the unmodified PU batch, which indicates that the incorporation of lignin in Bio-PUs leads to efficient crosslinking and the presence of less monomer in the final system. The incorporation of lignin results in an increased molecular weight of the lignin Bio-PU, which leads to the successful incorporation of lignin into the backbone of the polyurethane chain. Thus, the overall molecular weight of the adhesive is also increased [22]. Lignin-based, cross-linked PUs prepared from low- and medium-molecular-weight lignin fractions are characterized by high tensile strength and a high modulus of elasticity. However, their brittleness increases with the increased lignin content. The molecular weight of lignin can be altered by crosslinking with various solvents, which can tune the mechanical properties of the lignin-based PUs. Methanol fractionation of lignin increases the molecular weight (by removing the low-molecular-weight fraction), polydispersity, char yield, and glass transition temperature, and acetone fractionation improves the role of lignin as an antioxidant. Acetone-soluble lignin fractionation-based PU exhibits improved thermal stability due to crosslinking the lignin macromonomers [23].

The use of lignin as a Bio-PU pre-polymer follows two approaches: Lignin used directly without chemical modification or combined with other polymers, and using lignin after chemical modifications, such as esterification and etherification reactions to ensure hydroxyl groups react more easily [24]. Isolated lignin has aliphatic and hydroxyl phenolic groups, which provide excellent bonding to isocyanates. Most of the phenolic hydroxyls are linked to the adjacent phenylpropane units. In addition, chemical modifications such as methylolation and phenolation increase its chemical reactivity to formaldehyde and thus lignin can also act as a formaldehyde scavenger in conventional thermosetting formaldehyde-based adhesives and has unique characteristics with comprehensive strength and thermal stability because of the non-crystalline network structure [25–28].

Ramie (*Boehmeria nivea* (L.) Gaudich) is an abundant natural fiber that is easy to obtain, renewable, and environmentally sustainable as a resource material. Ramie fiber exhibits satisfactory strength and stiffness properties, low cost, and relatively high annual production [29]. Ramie fiber also has good decay resistance and the ramie-based composite products have twice the strength of regular wood fiber [30]. Ramie fiber has a tensile strength of around 95 MPa, which is higher than cotton and hemp [31]. Ramie fiber contains cellulose, hemicellulose, lignin, ash, and wax in different amounts, which cause differences in physical, mechanical, and thermal properties between ramie fiber and other cellulosic fibers. To obtain optimal composite characteristics, it is necessary to select the physical, mechanical, and thermal properties of the raw materials used [32]. Ramie fiber has been widely used in automotive, interior design, and furniture applications [33]. However, the main disadvantages of ramie fiber are its low thermal stability and poor flame retardancy, which limit the use of ramie as a raw material for functional textiles. Chemical treatments such as alkali, acetylation, and bleaching [34], as well as biological treatments have been performed to overcome those drawbacks [35]. In addition, the mechanical properties of ramie fiber often decrease due to the removal of cellulose, hemicellulose, and lignin after modifications.

One way to enhance both the thermal stability and mechanical properties of ramie fiber is by using the impregnation method. Impregnation is an important method in composite manufacturing. Different types of natural fibers have been impregnated with polypropylene, polylactic acid, and aluminum hydroxide [36,37]. The use of lignin derived from black liquor as polyols in the synthesis of bio-PU represents a sustainable approach to reduce waste from the kraft pulping process. The resulting bio-PU was used as an impregnation material aimed at improving the fiber thermal and mechanical characteristics. The impregnation of natural fibers can be simplified as a flow of viscous chemicals into porous materials. However, to the best of the authors' knowledge, no studies have reported the impregnation of ramie fiber to enhance its thermal and mechanical properties.

Therefore, the aim of this research work was to develop bio-PU resin derived from lignin for the impregnation of ramie fiber and investigate the properties of Bio-PU resin and the performance of impregnated ramie fiber, particularly its thermal stability and mechanical properties. The analyses used included functional group analysis using Fourier-transform infrared spectroscopy (FTIR), thermal analysis using Differential Scanning Calorimetry (DSC), thermal stability analysis using Thermogravimetric (TGA), and Pyrolysis-Gas-Chromatography-Mass-Spectroscopy (Py-GCMS) for determining lignin components, and the evaluation of tensile strength and the modulus of elasticity of ramie fiber using a universal testing machine (UTM).

2. Materials and Methods

2.1. Materials

Black liquor obtained from the kraft pulping process of *Acacia mangium* was provided by Tanjung Enim Lestari Pulp and Paper Company, Muara Enim, Indonesia (Figure 1). Standardized kraft lignin (guaiacyl lignin) from Sigma Aldrich (CAS No. 8068-05-1, Burlington, MA, USA) was used as a standard. Ramie fiber (*Boehmeria nivea* (L.) Gaudich) was obtained from Rabersa Company, Wonosobo, Indonesia. Hydrochloric acid (HCl 37%, analytical grade, Merck, Darmstadt, Germany), sulfuric acid (H_2SO_4 95–97%, analytical grade, Merck, Darmstadt, Germany), methanol (analytical grade, Merck, Darmstadt, Germany), acetone (analytical grade, Merck, Darmstadt, Germany), dioxane (1,4-dioxane, analytical grade, Merck, Darmstadt, Germany), and distilled water were used for isolation and fractionation of lignin. Polymeric 4,4-methane diphenyl diisocyanate (pMDI, ±31% NCO content), purchased from the company Anugerah Raya Kencana (ARK, Tangerang, Indonesia), and sodium hydroxide (NaOH 20%) were used to prepare the Bio-PU resin.

Figure 1. The materials used (**a**) black liquor; (**b**) pMDI; (**c**) ramie fibers.

2.2. Isolation and Fractionation of Lignin

Lignin was isolated via the acid precipitation method [38]. Approximately 200 g of black liquor and 2000 mL of distilled water were stuffed into a plastic jar. HCl 1 M was added while stirring to adjust the pH of black liquor from around 12 to 2. The solution was kept for 24 h at room temperature (25 °C) to separate the filtrate and residue. Further, the decantation process was carried out three times, and then the remaining sludge was refrigerated for 24 h. The precipitate was filtered with a Buchner funnel and kept in an oven for 24 h at 45 °C. Then the precipitate was powdered using a mortar and filtered with a size 60 mesh.

Lignin fractionation was carried out via a single-step fractionation using two different solvents, namely methanol (MeOH) and acetone (Ac) with some modifications [39]. The lignin sample was weighed and 20 g was mixed with 300 mL of solvent at a ratio of 1:15. The solution then was stirred with a magnetic stirrer for 24 h with a stirring speed of 200 rpm. The soluble and insoluble fractions were separated using a Buchner funnel and filter paper. After that, the dissolved fraction was concentrated with a rotary evaporator (Rotavapor® R-300, Buchi, Flawil, Switzerland). Furthermore, the lignin samples were dried for 24 h at 45 °C.

2.3. Lignin Characterization

Basic properties of lignin such as yield, moisture content, ash content, purity, and total phenolic hydroxyl groups were characterized using various analytical methods. The yield of lignin was calculated by dividing the weight of isolated lignin and black liquor. The moisture content of lignin was determined by drying lignin samples in an oven at 105 °C for 24 h.

The ash content of lignin was determined by heating the sample in a furnace for 6 h at a temperature of 525 ± 25 °C. Ash content of lignin was calculated according to the following equation [40].

$$\text{Ash content (\%)} = \frac{(C - A)}{B} \times 100\% \qquad (1)$$

Description:
A = weight of empty porcelain cup (g).
B = weight of lignin sample (g).
C = weight of oven-dried sample and porcelain cup (g).

The purity of isolated lignin was measured according to the standard procedure [41,42]. Approximately 0.3 g of lignin was added to a vial bottle. After that, 3 mL of H_2SO_4 72% was added and mixed with a magnetic stirrer at 150 rpm for 2 h. A blank solution was prepared by adding 3 mL of H_2SO_4 72% and 84 mL of distilled water. Furthermore, the sample was heated by autoclave for 1 h at a temperature of 121 °C. The solution was filtered using an IG3 filter glass (CTE33, IWAKI, Tohoku, Japan). The residue (acid-insoluble lignin) was put in the oven at 105 °C for 24 h. The filtrate was diluted in a test tube with a blank solution of 13 times dilution. The solution was stirred with a vortex, and its absorbance was measured using a UV-Vis spectrophotometer (UV-1800, Shimadzu, Kyoto, Japan) at a wavelength of 240 nm. Acid soluble lignin (ASL) and acid-insoluble lignin (AIL) were calculated using the following equations [41,42].

$$\text{ASL (\%)} = \frac{UVabs \times volume\ filtrate \times dilution}{\varepsilon \times A \times cuvette\ length} \times 100\% \qquad (2)$$

$$\text{AIL (\%)} = \text{AIR(\%)} - \text{Ash content (\%)} \qquad (3)$$

$$\text{Acid Insoluble Residue (AIR) (\%)} = \frac{C - B}{A} \times 100\% \qquad (4)$$

Description:
ε = absorptivity constant of biomass at specific wavelength (L/g·cm).
A = weight of sample without moisture content (g).
B = dry weight of IG3 filter glass (g).
C = dry weight of IG3 filter glass and AIL (g).

The total phenolic hydroxyl groups of lignin were determined according to the published methods [43]. The lignin sample was prepared by dissolving 20 mg of lignin in 10 mL of dioxane and 10 mL of NaOH 0.2 M. The solution was filtered using 0.45 μm nylon to remove insoluble particles. Further, 4 mL of the initial solution was diluted by adding 50 mL of solvents. Three different solvents were used: NaOH 0.2 M, pH 6 buffer, and pH 12 buffer. Dilution was carried out until the final concentration of the solution was 0.08 g/L. UV measurements were carried out using a UV-Vis spectrophotometer in the 200–600 nm range using a buffer solution of pH 6 as a reference. The maximum absorption occurred at the absorbance of 300 nm and 350 nm, which were used for calculations. The total phenolic hydroxyl group was calculated using the following equation [43]:

$$\text{Total OH (mmol g}^{-1}\text{)} = (0.425 \times A_{300nm}\ (\text{NaOH}) + 0.182 \times A_{350nm}\ (\text{NaOH})) \times a \qquad (5)$$

Description:
A = absorbance.
a = correction term (L/g·cm) = $1/(c \times l) \times 10/17$.
c = concentration of lignin solution (g/L).
l = path length (cm).

2.4. Preparation of Bio-Polyurethane Resin

Bio-polyurethane (Bio-PU) was prepared using fractionated lignin and pMDI at an NCO/OH mole ratio of 0.3. The ratio of NCO/OH was calculated by dividing the mole of NCO of pMDI and the mole of phenolic OH in lignin. The fractionated lignin was dissolved in a NaOH 20% solution with a ratio of 1:10 (w/v). A solution of pMDI in acetone (8% w/v) was added to the lignin fractions. The mixture was mechanically stirred with a stirring speed of 500 rpm for 30 min for the polymerization reaction. The obtained Bio-PU resin was stored in vials before the characterization and impregnation of ramie fibers. The lignin-based Bio-PU formation reaction scheme is shown in Figure 2.

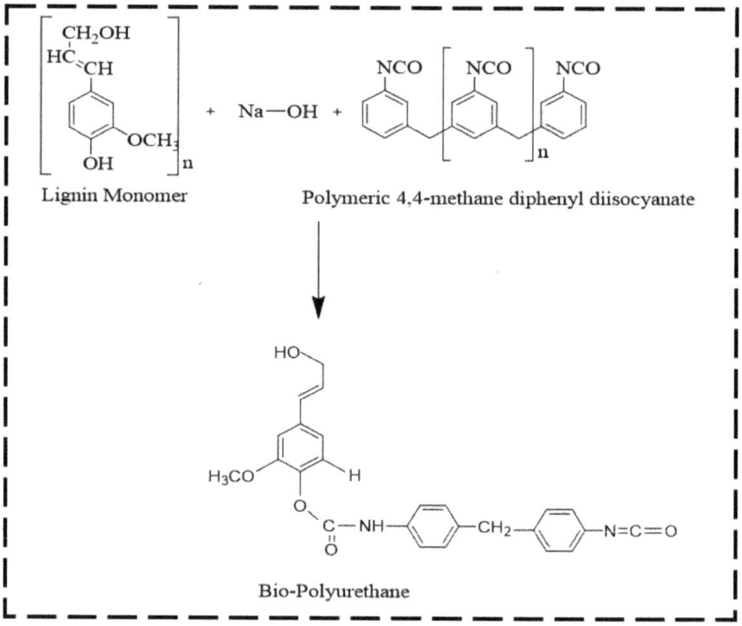

Figure 2. The possible reaction scheme of lignin-based Bio-PU resin.

2.5. Impregnation of Ramie Fiber with Bio-PU

The impregnation of ramie fibers was carried out in a 1 L vacuum chamber with a vacuum pump (VC0918SS, VacuumChambers.eu, Białystok, Poland) according to the published work [44]. The initial weight of ramie fiber was recorded before vacuum impregnation. Five grams of ramie fibers were immersed in 50 mL of Bio-PU resin and impregnated at $25 \pm 2\ ^\circ C$ under the pressure of 50 kPa for 30, 60, and 90 min. The impregnated fiber was then dried in an oven at $60\ ^\circ C$ for 24 h. Dried ramie fibers were weighed to determine the weight gain after impregnation. The weight gain (%) was calculated by dividing the mass of ramie fibers after impregnation by the initial mass of ramie fibers. The impregnated ramie fibers were then stored in a zip-lock plastic bag for further testing.

2.6. Characterization of Lignin-Based Bio-PU Resin

Functional groups of isolated lignin, fractionated lignin, and lignin-based Bio-PU resin were investigated using Fourier-transform infrared spectroscopy (FTIR) (SpectrumTwo, PerkinElmer, Waltham, MA, USA) by applying the universal attenuated total reflectance (UATR) method. The average accumulation was recorded as 16 scans at a resolution of 4 cm^{-1} with wavenumber in a range of 4000–400 cm^{-1} at $23 \pm 2\ ^\circ C$.

Thermal properties of isolated lignin, fractionated lignin, and lignin-based Bio-PU were analyzed using Differential Scanning Calorimetry (DSC4000, PerkinElmer, Waltham,

MA, USA). Around 4 mg of the sample was weighed in a standard aluminum pan (40 µL). The samples were heated in a nitrogen atmosphere with a flow rate of 20 mL/min and temperatures ranging from −50 to 350 °C with a heating rate of 10 °C/min. The Tp_1 and Tp_2 values were calculated automatically using Pyris 11 software (Version 11.1.1.0492, PerkinElmer, Shelton, CT, USA).

Thermogravimetric analysis (TGA) was performed using the TGA instrument (TGA 4000, Perkin Elmer, Waltham, MA, USA). Around 20 mg of the sample was weighed in a standard ceramic crucible and heated in a nitrogen atmosphere with a flow rate of 20 mL/min. The temperature used in heating ranged from 25 to 750 °C with a heating rate of 10 °C/min. The percent of weight loss, weight loss rate, and residue were analyzed with the help of Pyris software.

The analysis of lignin components was conducted using pyrolysis-gas-chromatography-mass-spectroscopy (Py-GCMS, Shimadzu, Kyoto, Japan). Lignin samples of 500–600 µg were put into the SF PYI-EC50F eco-cup, covered with glass wool, and Py-GCMS analyzed the samples. The eco-cup was pyrolyzed at 500 °C for 0.1 min using multi-shot pyrolysis (EGA/PY-3030D) connected to the GC/MS QP-2020 NX system (Shimadzu, Kyoto, Japan) and equipped with an SH-Rxi-5Sil column MS, which had a film thickness of 30 mm × 0.25 mm id 0.25 µm, 70 eV electrons, and helium as a carrier gas. The pressure used was 20.0 kPa (15.9 mL/min, column flow 0.61 mL/min). The temperature profile for GC was 50 °C, which was left for 1 min and then the temperature was increased to 280 °C with a heating rate of 5 °C/min, and the temperature was held at 280 °C for 13 min. Pyrolysis products were identified by comparing the retention time and mass spectrum data in the 2017 NIST LIBRARY program.

Rheological properties of lignin-based Bio-PU resin were investigated using a rotational rheometer (RheolabQC, AntonPaar, Graz, Austria). The measurement was conducted at room temperature using a concentric cylinder coupled with spindle No. 27. The viscosity and cohesion of lignin-based Bio-PU resin were obtained directly after the measurement.

2.7. Evaluation of Ramie Fibers Properties

FTIR, DSC, and TGA were performed as described in Section 2.6 to investigate the properties of ramie fibers before and after impregnation.

Micro confocal Raman hyperspectral imaging spectrometer (LabRAM HR Evolution, Horiba, Kyoto, Japan) was used to confirm the formation of urethane linkages on ramie fiber after impregnation. Ramie fiber before and after impregnation was put on the glass slide, and the image was captured using an NIR objective lens at 100-times magnification. The sample image then was bombarded with a laser at a wavelength of 785 nm using 600 grating. The measurement was performed at a Raman shift of 200–4000 cm^{-1}, an acquisition time of 10 s, and accumulations of 10 at room temperature.

The determination of mechanical properties of ramie fiber was carried out in accordance with the ASTM D3379-75 standard using a universal testing machine (UTM 5kN, Shimadzu, Kyoto, Japan) [45]. The specimen used is a single fiber separated by strand bonds. The specimen length was 20–30 mm, and the total fiber length was about three times the length of the specimen. The specimens were tested at a temperature of 23 °C. The tensile strength and elastic modulus of ramie fibers before and after impregnation were calculated according to the ASTM D3379-75 [45].

3. Results and Discussion

3.1. Basic Properties of Isolated Lignin

Black liquor used in this study was obtained by the kraft pulping process of the pulp and paper industry. The characteristics of black liquor and isolated lignin are shown in Table 1. The yield of isolated lignin obtained was 35.88%, which is lower than the one-stage lignin precipitation by Hermiati et al. [38], namely 45.76%. The yield of isolated lignin is affected by the lignin content in black liquor from the kraft pulping process and the precipitation process during isolation. A complete precipitation reaction will produce a higher lignin yield.

The moisture content of isolated lignin (L-isolated) was 5.07%. The moisture content of lignin was relatively low compared to the published work [46], which was 8.05%. Ash content is the residual content of combustion of inorganic materials found in lignin. The low ash content of 0.31% indicated that the three-stage decantation process to obtain the lignin in this study produced fewer impurities compared to other studies that used a one-stage decantation process and obtained an ash content of 8.25–19.19% [38,47].

Table 1. Characteristics of lignin isolated from black liquor.

Analysis Parameters	Value	References
Moisture Content of Black Liquor (%)	27.81 ± 1.11	10.00 [48]
Solid Content of Black Liquor (%)	76.79 ± 0.64	65.00–85.00 [49,50]
pH of Black Liquor	12.14 ± 0.02	12.00–13.00 [51,52]
Yield of Lignin (%)	35.88 ± 1.81	45.76 [38]
Moisture Content of Lignin (%)	5.07 ± 0.71	8.05 [46]
Ash Content of Lignin (%)	0.31 ± 0.19	8.25–19.19 [38,47]
Acid-insoluble Lignin (AIL) (%)	82.54 ± 0.96	53.08 [38]
Acid-soluble Lignin (ASL) (%)	12.77 ± 0.67	7.26 [38]
Purity Levels of Lignin (%)	95.32 ± 0.61	60.34 [38]

The determination of lignin purity is based on AIL and ASL. As presented in Table 1, the AIL content was 82.54% and ASL content was 12.77%. The high AIL and ASL values determined can be attributed to the three-stage decantation process during isolation. The decantation process involved washing with distilled water to remove impurities and other substances, thus increasing the total lignin generated from the isolation process. The level of purity of the isolated lignin was relatively high, namely 95.32%. This result was higher than the published works that used a similar source of black liquor and acid precipitation methods [38], calculating around 60.34% of lignin purity. The difference in the level of lignin purity can be attributed to the differences in mineral content in lignin isolates.

The yield of dissolved fractionation is shown in Table 2. The yield of lignin fractionated with MeOH (L-MeOH) was 71.25% and lignin fractionated with acetone (L-Ac) was 70.57%. The results are in agreement with the published work [53], where the highest solubility of fractionated technical lignin was observed in methanol and acetone solvents of 70–80%. The solubility of this lignin in organic solvents depends on the type of lignin, aliphatic hydroxyl number, and molecular weight [54].

Table 2. The yield and total phenolic hydroxyl group of isolated and fractionated lignin.

Type of Lignin	Yield of Fractionated Lignin (%)	Total OH Group
L-Standard	-	8.109
L-Isolated	-	7.968
L-MeOH	71.25	7.399
L-Ac	70.57	7.645

Total phenolic hydroxyl groups of fractionated lignin were determined using the UV method. The UV method is a fast and easy way of measuring phenolic hydroxyl groups by giving the total number of hydroxyl (OH) groups without any structural specifications [43]. The total phenolic OH group of L-Standard, L-Isolated, and fractionated lignin are shown in Table 2. The highest total OH group was found in the L-Standard (8.109%), and the lowest was obtained in the L-MeOH (7.399%). The isolated lignin contains high phenolic hydroxyl groups and a condensed structure due to the kraft pulping process [39]. The OH group value was further used in the formulation of lignin-based Bio-PU resin with an NCO/OH mole ratio of 0.3.

3.2. Characterizations of Lignin

3.2.1. FTIR Analysis

FTIR spectra of isolated lignin and fractionated lignin are shown in Figure 3. Standard lignin was used as a comparison. The isolated lignin and fractionated lignin were similar to standard lignin in terms of functional groups. However, there were different intensities at specific wavenumbers of isolated lignin and fractionated lignin.

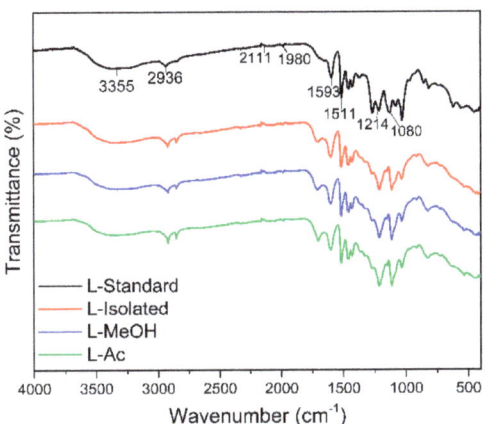

Figure 3. Typical FTIR spectra of lignin before and after fractionation.

The broad absorption band at 3550–3200 cm^{-1} is the O-H stretching vibration in phenolic and aliphatic O-H groups. The peaks were detected at 3355 cm^{-1} for standard lignin, 3347 cm^{-1} for isolated lignin, and 3327 cm^{-1}, and 3356.47 cm^{-1} for L-MeOH and L-Ac, respectively. The wavenumber at 2936–2917 cm^{-1} shows typical C-H stretching of CH_3 and CH_2, 1710–1698 cm^{-1} was assigned to C=O stretching from conjugate acid, 1601–1593 cm^{-1} was attributed to C=O stretching vibrations from skeletal aromatic, and 1514–1511 cm^{-1} represented the C-C stretching of the aromatic ring. The absorption bands at 1605–1600 cm^{-1} and 1515–1505 cm^{-1} were assigned to the aromatic ring vibration of the phenyl-propane (C9) skeleton [47]. The peaks at 1470–1460 cm^{-1} indicated the presence of C-H deformation (asymmetrical) of methyl, methylene, and methoxyl groups, which indicated the aromatic structure of lignin did not change after the isolation and fractionation process.

Standard lignin showed strong absorption bands at 1269 cm^{-1} and 1210–1220 cm^{-1}, assigned to the C-O group of guaiacyl, phenolic O-H, and ether in syringyl and guaiacyl. Meanwhile, isolated lignin and fractionated lignin have a strong absorption band at 1080–1030 cm^{-1}, indicating a C-O group in syringyl and guaiacyl. It is suspected that the standard lignin used is a guaiacyl lignin derived from softwood, which has a higher intensity at a wavenumber of 1269 cm^{-1}. Isolated lignin and fractionated lignin were thought to be syringyl-guaiacyl lignin derived from hardwood and have a higher intensity at a wavenumber of 1030 cm^{-1} [55]. This study showed that isolated lignin from black liquor and fractionated lignin did not damage the structure of lignin.

3.2.2. DSC Analysis

Thermal properties of isolated and fractionated lignin were analyzed using DSC analysis. Standard lignin was used as a comparison. In Figure 4, two endothermic peaks were detected during the heating process of lignin. The initial peak endothermic reaction (Tp_1) appeared at 56 °C for L-standard and 73 °C for L-isolated. Meanwhile, Tp_1 for fractionated lignin was in the range of 63–65 °C. The initial endothermic reaction is the beginning of the process to evaporate the water contained in lignin. The higher Tp_1 of isolated and fractionated lignin indicated that lignin had more water than standard lignin. The second endothermic reaction (Tp_2) is the final part of the endothermic process that changes the lignin structure and reduces stiffness (plasticization). This reaction is also known as the glass transition (Tg), which is the inflection point of the heat flow change of the heating cycle [56]. The Tg of lignin is generally in the range of 100–180 °C, and Tg from standard lignin, isolated lignin, and fractionated lignin were in the range of 136–158 °C.

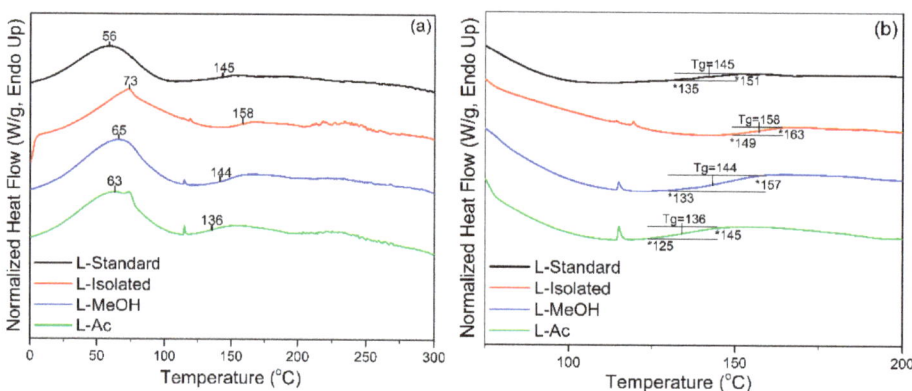

Figure 4. DSC thermograms of lignin before and after fractionation: (**a**) full thermogram, (**b**) enlarged thermogram.

3.2.3. TGA Analysis

The thermal degradation of lignin was determined by TGA-DTG analysis. The results of the TGA-DTG analysis are shown in Figure 5. There are three stages of lignin degradation based on the TGA-DTG analysis. The first stage occurs at a temperature of 25–100 °C, which causes lignin to lose weight due to water evaporation or reduction in moisture. At this stage, the derivative weight loss was around 0.2–1.0%/°C. The next step is weight reduction due to carbohydrate degradation, which occurs at a temperature of 120–250 °C. In this state, lignin lost weight up to 10% with a derivative weight loss of around 0.5%/°C. The main lignin degradation occurred at the broadest temperature range from 200 to 480 °C. Based on the DTG curve, the most significant lignin degradation occurs at a temperature range of 370 °C with a derivative weight loss reaching an average of 2.5%/°C.

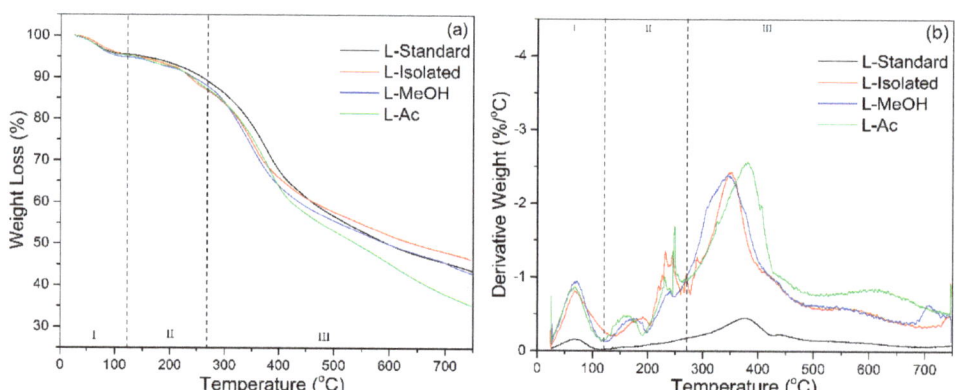

Figure 5. Thermal behavior of lignin before and after fractionation: (**a**) TGA, (**b**) DTG.

Lignin loses ~50% of its weight in the temperature interval of 400–650 °C (Table 3). The heterogeneous structure of lignins with a non-repetitive combination of carbon–carbon and ether linkages and with a broad molecular mass distribution results in the facile release of considerable amounts of volatile products. At temperatures above 500 °C, lignin degradation occurs, associated with the decomposition of the aromatic ring. At a temperature of 750 °C, the final percentage of residual lignin combustion was in the range of 31–46%. Isolated lignin had the most percentage of lignin residue (46.25%) and L-Ac had the smallest residual percentage (35.29%). Isolated lignin has better thermal stability compared to fractionated lignin. Fractionation of lignin broke down lignin macromolecules into smaller molecules, which resulted in lower thermal stability. It has been reported that lignin is more thermally stable, so it can be used as a synthesis agent for polyurethane and phenol-formaldehyde resin in high-temperature applications [57].

Table 3. Degradation temperature and weight loss of lignin detected by TGA.

Type of Lignin	$T_{WL10\%}$ (°C)	$T_{WL25\%}$ (°C)	$T_{WL50\%}$ (°C)	WL (%)	Residue (%)
L-standard	259	366	597	56.49	43.51
L-isolated	235	347	644	53.75	46.25
L-MeOH	242	341	596	57.06	42.94
L-Ac	239	351	540	64.71	35.29

3.2.4. Py-GCMS Analysis

Lignin is a complex polymer synthesized from three hydroxyl alcohols that differ in degree of methoxylation, namely p-coumaryl, coniferyl, and synapyl alcohols. Each monolignol produces a different type of lignin unit called p-hydroxyphenyl (H), guaiacyl (G), and syringyl (S). Pyrolysis gas chromatography/mass spectrometry (Py-GCMS) was performed as analysis to determine the lignin composition contained in isolated lignin and fractionated lignin. Pyrolysis is based on thermal fragmentation at high temperatures up to 500 °C and in an oxygen-free environment. The results of the Py-GCMS chromatogram of isolated lignin and fractionated lignin are shown in Figure 6. The results showed that isolated lignin and fractionated lignin produced more syringyl than guaiacyl and hydroxyphenyl, which were recorded at retention times of 21 min to 32.5 min. Meanwhile, standard lignin produced more guaiacyl, which became visible at retention times of 12.5 min to 28.5 min (Figure 6). Based on the result, the S/G ratio of isolated lignin is 0.96. Meanwhile, the fractionated lignin had a higher S/G ratio, which was 0.99 for L-MeOH and 1.01 for L-Ac. A similar trend was found for the S/GH ratio, where fractionated lignin had a greater S/G/H ratio than the isolated lignin (Table 4). The prominent peaks in the lignin chromatogram obtained were Guaiacol (G1), Syringol (S1), Guaiacol. 4-methyl (G2), Syringol-4-methyl (S2), and Guaiacol. 4-vinyl (G5) [58].

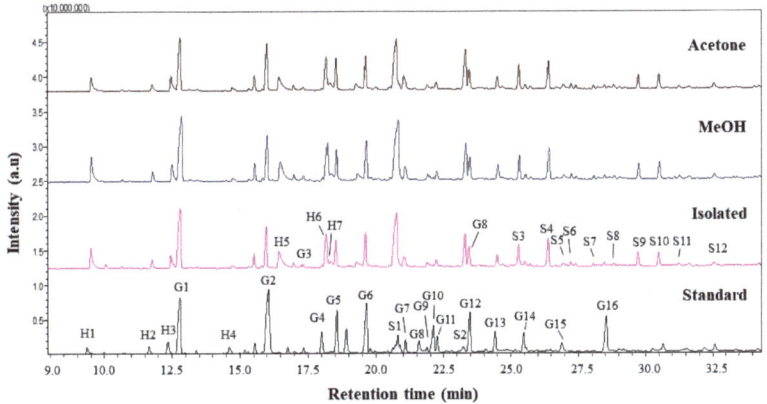

Figure 6. Py-GCMS chromatograms of lignin before and after fractionation.

Table 4. Pyrolysis products of lignin assessed by Py-GCMS.

No	RT (min)	Pyrolysis Product	Origin	Lignin-Isolated	Fractionated Lignin	
					MeOH	Ac
1	9.50	Phenol	H	3.93	4.17	2.85
2	11.76	Phenol. 2-methyl-	H	1.22	1.35	1.35
3	12.46	Phenol. 4 methyl	H	2.7	2.9	3.13
4	12.80	Guaiacol	G	14.24	14.87	10.86
5	15.97	Guaiacol. 4-methyl	G	6.79	6.66	8.7
6	16.44	Catechol	H	4.65	5.04	4.07
7	18.16	Catechol. 3-methoxy	H	7.28	6.58	6.72
8	18.30	Catechol. 4 methyl	H	1.44	0.2	1.84
9	19.62	Guaiacol. 4-vinyl	G	5.58	5.61	5.23
10	20.76	Syringol	S	16.77	17.5	14.38
11	23.30	Syringol-4-methyl	S	6.24	6.3	8.36
		LH (hydroxypenhyl)		21.22	20.24	19.96
		LG (guaiacyl)		39.84	39.94	39.51
		LS (syringyl)		38.52	39.44	39.9
		S/G		0.9669	0.9875	1.0099
		S/G/H		0.0456	0.0488	0.0506

3.3. Properties of Bio-Polyurethane Resin

In this study, the Bio-PU resin was designed as an impregnation material to improve the characteristics of ramie fiber. Therefore, the Bio-PU resin must have a low viscosity to be impregnated into the ramie fiber [59]. The viscosity and cohesion strength of Bio-PU resin was measured at 25 °C (Figure 7). The lowest viscosity value of 77.02 mPa·s was determined for Bio-PU L-Isolated, followed by Bio-PU L-MeOH and Bio-PU L-Ac, which had values of 100.71 and 223.58 mPa·s, respectively (Figure 7). The cohesion strength followed the results of viscosity, where the cohesion strength increased with greater viscosity of Bio-PU resin. Lower viscosity can increase the ability of Bio-PU resin to be impregnated and absorbed by ramie fiber. Viscosity is related to hydrogen bonds between isocyanate and polyol molecules (NCO/OH molar ratio). An increase in the NCO/OH ratio will result in more urethane bonds forming hard segments in Bio-PU and produce high viscosity [59,60].

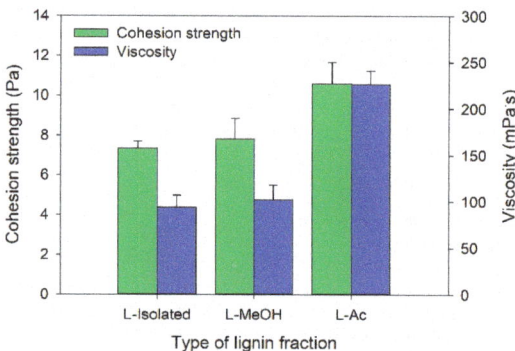

Figure 7. Viscosity and cohesion strength of Bio-PU from fractionated lignin.

The functional groups of Bio-PU resin are shown in Figure 8. A typical strong peak at the wavenumber of 2263 cm^{-1} indicated an isocyanate group (N=C=O) of pMDI. In the Bio-PU resin (L-isolated, L-MeOH, and L-Ac), no absorption of the N=C=O group was found, indicating that the N=C=O group reacted completely with the OH group of lignin during the preparation of PU resin. It has been reported that the isocyanate group derived from pMDI will react with the hydroxyl group of the polyol [61]. The reaction between fractionated lignin and pMDI in the formation of Bio-PU resin produces a specific functional group, namely the urethane group (R-NH-C=O-R) [62].

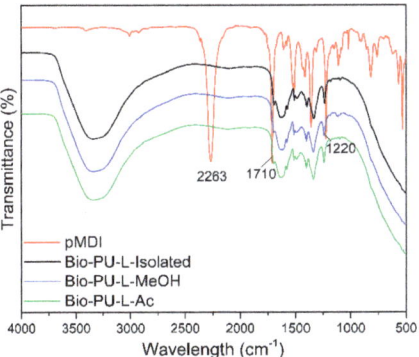

Figure 8. ATR-FTIR spectra of Bio-PU resin derived from fractionated lignin.

Bio-PU resin showed broad peaks at wavenumbers 3400–3200 cm^{-1}, indicating the formation of N-H groups in the aliphatic primary amine structure. In addition, this broad peak also overlapped with the -OH groups from the lignin sample and aqueous NaOH for lignin dissolution. The urethane bond formed from the lignin phenolic OH group and isocyanate (NCO group) from pMDI [63]. Another characteristic band in the formation of the Bio-PU resin at 1710–1685 cm^{-1} revealed the formation of C=O stretching from the urethane bond (R-NH-C=O-R), and the wavenumber of 1250–1020 cm^{-1} showed C-N stretching. It is indicated that the Bio-PU resin was successfully prepared from fractionated lignin (L-MeOH and L-Ac) as an alternative polyol.

Thermal properties of the Bio-PU resin were determined based on the results of the DSC analysis shown in Figure 9. The DSC thermograms show three-step endothermic reactions in the Bio-PU resin. The Tp_1 of Bio-PU (L-Isolated, L-MeOH, and L-Ac) ranged 56–63 °C, the Tp_2 ranged 103–109 °C, and the Tp_3 or Tg ranged 137–140 °C. The Tp_1 corresponded to the initial moisture evaporation in Bio-PU resin and the evaporation peak at Tp_2. The Tg value represents the polymer transition at a specific temperature [64]. The Tg of Bio-PU L-MeOH was 140 °C; meanwhile, the Tg of Bio-PU L-isolated and L-Ac was 137 °C. High Tg produces hard segments in Bio-PU due to the formation of urethane. Inter-urethane hydrogen bonds play an essential role in the stability of Bio-PU resin. The structure of the hard segment is more influential in thermal degradation than the soft segment [65].

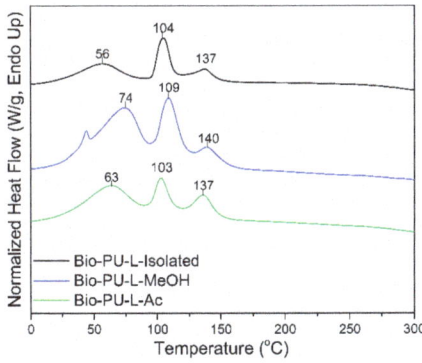

Figure 9. DSC thermograms of Bio-PU resin derived from fractionated lignin.

Thermogravimetric analysis was performed to evaluate the thermal degradation of lignin-based Bio-PU resin. TGA-DTG thermogram is shown in Figure 10. The initial weight loss occurred at a temperature of 60–110 °C, caused by water evaporation and evaporation of chemicals in unreacted lignin during the polymerization process [44,61]. Lignin-based Bio-PU lost 10% of the initial weight with a derivative weight loss of 1.5%/°C at a temperature of 80 °C. There was significant weight loss at temperatures above 160 °C up to 300 °C. The degradation peak occurred at a temperature of 250 °C, which indicated the initial decomposition of the urethane bond with a derivative weight loss of 2.0%/°C. Further degradation occurred at ~350 °C, which is oxidative degradation by the urethane group with a derivative weight loss of 0.75%/°C [66]. At a temperature of 450–600 °C, both the aromatic ring decomposition of lignin and the primary oxidative decomposition of lignin occurred [66,67].

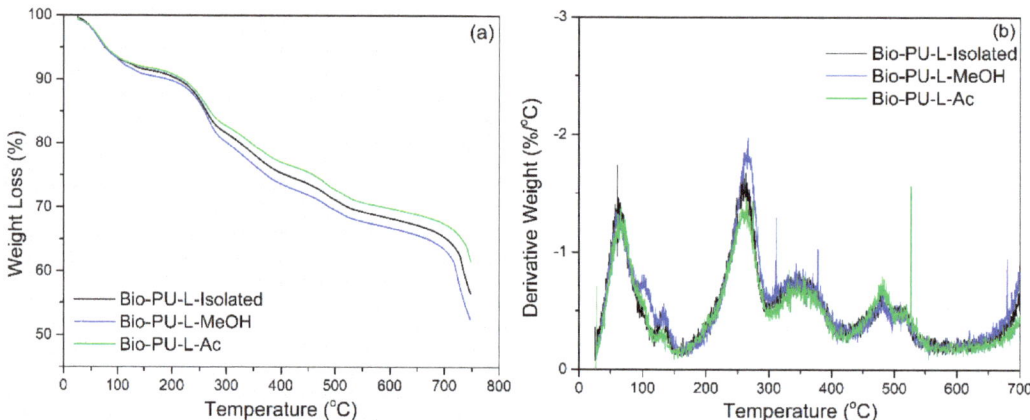

Figure 10. Thermal degradation of Bio-PU resin derived from lignin: (**a**) TGA, (**b**) DTG.

The weight residue after heating at 750 °C of each Bio-PU resin (L-Isolated, L-MeOH, and L-Ac) was in the range of 52–61% (Figure 10). Bio-PU L-Ac had the highest weight residue of 61.55 and bio-PU L-MeOH had the lowest weight residue of 52.76%. The bio-PU L-Ac showed lower thermal degradation compared to Bio-PU from L-Isolated and L-MeOH due to crosslinking the lignin macromonomers. The weight loss of Bio-PU from each fractionation of lignin was 25% at 300–400 °C during degradation of the urethane group. The weight loss of the fractionated lignin-based Bio-PU resin in this study was not more than 50%. Meanwhile, rigid or semi-rigid polyurethane resin has a percentage of weight loss at 340–525 °C of 62–75% [68]. This study showed that lignin fractionation using Acetone (Ac) could produce lignin-based Bio-PU resin with lower thermal degradation than isolated lignin.

3.4. Properties of Ramie Fiber Impregnated with Bio-PU Resin

The impregnation of ramie fiber with lignin-based Bio-PU was aimed at improving ramie fiber's thermal stability as a functional material. The weight gain of ramie fiber after impregnation was calculated as a way to investigate the influence of the impregnation process on ramie fiber. The weight gain of ramie fiber after impregnation is shown in Table 5. In general, the weight gain increased with longer impregnation time. Impregnation for 90 min resulted in more significant weight gain for each type of lignin-based Bio-PU used, followed by impregnation times of 60 and 30 min. Ramie impregnated with Bio-PU derived from L-isolated had the highest weight gain compared to Bio-PU L-MeOH and L-Ac. This could be due to the viscosity of Bio-PU L-isolated being lower than that of L-MeOH and L-Ac (Figure 6). The lower viscosity of Bio-PU will make it easier for the solution to be impregnated into the fiber. Ramie fiber has a fiber diameter of 40–60 μm [69].

This could facilitate the absorption of the Bio-PU solution and increase weight gain after Bio-PU impregnation into the ramie fiber.

Table 5. Weight gain of ramie fiber after impregnation with lignin-based Bio-PU at 30, 60, and 90 min.

Type	Weight Gain (%)		
	30 min	60 min	90 min
Ramie L-Isolated	12.38 ± 1.69	15.17 ± 0.36	15.93 ± 2.43
Ramie L-MEOH	6.25 ± 1.10	7.21 ± 3.17	8.62 ± 1.09
Ramie L-Ac	6.68 ± 0.74	8.26 ± 0.06	9.07 ± 0.62

The FTIR spectra of the impregnated ramie fiber investigated in this study are shown in Figure 11. In non-impregnated ramie fiber, there were six prominent peaks. Firstly, a wavenumber of 3330 cm^{-1} represented OH groups from cellulose, hemicellulose, and lignin that arrange the ramie fiber [70]. A wavenumber of 2897 cm^{-1} described C-H vibration carbohydrates [44]. A wavenumber of 1740 cm^{-1} indicated the presence of C=O vibration in the ester, carboxylate groups, and hemicellulose of ramie fiber [70–72]. Wavenumbers of 1607 cm^{-1}, 1424 cm^{-1}, and 1024 cm^{-1} showed the C-H stretching of carbohydrates, C-O stretching of cellulose, C-O vibrations, and O-H vibrations of cellulose, respectively [44,72].

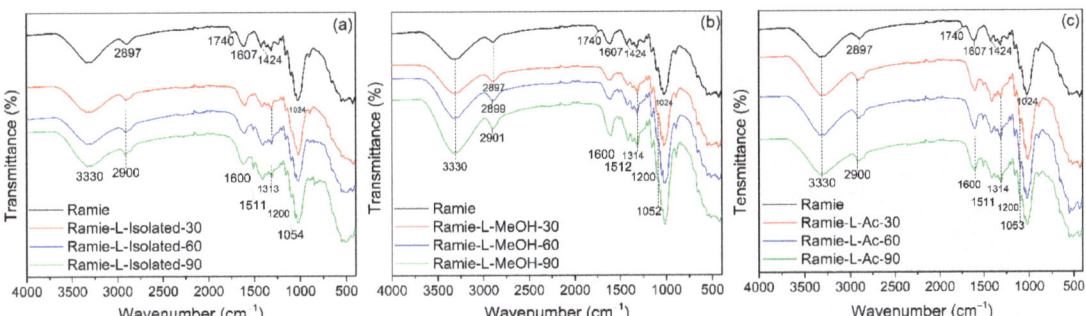

Figure 11. FTIR Spectra of impregnated ramie fibers: (**a**) Ramie L-Isolated, (**b**) Ramie L-MeOH, and (**c**) Ramie L-Ac.

The functional groups of ramie fiber after impregnation were also observed using FTIR. A peak similar to the original ramie at a wavenumber of 3330 cm^{-1} indicated O-H vibration. The peak of ramie fiber at 2897 cm^{-1} shifted to 2900 cm^{-1}, which showed N-H bending and -CH$_2$ stretching of bio-PU in impregnated ramie fiber with Bio-PU resins. The wavenumber of 1740 cm^{-1} was not found in ramie fiber after impregnation. It can be assumed that there was a reaction between ramie fiber and Bio-PU resins, and the formation of a peak at 1600 cm^{-1} indicates the formation of C=O urethane from Bio-PU [44]. The wavenumber of 1515–1310 cm^{-1} represented the C-N stretching group from primary and secondary amides of bio-PU resin in impregnated ramie fiber. Furthermore, the wavenumbers 1200 cm^{-1} and 1050 cm^{-1} showed C=O vibrations of Bio-PU and C-O-C ether linkages, respectively [19], which demonstrates that lignin-based Bio-PU successfully impregnated the ramie fiber.

Micro Confocal Raman analysis was performed to investigate the formation of urethane linkages in ramie fiber after impregnation with lignin-based Bio-PU resin. As depicted in Figure 12a, the original ramie fiber had a typical broad peak at 1350–1400 cm^{-1}, which indicated the presence of cellulosic materials. After impregnation with Bio-PU resin, the Raman spectra of ramie fiber remarkably changed due to the presence of a strong peak of urethane linkages at 1614 cm^{-1}. This result confirmed that Bio-PU resin impregnated and covered the ramie fiber. The images of ramie fiber before and after impregnation are presented in Figure 12b,c. The results showed that impregnation of Bio-PU into ramie fiber altered the color of ramie fiber to a light brown. Based on the result of FTIR and Micro Confocal Raman Hyperspectral Spectroscopies, the possible reaction of ramie fiber with lignin-based Bio-PU resin is illustrated in Figure 13.

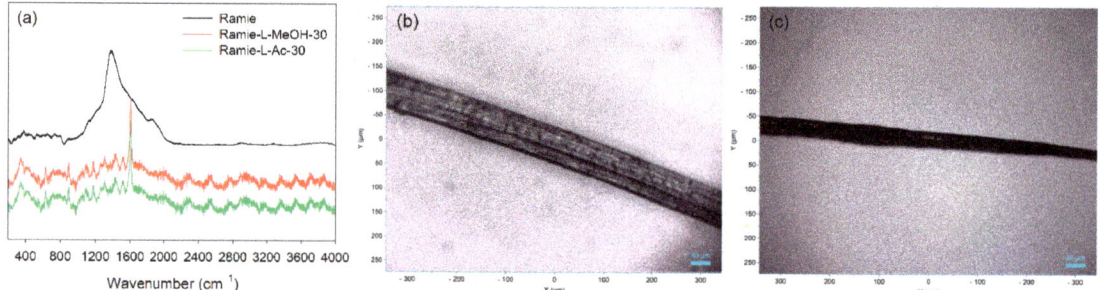

Figure 12. Micro Confocal Raman analysis of ramie fiber: (**a**) Typical Raman spectra of ramie fiber, (**b**) image of ramie fiber before impregnation at 10× magnification, and (**c**) image of ramie fiber after impregnation at 10× magnification.

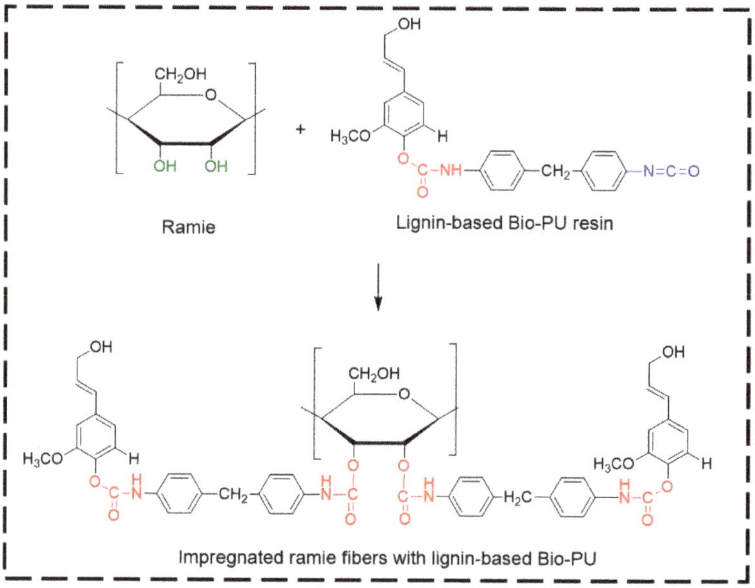

Figure 13. The possible scheme reaction of ramie fibers with lignin-based Bio-PU resin.

DSC analysis detected two thermal events on ramie fiber before and after impregnation, namely an endothermic reaction and an exothermic reaction (Figure 14). The endothermic reaction is a heat absorption reaction by the ramie fiber, while the exothermic reaction is a heat release reaction by the ramie fiber [73]. The peak of the endothermic reaction (Tp_1) in the original ramie fiber occurred at a temperature of 46 °C, which is the process of evaporation of water. In general, impregnation resulted in a slightly higher Tp_1 value. Ramie fiber impregnated with L-Isolated and L-Ac had a lower Tp_1 value than that of ramie impregnated with L-MeOH. The difference in Tp_1 value between natural ramie and impregnated ramie is influenced by differences in fiber moisture content. The moisture content of natural ramie is 5.15%, and the moisture content of impregnated ramie is between 6 and 10%. No significant influence of increasing the impregnation time on the Tp_1 value of ramie impregnated with lignin-based Bio-PU resin was observed.

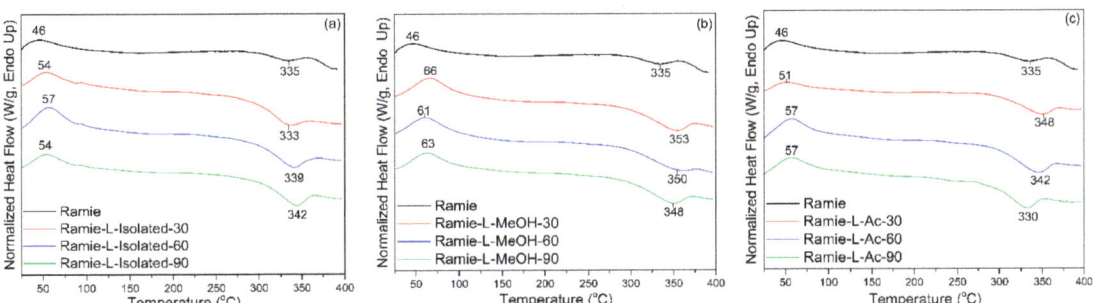

Figure 14. DSC analysis of impregnated ramie fibers: (**a**) Ramie L-Isolated, (**b**) Ramie L-MeOH, and (**c**) Ramie L-Ac.

The exothermic reaction in ramie fiber indicates the formation of solid residues due to hemicellulose decomposition and lignin degradation [74]. The peak of the exothermic reaction (Tp_2) generally occurred after the heating temperature reached 300 °C. In non-treated ramie, the Tp_2 value was around 335 °C. Impregnation resulted in a slightly higher Tp_2 value. Ramie fiber impregnated with L-Isolated and L-Ac had a lower Tp_2 value than that of ramie impregnated with L-MeOH. The increase in Tp_2 value in impregnated ramie compared to original ramie indicates that ramie has more heat-resistant properties after impregnation. Ramie fiber impregnated at 90 min for all types of Bio-PU generally had an increasing Tp_2 value compared to original ramie.

Thermal degradation of ramie fiber before and after impregnation with lignin-based Bio-PU resin was monitored using TGA-DTG analysis. As displayed in Figure 15, the initial stage of the degradation process occurred at a temperature of 25–100 °C, which is attributed to the release of water contained in ramie fiber and the degradation of low-molecular-weight components. At this stage, the ramie fiber lost 5% weight with a derivative weight loss of around 1.0–2.0%/°C. Significant weight loss was observed after heating at 250–450 °C, which caused the ramie fiber to experience an extreme weight loss of around 75% at a derivative weight loss of 13%/°C. The results showed that the impregnation of ramie fiber with Bio-PU remarkably enhanced the thermal stability by reducing the weight loss value to 50% with a lower derivative weight loss of 6.0–7.0%/°C. At a temperature of 200–400 °C, a complex reaction occurs from the degradation of the constituent components of lignocellulose fibers, such as cellulose, hemicellulose, and lignin [75]. Hemicellulose begins to degrade at a temperature of 200–290 °C, at a temperature of 240–350 °C, cellulose degradation occurs, and lignin degradation begins at a temperature of 280–500 °C. Several related studies also stated that the degradation of lignin, cellulose, and hemicellulose in ramie fiber occurred at a temperature of 300–400 °C, whereas at a temperature of 320–400 °C, the cellulose glycosidic bond was broken, and the decomposition of the lignin structure by a high molecular weight followed at a temperature of 360–750 °C [44,72,76].

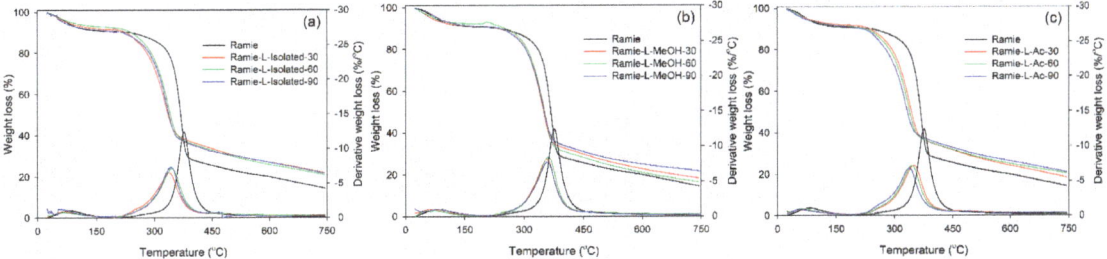

Figure 15. Thermal stability of impregnated ramie fibers determined with TGA-DTG: (**a**) Ramie L-Isolated, (**b**) Ramie L-MeOH, and (**c**) Ramie L-Ac.

The impregnated ramie fiber lost 75% of its weight at a temperature of 550–660 °C, while the non-impregnated ramie fiber experienced a similar weight loss at 469 °C. This can be attributed to the occurrence of lignin degradation at that temperature. Maximum decomposition occurred at a temperature of 300–400 °C, and has also been noted that lignin degradation in ramie fiber is the limit for determining thermal stability [76]. This study showed that lignin-based Bio-PU can increase the thermal stability of ramie fiber so that it is better than natural ramie. The impregnation time affects the thermal degradation of the ramie fiber remarkably. The longer the impregnation time, the lower the thermal degradation of the impregnated fiber. The weight residue of the original ramie fiber was 14.3%. Meanwhile, impregnated ramie has a higher residue weight, at 18.4–21.7%. An impregnation time of 90 min produced lower thermal degradation of ramie fiber with greater weight residue than those at 30 and 60 min. The lowest thermal degradation was obtained in both ramie fiber impregnated with L-MeOH and L-Ac for 90 min with an average weight residue of around 21.7%.

Mechanical properties are essential characteristics for ramie fiber as a textile and composites material. A graphical representation of the stress—strain curves of ramie fiber impregnated with different lignin-based Bio-PU resins for different impregnation times is presented in Figure 16. The results showed that increasing the impregnation time generally enhanced the maximum stress. Ramie impregnated with Bio-PU L-Ac showed the highest maximum stress after 90 min of impregnation, followed by Bio-PU L-isolated and L-MeOH. The lowest stress was obtained in the original ramie fiber. This indicated that the impregnation of Bio-PU resin into ramie fiber could enhance the mechanical properties of the fiber. The fiber deformation can be divided into the following three stages: (i) A linear part related to the deformation of the fiber cell wall; (ii) a non-linear part interpreted as a thicker cell wall deformation (S2) and known as elasto-visco-plastic; and (iii) a final linear section, which is the elastic response aligned with the tensile strain [33].

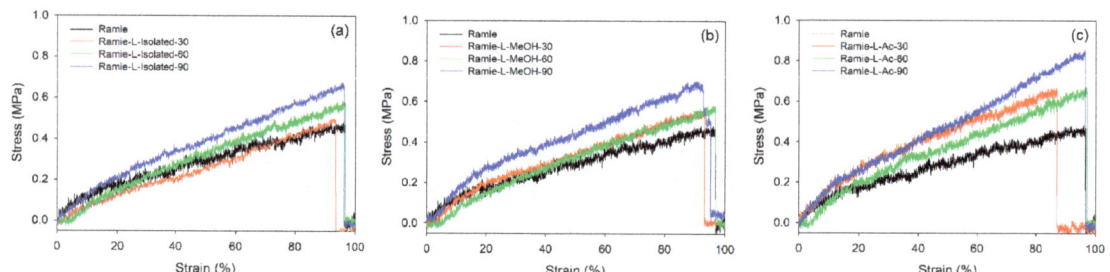

Figure 16. Typical stress–strain curve of ramie fibers before and after impregnation: (**a**) Ramie L-Isolated, (**b**) Ramie L-MeOH, and (**c**) Ramie L-Ac.

The values obtained for the modulus of elasticity (MOE) and tensile strength of non-impregnated and impregnated ramie fibers are presented in Table 6. In general, the mechanical properties of impregnated ramie fibers increased compared to the original ramie. The tensile strength of the original ramie was 397.72 MPa with an MOE value of 10.45 GPa. The tensile strength of impregnated ramie was in the range of 397.72–648.48 MPa with MOE values ranging from 13.06 to 31.10 GPa. Increasing the impregnation time generally increased the elastic modulus and tensile strength of ramie fiber. Ramie fiber impregnated with Bio-PU L-isolated for 90 min resulted in the highest MOE value of 31.10 GPa. Meanwhile, ramie impregnated with Bio-PU L-Ac for 90 min had the highest tensile strength of around 577.61 MPa. Several factors besides chemical modification affected the mechanical properties of ramie fiber, namely parameters related to relative environmental humidity, fiber length, fiber microstructure, moisture content and drying, and fiber diameter [77,78].

Table 6. Modulus of elasticity (MOE) and tensile strength of ramie and impregnated ramie fibers at different impregnation times.

Type	Impregnation Time (min)	Modulus of Elasticity (GPa)	Std. Dev	Tensile Strength (MPa)	Std. Dev
Ramie	-	10.45	0.2402	397.72	58.26
Ramie-L-isolated	30	15.23	0.4169	441.19	226.37
	60	20.35	0.6712	497.22	91.62
	90	31.10	0.6705	574.11	133.97
Ramie-L-MeOH	30	13.06	0.4446	447.06	58.59
	60	11.98	0.2780	406.71	175.64
	90	17.86	0.5629	461.32	106.13
Ramie-L-Ac	30	16.78	0.2554	523.38	80.12
	60	19.81	0.2923	547.66	91.37
	90	21.99	0.5288	577.61	68.87

4. Conclusions

This study demonstrated the potential use of lignin derived from black liquor as a pre-polymer of bio-polyurethane (Bio-PU) resin and its application for the modification of ramie fiber. The isolated lignin from black liquor was fractionated using MeOH and Ac. The isolated and fractionated lignin then reacted with pMDI to form Bio-PU resin with an NCO/OH ratio of 0.3. FTIR results showed that the reaction between -OH of lignin and -NCO of pMDI formed an absorption band at a wavenumber of 1605 cm^{-1}, which was urethane linkages (R-NH-C=O-R) of Bio-PU resin. The Bio-PU L-Isolated has a lower viscosity than fractionated lignin. The thermal properties and thermal stability of Bio-PU L-Ac were better than L-isolated and L-MeOH. Impregnation of Bio-PU into ramie fiber resulted in weight gain varying from 6% to 15%, where the value increased with longer impregnation time. The reaction between Bio-PU and ramie fiber formed C=O of the urethane group as confirmed by FTIR and Micro Confocal Raman Spectroscopies. This resulted in greater thermal properties and stability of ramie fiber after impregnation with the weight residue reaching 21.7%. The mechanical properties of ramie fiber also increased after impregnation with lignin-based Bio-PU, resulting in a modulus of elasticity of around 31 GPa for Ramie-L-isolated and tensile strength of around 577 MPa for Ramie-L-Ac. This study showed that lignin-based Bio-PU resin derived from isolated lignin and fractionated L-Ac could be used as a pre-polymer of Bio-PU resin for the modification of ramie fiber via impregnation. The enhanced thermal stability and mechanical properties of impregnated ramie fiber could increase the future potential for greater industrial application of ramie fiber as a sustainable and functional material.

Author Contributions: Methodology, S.O.H., M.A.R.L. and R.K.S.; validation, S.O.H., M.A.R.L. and R.K.S.; investigation, S.O.H., M.A.R.L. and R.P.B.L.; writing—original draft preparation S.O.H., M.A.R.L., R.K.S., P.A. and A.H.I.; writing—review and editing, M.A.R.L., R.K.S., P.A. and M.G.; visualization, M.A.R.L., P.A., V.S. and M.G.; supervision, M.A.R.L. and R.K.S. All authors have read and agreed to the published version of the manuscript.

Funding: This work was supported by the research grant No. 172/E1/PRN/2020 (Pengembangan Serat Rami Tahan Api Terimpregnasi Poliuretan Berbasis Lignin dan Tanin) from the Ministry of Research and Technology of the Indonesia/National Research and Innovation Agency. This research was also supported by project No. НИС-Б-1145/04.2021 "Development, Properties and Application of Eco-Friendly Wood-Based Composites" carried out at the University of Forestry, Sofia, Bulgaria.

Institutional Review Board Statement: Not applicable.

Informed Consent Statement: Not applicable.

Data Availability Statement: The data presented in this study are available on request from the corresponding author.

Acknowledgments: The authors are grateful to the Integrated Laboratory of Bio-products (iLaB), National Research and Innovation Agency of Indonesia for the facilities support. This publication was also supported by the Slovak Research and Development Agency under contracts No. APVV-18-0378 and APVV-19-0269.

Conflicts of Interest: The authors declare no conflict of interest.

References

1. Hwang, J.Z.; Wang, S.C.; Chen, P.C.; Huang, C.Y.; Yeh, J.T.; Chen, K.N. A new UV-curable PU resin obtained through a nonisocyanate process and used as a hydrophilic textile treatment. *J. Polym. Res.* **2012**, *19*, 9900. [CrossRef]
2. Zia, K.M.; Anjum, S.; Zuber, M.; Mujahid, M.; Jamil, T. Synthesis and molecular characterization of chitosan based polyurethane elastomers using aromatic diisocyanate. *Int. J. Biol. Macromol.* **2014**, *66*, 26–32. [CrossRef]
3. Das, A.; Mahanwar, P. A brief discussion on advances in polyurethane applications. *Adv. Ind. Eng. Polym. Res.* **2020**, *3*, 93–101. [CrossRef]
4. Chen, W.-J.; Wang, S.-C.; Chen, P.-C.; Chen, T.-W.; Chen, K.-N. Hybridization of aqueous PU/epoxy resin via a dual self-curing process. *J. Appl. Polym. Sci.* **2008**, *110*, 147–155. [CrossRef]
5. Wang, S.-C.; Chen, P.-C.; Yeh, J.-T.; Chen, K.-N. Curing reaction of amino-terminated aqueous-based polyurethane dispersions with triglycidyl-containing compound. *J. Appl. Polym. Sci.* **2008**, *110*, 725–731. [CrossRef]
6. Strakowska, A.; Członka, S.; Kairyte, A. Rigid polyurethane foams reinforced with poss-impregnated sugar beet pulp filler. *Materials* **2020**, *13*, 5493. [CrossRef]
7. Uram, K.; Kurańska, M.; Andrzejewski, J.; Prociak, A. Rigid Polyurethane Foams Modified with Biochar. *Materials* **2021**, *14*, 5616. [CrossRef] [PubMed]
8. Gama, N.V.; Ferreira, A.; Barros-Timmons, A. Polyurethane foams: Past, present, and future. *Materials* **2018**, *11*, 1841. [CrossRef]
9. Kurańska, M.; Pinto, J.A.; Salach, K.; Barreiro, M.F.; Prociak, A. Synthesis of thermal insulating polyurethane foams from lignin and rapeseed based polyols: A comparative study. *Ind. Crops Prod.* **2020**, *143*, 111882. [CrossRef]
10. Hu, S.; Luo, X.; Li, Y. Polyols and polyurethanes from the liquefaction of lignocellulosic biomass. *ChemSusChem* **2014**, *7*, 66–72. [CrossRef]
11. Petrovic, Z.S. Polyurethanes from vegetable oils. *Polym. Rev.* **2008**, *48*, 109–155. [CrossRef]
12. Pfister, D.P.; Xia, Y.; Larock, R.C. Recent advances in vegetable oil-based polyurethanes. *ChemSusChem* **2011**, *4*, 703–717. [CrossRef] [PubMed]
13. Lligadas, G.; Ronda, J.C.; Galiá, M.; Cádiz, V. Plant oils as platform chemicals for polyurethane synthesis: Current state-of-the-art. *Biomacromolecules* **2010**, *11*, 2825–2835. [CrossRef] [PubMed]
14. Ge, X.; Chang, C.; Zhang, L.; Cui, S.; Luo, X.; Hu, S.; Qin, Y.; Li, Y. *Conversion of Lignocellulosic Biomass into Platform Chemicals for Biobased Polyurethane Application*, 1st ed.; Elsevier Inc.: Amsterdam, The Netherlands, 2018; Volume 3.
15. Thébault, M.; Pizzi, A.; Essawy, H.A.; Barhoum, A.; Van Assche, G. Isocyanate free condensed tannin-based polyurethanes. *Eur. Polym. J.* **2015**, *67*, 513–526. [CrossRef]
16. Aristri, M.A.; Lubis, M.A.R.; Iswanto, A.H.; Fatriasari, W.; Sari, R.K.; Antov, P.; Gajtanska, M.; Papadopoulos, A.N.; Pizzi, A. Bio-Based Polyurethane Resins Derived from Tannin: Source, Synthesis, Characterisation, and Application. *Forests* **2021**, *12*, 1516. [CrossRef]
17. Watanabe, M.; Kanaguri, Y.; Smith, R.L. Hydrothermal separation of lignin from bark of Japanese cedar. *J. Supercrit. Fluids* **2018**, *133*, 696–703. [CrossRef]
18. Tudor, E.M.; Barbu, M.C.; Petutschnigg, A.; Réh, R.; Krišťák, Ľ. Analysis of larch-bark capacity for formaldehyde removal in wood adhesives. *Int. J. Environ. Res. Public Health* **2020**, *17*, 764. [CrossRef] [PubMed]
19. Cateto, C.A.; Barreiro, M.F.; Rodrigues, A.E. Monitoring of lignin-based polyurethane synthesis by FTIR-ATR. *Ind. Crops Prod.* **2008**, *27*, 168–174. [CrossRef]
20. Bajwa, D.S.; Pourhashem, G.; Ullah, A.H.; Bajwa, S.G. A concise review of current lignin production, applications, products and their environment impact. *Ind. Crops Prod.* **2019**, *139*, 111526. [CrossRef]
21. Mandlekar, N.; Cayla, A.; Rault, F.; Giraud, S.; Salaün, F.; Malucelli, G.; Guan, J.-P. An Overview on the Use of Lignin and Its Derivatives in Fire Retardant Polymer Systems. In *Lignin—Trends and Applications*; InTech: London, UK, 2018; pp. 207–231.
22. Zhang, H.; Bai, Y.; Yu, B.; Liu, X.; Chen, F. A practicable process for lignin color reduction: Fractionation of lignin using methanol/water as a solvent. *Green Chem.* **2017**, *19*, 5152–5162. [CrossRef]
23. Evtuguin, D.V.; Andreolety, J.P.; Gandini, A. Polyurethanes based on oxygen-organosolv lignin. *Eur. Polym. J.* **1998**, *34*, 1163–1169. [CrossRef]
24. Aristri, M.A.; Lubis, M.A.R.; Yadav, S.M.; Antov, P.; Papadopoulos, A.N.; Pizzi, A.; Fatriasari, W.; Ismayati, M.; Iswanto, A.H. Recent Developments in Lignin- and Tannin-Based Non-Isocyanate Polyurethane Resins for Wood Adhesives—A Review. *Appl. Sci.* **2021**, *11*, 4242. [CrossRef]
25. Thring, R.W.; Vanderlaan, M.N.; Griffin, S.L. Polyurethanes from Alcell® lignin. *Biomass Bioenergy* **1997**, *13*, 125–132. [CrossRef]
26. Antov, P.; Savov, V.; Trichkov, N.; Krišťák, Ľ.; Réh, R.; Papadopoulos, A.N.; Taghiyari, H.R.; Pizzi, A.; Kunecová, D.; Pachikova, M. Properties of high-density fiberboard bonded with urea–formaldehyde resin and ammonium lignosulfonate as a bio-based additive. *Polymers* **2021**, *13*, 2775. [CrossRef]

27. Alzagameem, A.; El Khaldi-Hansen, B.; Büchner, D.; Larkins, M.; Kamm, B.; Witzleben, S.; Schulze, M. Lignocellulosic biomass as source for lignin-based environmentally benign antioxidants. *Molecules* **2018**, *23*, 2664. [CrossRef] [PubMed]
28. Réh, R.; Krišťák, Ľ.; Sedliačik, J.; Bekhta, P.; Božiková, M.; Kunecová, D.; Vozárová, V.; Tudor, E.M.; Antov, P.; Savov, V. Utilization of birch bark as an eco-friendly filler in urea-formaldehyde adhesives for plywood manufacturing. *Polymers* **2021**, *13*, 511. [CrossRef]
29. Bachtiar, E.V.; Kurkowiak, K.; Yan, L.; Kasal, B.; Kolb, T. Thermal stability, fire performance, and mechanical properties of natural fibre fabric-reinforced polymer composites with different fire retardants. *Polymers* **2019**, *11*, 699. [CrossRef]
30. Twite-Kabamba, E.; Mechraoui, A.; Rodrigue, D. Rheological properties of polypropylene/hemp fiber composites. *Polym. Compos.* **2009**, *30*, 1401–1407. [CrossRef]
31. Rehman, M.; Gang, D.; Liu, Q.; Chen, Y.; Wang, B.; Peng, D.; Liu, L. Ramie, a multipurpose crop: Potential applications, constraints and improvement strategies. *Ind. Crops Prod.* **2019**, *137*, 300–307. [CrossRef]
32. Mulyawan, A.S.; Sana, A.W.; Kaelani, Z. Identifikasi Sifat Fisik Dan Sifat Termal Serat-Serat Selulosa Untuk Pembuatan Komposit. *Arena Tekst.* **2015**, *30*, 75–82. [CrossRef]
33. Yan, L.; Chouw, N.; Jayaraman, K. Flax fibre and its composites—A review. *Compos. Part B Eng.* **2014**, *56*, 296–317. [CrossRef]
34. Yuan, J.M.; Feng, Y.R.; He, L.P. Effect of thermal treatment on properties of ramie fibers. *Polym. Degrad. Stab.* **2016**, *133*, 303–311. [CrossRef]
35. Kalia, S.; Sheoran, R. Modification of ramie fibers using microwaveassisted grafting and cellulase enzyme-assisted biopolishing: A comparative study of morphology, thermal stability, and crystallinity. *Int. J. Polym. Anal. Charact.* **2011**, *16*, 307–318. [CrossRef]
36. Liu, X.; Dai, G. Impregnation of thermoplastic resin in jute fiber mat. *Front. Chem. Eng. China* **2008**, *2*, 145–149. [CrossRef]
37. Xia, C.; Zhang, S.; Shi, S.Q.; Cai, L.; Huang, J. Property enhancement of kenaf fiber reinforced composites by in situ aluminum hydroxide impregnation. *Ind. Crop. Prod.* **2016**, *79*, 131–136. [CrossRef]
38. Hermiati, E.; Risanto, L.; Lubis, M.A.R.; Laksana, R.P.B.; Dewi, A.R. Chemical characterization of lignin from kraft pulping black liquor of Acacia mangium. *AIP Conf. Proc.* **2017**, *1803*, 020005. [CrossRef]
39. Ponnuchamy, V.; Gordobil, O.; Diaz, R.H.; Sandak, A.; Sandak, J. Fractionation of lignin using organic solvents: A combined experimental and theoretical study. *Int. J. Biol. Macromol.* **2021**, *168*, 792–805. [CrossRef] [PubMed]
40. TAPPI. *T211 Ash in Wood, Pulp, Paper and Paperboard: Combustion at 525 °C*; TAPPI Standard Test Methods; TAPPI: Peachtree Corners, GA, USA, 2007; Volume T211 om-02, p. 5.
41. Templeton, D.; Ehrman, T. *Determination of Acid-Insoluble Lignin in Biomass—LAP-003*; NREL—National Renewable Energy Laboratory: Golden, CO, USA, 1995; p. 14.
42. Ehrman, T. *Determination of Acid-Soluble Lignin in Biomass—LAP-004*; NREL—National Renewable Energy Laboratory: Golden, CO, USA, 1996; Volume LAP-004, p. 8.
43. Serrano, L.; Esakkimuthu, E.S.; Marlin, N.; Brochier-Salon, M.C.; Mortha, G.; Bertaud, F. Fast, Easy, and Economical Quantification of Lignin Phenolic Hydroxyl Groups: Comparison with Classical Techniques. *Energy Fuels* **2018**, *32*, 5969–5977. [CrossRef]
44. Aristri, M.A.; Lubis, M.A.R.; Laksana, R.P.B.; Fatriasari, W.; Ismayati, M.; Wulandari, A.P.; Ridho, M.R. Bio-Polyurethane Resins Derived from Liquid Fractions of Lignin for the Modification of Ramie Fibers. *J. Sylva Lestari* **2021**, *9*, 223–238. [CrossRef]
45. ASTM. *ASTM D 3379–75 Standard Test Method for Tensile Strength and Young's Modulus for High-Modulus Single-Filament Materials*; ASTM International: Washington, DC, USA, 2000; Volume 75, pp. 1–5.
46. Sameni, J.; Krigstin, S.; Derval dos Santos, R.; Leao, A.; Sain, M. Thermal characteristics of lignin residue from industrial processes. *BioResources* **2014**, *9*, 725–737. [CrossRef]
47. Lubis, M.A.R.; Dewi, A.R.; Risanto, L.; Zaini, L.H.; Hermiati, E. Isolation and Characterization of Lignin from Alkaline Pretreatment Black Liquor of Oil Palm Empty Fruit Bunch and Sugarcane Bagasse. In Proceedings of the ASEAN COSAT 2014, Bogor, Indonesia, 18–19 August 2014; pp. 483–491.
48. Gordobil, O.; Herrera, R.; Poohphajai, F.; Sandak, J.; Sandak, A. Impact of drying process on kraft lignin: Lignin-water interaction mechanism study by 2D NIR correlation spectroscopy. *J. Mater. Res. Technol.* **2021**, *12*, 159–169. [CrossRef]
49. Cardoso, M.; de Oliveira, É.D.; Passos, M.L. Chemical composition and physical properties of black liquors and their effects on liquor recovery operation in Brazilian pulp mills. *Fuel* **2009**, *88*, 756–763. [CrossRef]
50. Nikolskaya, E.; Janhunen, P.; Haapalainen, M.; Hiltunen, Y. Solids Content of Black Liquor Measured by Online. *Appl. Sci.* **2019**, *9*, 2169. [CrossRef]
51. Yotwadee, H.; Duangduen, A.; Viboon, S. Lignin isolation from black liquor for wastewater quality improvement and bio-material recovery. *Int. J. Environ. Sci. Dev.* **2020**, *11*, 365–371. [CrossRef]
52. Jusuf, P.G.; Purwono, S.; Tawfiequrahman, A. Permodelan Ekstraksi Lignin Mentah dari Black Liquor dengan Metode Asidifikasi pada pH Rendah. In Proceedings of the Seminar Nasional Teknik Kimia "Kejuangan", Jakarta, Indonesia, 25 April 2019; pp. 1–5.
53. Sadeghifar, H.; Sadeghifar, H.; Ragauskas, A.; Ragauskas, A.; Ragauskas, A.; Ragauskas, A. Perspective on Technical Lignin Fractionation. *ACS Sustain. Chem. Eng.* **2020**, *8*, 8086–8101. [CrossRef]
54. Sameni, J.; Krigstin, S.; Sain, M. Solubility of Lignin and Acetylated Lignin in Organic Solvents. *BioResources* **2017**, *12*, 1548–1565. [CrossRef]
55. Kubo, S.; Kadla, J.F. Hydrogen bonding in lignin: A fourier transform infrared model compound study. *Biomacromolecules* **2005**, *6*, 2815–2821. [CrossRef]
56. Li, H.; McDonald, A.G. Fractionation and characterization of industrial lignins. *Ind. Crops Prod.* **2014**, *62*, 67–76. [CrossRef]

57. Buranov, A.U.; Ross, K.A.; Mazza, G. Isolation and characterization of lignins extracted from flax shives using pressurized aqueous ethanol. *Bioresour. Technol.* **2010**, *101*, 7446–7455. [CrossRef]
58. Lucejko, J.J.; Tamburini, D.; Modugno, F.; Ribechini, E.; Colombini, M.P. Analytical pyrolysis and mass spectrometry to characterise lignin in archaeological wood. *Appl. Sci.* **2021**, *11*, 240. [CrossRef]
59. Gogoi, R.; Alam, M.; Khandal, R. Effect of increasing NCO/OH molar ratio on the physicomechanical and thermal properties of isocyanate terminated polyurethane prepolymer. *Int. J. Basic Appl. Sci.* **2014**, *3*, 118–123. [CrossRef]
60. Restasari, A.; Ardianingsih, R.; Abdillah, L.H.; Hartaya, K. Effects of Toluene Diisocyanate's Chemical Structure on Polyurethane's Viscosity and Mechanical Properties for Propellant. In Proceedings of the International Seminar on Aerospace Science and Technology, Kuta, Indonesia, 27–29 October 2015; pp. 59–67.
61. Gharib, J.; Pang, S.; Holland, D. Synthesis and characterisation of polyurethane made from pyrolysis bio-oil of pine wood. *Eur. Polym. J.* **2020**, *133*, 109725. [CrossRef]
62. Lubis, M.A.R.; Park, B.D.; Lee, S.M. Microencapsulation of polymeric isocyanate for the modification of urea-formaldehyde resins. *Int. J. Adhes. Adhes.* **2020**, *100*, 102599. [CrossRef]
63. Luo, S.; Gao, L.; Guo, W. Effect of incorporation of lignin as bio-polyol on the performance of rigid lightweight wood–polyurethane composite foams. *J. Wood Sci.* **2020**, *66*, 23. [CrossRef]
64. Chen, Y.; Zhang, H.; Zhu, Z.; Fu, S. *High-Value Utilization of Hydroxymethylated Lignin in Polyurethane Adhesives*; Elsevier B.V.: Amsterdam, The Netherlands, 2020; Volume 152, ISBN 1343036402.
65. Amado, J.C.Q. Thermal Resistance Properties of Polyurethanes and Its Composites. In *Thermosoftening Plastics*; InTech: London, UK, 2019; pp. 1–13.
66. Wang, Y.-Y.; Wyman, C.E.; Cai, C.M.; Ragauskas, A.J. Lignin-Based Polyurethanes from Unmodified Kraft Lignin Fractionated by Sequential Precipitation. *ACS Appl. Polym. Mater.* **2019**, *1*, 1672–1679. [CrossRef]
67. Tavares, L.; Stilhano, C.R.; Boas, V. Bio-Based polyurethane prepared from Kraft lignin and modified castor oil. *Express Polym. Lett.* **2016**, *10*, 927–940. [CrossRef]
68. Trovati, G.; Sanches, E.A.; Neto, S.C.; Mascarenhas, Y.P.; Chierice, G.O. Characterization of Polyurethane Resins by FTIR, TGA and XRD. *J. Appl. Polym. Sci.* **2009**, *115*, 263–268. [CrossRef]
69. Novarini, E.; Sukardan, M.D. Potensi Serat Rami (Boehmeria Nivea S. Gaud) Sebagai Bahan Baku Industri Tekstil Dan Produk Tekstil Dan Tekstil Teknik. *Arena Tekst.* **2015**, *30*, 113–122. [CrossRef]
70. Bevitori, A.B.; da Silva, I.L.A.; Rohen, L.A.; Margem, F.M.; de Moraes, Y.M.; Monteiro, S.N. Evaluation of Ramie Fibers Component by Infrared Spectroscopy. In Proceedings of the 21st CBECIMAT—Congresso Brasileiro de Engenharia e Ciência dos Materiais, Cuiabá, Brazil, 9–13 November 2014; pp. 2109–2116.
71. Simonassi, N.T.; Pereira, A.C.; Monteiro, S.N.; Muylaert, F.; De Deus, J.F.; Fontes, C.M.; Drelich, J.; Vermelha, P.; De Janeiro, R.; California, P. Reinforcement of Polyester with Renewable Ramie Fibers. *Mater. Res.* **2017**, *20*, 51–59. [CrossRef]
72. Kandimalla, R.; Kalita, S.; Choudhury, B.; Devi, D.; Kalita, D.; Kalita, K.; Dash, S.; Kotoky, J. Fiber from ramie plant (Boehmeria nivea): A novel suture biomaterial. *Mater. Sci. Eng. C* **2016**, *62*, 816–822. [CrossRef]
73. Shahinur, S.; Hasan, M.; Ahsan, Q.; Haider, J. Effect of Chemical Treatment on Thermal Properties of Jute Fiber Used in Polymer Composites. *J. Compos. Sci.* **2020**, *4*, 132. [CrossRef]
74. Shahinur, S.; Hasan, M.; Ahsan, Q.; Saha, D.K.; Islam, M.S. Characterization on the Properties of Jute Fiber at Different Portions. *Int. J. Polym. Sci.* **2015**, *2015*, 262348. [CrossRef]
75. Tomczak, F.; Satyanarayana, K.G.; Sydenstricker, T.H.D. Studies on lignocellulosic fibers of Brazil: Part III—Morphology fibers and properties of Brazilian curaua. *Compos. Part. A* **2007**, *38*, 2227–2236. [CrossRef]
76. Bevitori, A.B.; Margem, F.M.; Carreiro, R.S.; Monteiro, S.N.; Calado, V. Thermal Caracterization Behavior of Epoxy Composites Reinforced Ramie Fibers. In Proceedings of the 67th ABM International Congress, Rio de Janeiro, Brazil, 31 July–3 August 2012; pp. 473–480.
77. Charlet, K.; Jernot, J.P.; Breard, J.; Gomina, M. Scattering of morphological and mechanical properties of flax fibres. *Ind. Crops Prod.* **2010**, *32*, 220–224. [CrossRef]
78. Le Duigou, A.; Bourmaud, A.; Balnois, E.; Davies, P.; Baley, C. Improving the interfacial properties between flax fibres and PLLA by a water fibre treatment and drying cycle. *Ind. Crops Prod.* **2012**, *39*, 31–39. [CrossRef]

Article

Prediction of Mechanical Properties of Artificially Weathered Wood by Color Change and Machine Learning

Vahid Nasir [1], Hamidreza Fathi [2], Arezoo Fallah [2,*], Siavash Kazemirad [2,*], Farrokh Sassani [1] and Petar Antov [3,*]

[1] Department of Mechanical Engineering, The University of British Columbia (UBC), Vancouver, BC 2054-6250, Canada; vahid.nasir@alumni.ubc.ca (V.N.); sassani@mech.ubc.ca (F.S.)
[2] School of Mechanical Engineering, Iran University of Science and Technology, Tehran 16846-13114, Iran; hrfathi93@gmail.com
[3] Department of Mechanical Wood Technology, Faculty of Forest Industry, University of Forestry, 1797 Sofia, Bulgaria
* Correspondence: arezoofallah712@gmail.com (A.F.); skazemirad@iust.ac.ir (S.K.); p.antov@ltu.bg (P.A.)

Abstract: Color parameters were used in this study to develop a machine learning model for predicting the mechanical properties of artificially weathered fir, alder, oak, and poplar wood. A CIELAB color measuring system was employed to study the color changes in wood samples. The color parameters were fed into a decision tree model for predicting the MOE and MOR values of the wood samples. The results indicated a reduction in the mechanical properties of the samples, where fir and alder were the most and least degraded wood under weathering conditions, respectively. The mechanical degradation was correlated with the color change, where the most resistant wood to color change exhibited less reduction in the mechanical properties. The predictive machine learning model estimated the MOE and MOR values with a maximum R^2 of 0.87 and 0.88, respectively. Thus, variations in the color parameters of wood can be considered informative features linked to the mechanical properties of small-sized and clear wood. Further research could study the effectiveness of the model when analyzing large-sized timber.

Keywords: wood characterization; mechanical properties; photodegradation; artificial weathering; color change; ultraviolet radiation; machine learning

1. Introduction

Nondestructive evaluation (NDE) of wood is crucial for monitoring purposes and timber grading [1,2], especially when the wood is used in load-carrying applications. The characterization of the mechanical properties of wood, including the modulus of elasticity (MOE) and modulus of rupture (MOR), can be performed using tensile, compression, and bending static tests [3]. However, these methods are costly and time-consuming, and are not suitable for in situ characterization and monitoring purposes. While visual strength grading is being practiced in some industrial applications [4], fast and reliable assessment of timber and wood-based materials that accounts for material variability and anisotropic properties and natural defects requires the application of NDE methods.

Near-infrared (NIR) spectroscopy is one of the most commonly used NDE methods for wood characterization. NIR spectroscopy is sensitive to changes in the chemical composition of wood [5] and has been used for wood classification and the estimation of different wood properties [6–10]. Wave propagation-based methods have also been employed to estimate timber MOE [11,12]. For example, a wave is generated in wood using an impact or piezoelectric actuators, respectively, in the stress wave or ultrasonic wave methods. The propagated wave velocity, and consequently the wood dynamic MOE, are typically calculated in wave propagation techniques using the time-of-flight method. The prediction of the mechanical properties [13,14] and the detection of internal check formation [15,16] in weathered thermally modified timber have been performed using the ultrasonic wave method. The physical and mechanical properties of thermally modified wood have also been predicted using the stress wave method [17]. Recently, the elastic and viscoelastic properties of wood have been characterized using the Lamb wave propagation method [18,19]. Another readily available NDE method for wood classification and characterization is related to measuring the surface color of wood.

The surface color is affected when the wood is used in applications that cause changes to its chemical composition, such as through thermal treatment, or ultraviolet (UV) or laser irradiation [20,21]. The color change has been shown to be a quality indicator for thermally modified timber (TMT) [22,23]. It has been reported that color change can be linked to the intensity of thermal treatment [24–29] and the mechanical properties of TMT [30]. The correlation between the color change and the pressure treatment of wood has also been reported in the literature [31]. Color measurement has also been employed for the classification [32] and characterization [33,34] of thermally modified timber. The color is also an important feature to be studied during the wood weathering since the weathering of wood causes photodegradation. The CIELAB color measuring system has been widely employed to study the color change in wood under UV radiation [35–42]. The color was considered as an informative parameter to monitor the wood photodegradation [43]. Additionally, the infrared spectroscopy analysis of wood has been used to assess the change in wood chemical composition under weathering [44–48]. The critical role of color change during wood weathering may suggest further investigations to see if the color change can be used as a quality control tool for monitoring the in-service weathered wood. One of these monitoring tasks is to assess the feasibility of using the color change for predicting the MOE and MOR of weathered wood.

The mechanical degradation of wood under weathering depends on different factors. These factors include but are not limited to the type of wood, weathering condition (UV radiation, temperature, humidity, rain, etc.), thickness of wood under weathering, and conditioning parameters. It has been reported that UV and solar irradiation result in the degradation of the mechanical properties in thin wood strips [49–51]. Accelerated UV exposure can result in a 20–40% drop in the strength of wood strips [52]. The mechanical degradation in wood-based composites under weathering conditions with a panel thickness of 8 mm and 12.5 mm has also been reported in the literature [53,54]. Weathering has also been reported to cause a reduction in the impact bending properties of spruce, fir, and oak wood with a board thickness of 20 mm [55]. The impact of weathering on the mechanical properties of full-size timber has also been investigated [56]. van Blokland et al. [14] studied the impact of natural weathering on the mechanical properties of Norway spruce timber. The timbers were conditioned after the weathering and mechanically tested. They reported that the bending strength of the control and thermally treated spruce were reduced after weathering by 6% and 9%, respectively. A 4% reduction in the MOE of both types of timbers was also reported.

Apart from the size of the wood cross-section, the type of wood species, weathering situations, and conditioning also impact the mechanical properties of weathered wood. Tomak et al. [57] showed that the reduction in the mechanical properties under weathering is significantly dependent on the type of wood species. They reported that while the MOR of Ash wood decreased by 18% after 24 months of weathering, Iroko wood experienced a 40% MOR reduction in the same time period. The mechanical degradation of softwood and hardwood under weathering can also be significantly different [57]. The impact of weathering is greatly affected by the environmental conditions. In the case of artificial weathering, for example, the impact of UV radiation is affected by the moisture content (MC), relative humidity (RH), and temperature conditions. Timar et al. [36] reported that the combination of UV radiation and temperature could cause lignin degradation, while exposure of wood to temperature alone did not affect the lignin. Persze and Tolvaj [58] also reported that the photodegradation of wood is affected by temperature. Therefore, while UV radiation alone is mainly known to be a phenomenon causing surface damage, its impact on the wood can be aggravated when combined with harsh environmental conditions. To study the effect of UV radiation, the exposed samples are conditioned after the weathering test. This is useful to understand the mechanism of damage caused; however, some researchers did not condition their samples after the weathering and before mechanical testing to simulate the practical situations [56]. For example, Boonstra et al. [56] showed that, compared to the conditioned samples after weathering, the weathered samples that were directly tested without conditioning showed higher reduction in the MOE and MOR. While such an approach assesses the combined effect of different parameters (UV radiation, temperature, MC change, etc.) on the mechanical properties of wood under weathering, this is more aligned with real situations, where in situ monitoring of in-service weathered wood structures is intended.

The above discussion shows that it is hard to generalize the mechanical behavior of wood under weathering conditions, since it depends on many factors. The mechanical behavior of the weathered wood depends on the size of wood cross-section, type of wood species, and range and intensity of environmental parameters such as the temperature and conditioning situations. Therefore, finding an NDE method for monitoring the mechanical behavior of weathered wood becomes a crucial quality control task. The MOE and MOR of defect-free wood subjected to UV radiation were recently estimated using the Lamb wave propagation method [59,60]. Yet, the color parameters of wood have not been used to make an intelligent monitoring model to predict the MOE and MOR of degraded wood under weathering. The use of color parameters to monitor the mechanical properties of photodegraded wood may offer a fast, cost-effective, and reliable complement to wave propagation-based methods.

The aim of this research work was to link the color change of weathered wood to its mechanical properties and develop a machine learning-based model for monitoring the MOE and MOR of weathered wood. The weathering condition consisted of UV radiation at 40 °C. This is a very common temperature in regions with hot climate, such as Iran. The weathered samples were not conditioned after the weathering to better simulate the practical conditions. Thus, the weathering impacted the samples not only due to UV radiation, but also as a result of the temperature conditions. Since the objective of this study was to develop an NDE tool for the in situ monitoring of weathered wood and the evaluation of the combined effect of all parameters involved during the weathering on its mechanical properties, finding the UV penetration depth or separating the share of different playing factors on the mechanical degradation of wood was not within the scope of this study.

2. Materials and Methods
2.1. Sample Preparation

In this study, twenty samples of poplar (*Populus euroamerican*), alder (*Alnus glutinosa*), oak (*Quercus spp.*), and fir (*Abies alba*) were prepared, resulting in a total of eighty samples. The samples were clear, with no types of defects such as knots. The dimensions of the flat-sawn samples in the radial, tangential, and longitudinal directions were 20 mm × 20 mm × 300 mm, respectively, according to ISO-13061-3 and -4 standard methods [61,62]. The samples were divided into groups of four specimens from each wood species. One group was considered as the control and was not exposed to artificial weathering. Other groups were exposed to weathering for different time periods. Considerations were made to ensure that there was no significant variation of wood initial properties between the different groups according to the procedure explained in [60] based on guided wave propagation method. Figure 1 illustrates the experimental procedure employed in this study.

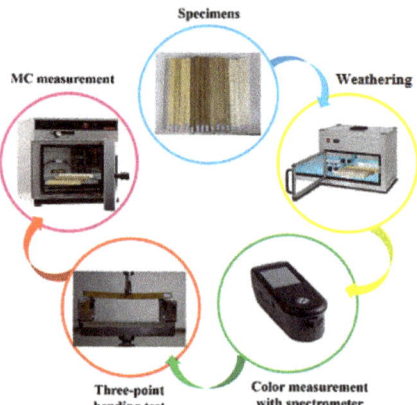

Figure 1. The experimental procedure employed in this study.

2.2. Weathering Test

In the present study, the wood samples were placed in a weathering chamber. Weathering experiments were conducted immediately after the sample preparation. Samples of each group from each wood species were exposed to UV radiation for 24, 100, 150, and 240 h, respectively. The chamber had two lamps (OSRAM HQE-40 Hg, Munich, Germany) with a length, diameter, and a spectrum of 90 mm, 10 mm, and 240–570 nm, respectively, with a radiated power in the wavelength range of 315–400 nm. The wood samples were placed at a distance of 500 mm from the UV light source. Before the exposure to UV radiation, the samples were placed in a conditioning room with a relative humidity of 65% and a temperature of 20 °C to reach the equilibrium moisture content of 12% (standard deviation = 0.33). During the experiment, the humidity and temperature inside the chamber were controlled to be 55–60% and 40 °C, respectively. One of the longitudinal-tangential (L-T) surfaces of the wood samples was subjected to the UV radiation and then the color measurements and mechanical bending tests were immediately conducted without any further conditioning. As already explained in the Introduction section, such a method can better simulate the practical conditions, especially during the in situ monitoring tasks. The weathering conditions can degrade wood through both the UV effect and the temperature condition resulting in a change in the MC. This study does not focus on the pure effect of UV radiation on the degradation of wood samples. Mechanisms of degradation under weathering, especially UV radiation, has been discussed in the literature [36,49,58,63–65]. Thus, instead of analyzing the role of different governing factors in wood degradation, this study aims to evaluate the feasibility of developing monitoring models for predicting the mechanical degradation of weathered wood.

2.3. Measurement of Color Parameters

The color of the wood samples was evaluated using the CIELAB color measuring system. The color measurement was performed on all wood samples before and after the weathering experiment. The color measurement was performed on three locations in the center and close to the ends of the samples (30 mm away). The color measurement was carried out on the degraded surface of the wood samples and performed using a spectrophotometer (Model #CM-2600d, Konica Minolta Inc., Tokyo, Japan) with a D65 illuminant, a 10° standard observer, and a sensor head of 6-mm (ASTM D2244-16 standard [66]). Once the color coordinates (L, a, and b) were measured before and after the weathering tests, the difference in the lightness (ΔL) and the chromatic coordinates (Δb and

Δa) were calculated. The total color change (ΔE) was obtained for each wood sample using the following equation:

$$\Delta E = \left[(\Delta L)^2 + (\Delta a)^2 + (\Delta b)^2 \right]^{1/2} \qquad (1)$$

The color measurement was also performed on the samples that were not placed into the weathering chamber.

2.4. Mechanical Properties

The three-point bending tests were performed using a STM-1 50 testing machine (Santam Engineering Desgin Co. Ltd., Tehran, Iran). The crosshead speed was set to 1 mm/min (ISO 13061-3 and -4 [61,62]). The bending tests were performed on the longitudinal-tangential surface of the samples where the degraded plane was placed under tension. The MOE and MOR of the wood samples were calculated based on the ISO 13061-3 and -4 from the mechanical bending tests.

2.5. MC Measurements

The measurement of MC and density after weathering was done based on the ISO-13061-1 and -2 standard methods [67,68] from the cookie samples (50 mm × 20 mm × 20 mm [L, T, R]) cut from the vicinity of the wood samples. The average MC of the samples was about 12% before the weathering experiment. For this purpose, separate cookies with the same size were prepared before the preparation of the final weathering samples (20 mm × 20 mm × 300 mm). The samples prepared for MC measurements were weighed (m_1) and then placed in an oven at the temperature of 101 °C for 24 h. The oven-dried weights (m_2) of the samples were then measured and the MC was calculated via:

$$W\,(\%) = \frac{m_1 - m_2}{m_2} \times 100 \qquad (2)$$

where W, m_2, and m_1 are the MC, the dry mass and the wet mass of the wood samples, respectively.

2.6. Statistical Analysis

The data of color change were analyzed with a one-way ANOVA in Minitab 19 (AppOnFly s.r.o., Prague, Czech Republic). For each wood species, the significance between the color changes of the samples weathered at different time periods was analyzed using Tukey's comparison test. Having a low number of replications is a limitation of such a study. However, it should be noted that the main aim of this research was to monitor the weathered wood samples through machine learning modeling, which does not necessarily depend on the requirements for statistical analysis.

2.7. Machine Learning Analysis

Figure 2 illustrates the flowchart of the adopted methodology in this study to estimate the mechanical properties of weathered wood through machine learning (ML) modeling. Different ML models and artificial neural networks (ANNs) have been employed for damage/defect detection [69,70], prediction of the material's properties [71,72], and process condition monitoring [73–76]. Decision tree regression modeling was used in this study for the prediction of the mechanical properties of wood samples. While ANNs are challenging to interpret, the outputs of decision tree models can be comprehended easier. Furthermore, the significance of predictor variable can be identified in the model, which helps to study the relationships between the features and take care of the redundant features with relatively lower importance [77]. The current study dealt with a small dataset; however, decision tree modeling has been successfully employed in the literature to analyze small datasets [15,59,78] when predicting the mechanical properties or checks formation in weathered wood samples. The classification and regression trees (CART) algorithm [79] was used in this study. The decision tree development process included tree growing and pruning phases. In the tree growing phase, the node splitting criteria were based on the highest contribution to lowering the least squared error during the model training. The optimal tree was chosen in the pruning phase as the smallest tree that has an R^2 within one standard error of the tree representing the highest R^2 obtained during the model validation. The 5-fold cross-validation method was employed in this study. The details of the developed decision tree model were similar to those chosen in [59] and based on the discussion provided in [80].

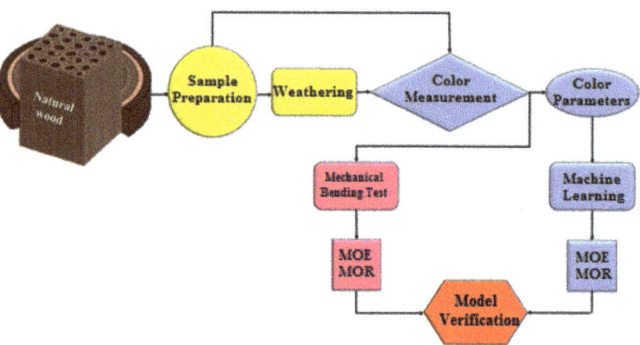

Figure 2. Flowchart of the proposed methodology to monitor the mechanical properties of the weathered wood by color measurement.

The contribution of each feature (either as a primary splitters or surrogate one) in improving the performance of the model was studied based on minimizing the least squared error. The most important feature causes the highest improvement to the model and the other features are relatively ranked, accordingly. Additionally, clustering analysis was employed to observe the level of similarity and find the common characteristics between the color parameters, MOE and MOR, using agglomerative hierarchical clustering as explained by Fathi et al. [81].

3. Results

3.1. Mechanical Degradation under Artificial Weathering

The MC and density of the wood samples decreased under weathering. Following 240 h of exposure, fir, alder, oak, and poplar wood experienced a 4.9%, 3.2%, 4.7%, and 8.1% reduction in the density, respectively. The initial MC of fir (12.1%), alder (11.6%), oak (12.4%), and poplar (12.1%) were also reduced to 6.3%, 5.3%, 5.8%, 5.3%, respectively, after 240 h of weathering. Ouadou et al. [49] reported that 120 h of weathering resulted in the mass loss due to the reduction of the MC. Figure 3 shows the impact of weathering on the MOE, MOR, and deflection to failure of different wood samples. It shows that the MOE and MOR decreased with increasing weathering duration. Overall, alder was shown to be the most resilient wood with an 11.5% and 11% reduction in its MOE and MOR after 240 h of weathering. Fir also exhibited the highest degradation, with a 21.5% and 17.5% reduction in its MOE and MOR after 240 h of weathering. The mechanical degradation under ultraviolet conditions was discussed to affect the microstructure of wood and its effect on the chemical composition and results in degradation of lignin [49]. Additionally, Persze and Tolvaj [58] explained that the higher temperatures increase the wood degradation under ultraviolet conditions. The mass loss of wood may also occur during the weathering [49]. Boonstra et al. [56] reported that wood specimens conditioned after weathering showed lower reduction in the mechanical properties. Thus, the mechanical degradation observed in Figure 3 is affected by the combined impact of the UV radiation and temperature condition resulting in mass loss due to reduction in the MC.

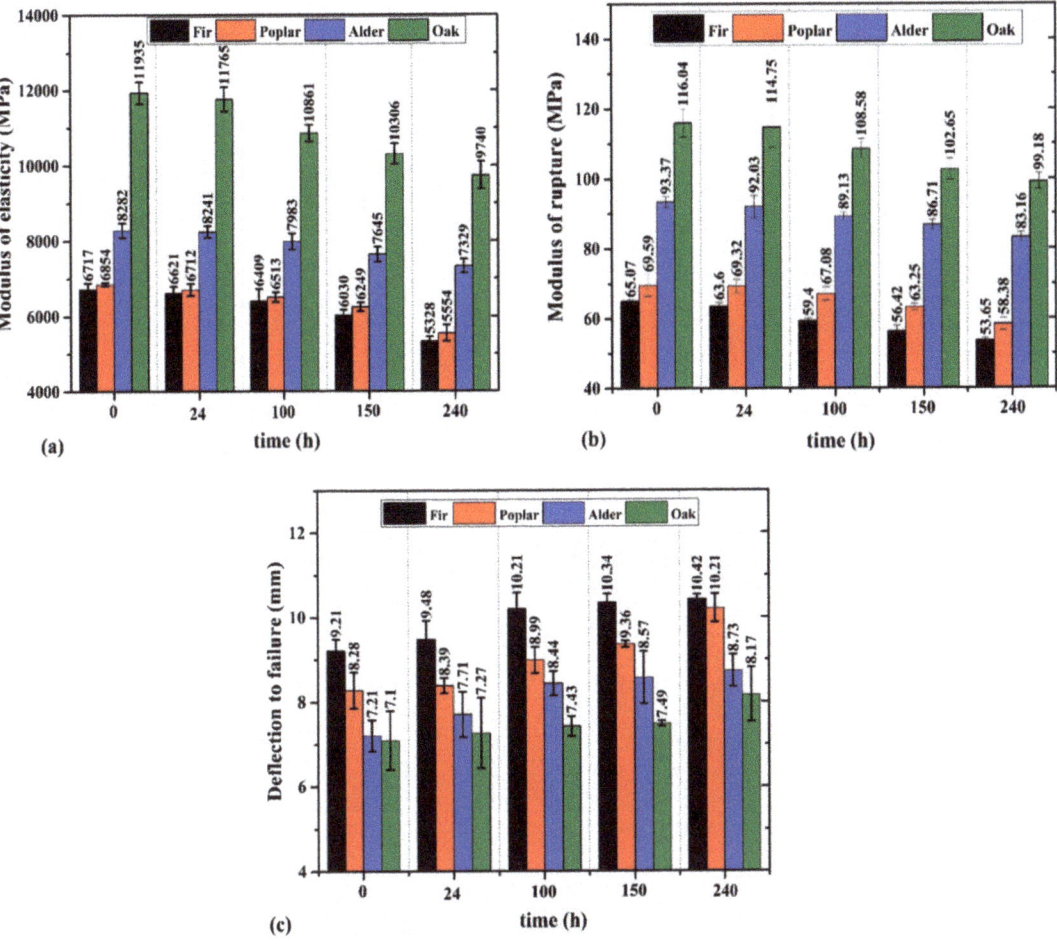

Figure 3. The variation of MOE (**a**), MOR (**b**), and deflection to failure (**c**) in weathered wood samples.

Despite the reduction in the MOE and MOR, weathering led to an increase in the deflection to failure in all wood species. This increment was 13%, 23%, 21%, and 15% for fir, poplar, alder, and oak wood after 240 h of exposure. This observation may be linked to the impact of ultraviolet condition on enhancing the viscoelasticity of wood. Fathi et al. [60] showed that the loss modulus and loss factor of the wood samples tested in this study increased under ultraviolet condition, proving that wood exhibits more viscoelasticity when exposed to UV degradation. This may result in an increase in the deflection to failure in the degraded samples. Feist [64] reported the leaching and plasticizing effects of water during the weathering process. This plasticizing effect can also facilitate the enlargement of micro-checks. Figure 4 shows the load-deflection diagram of the wood samples after different weathering time periods. It shows that the degradation resulted in having a larger deflection at certain load levels. It was also indicated that after 150 h or 240 h of exposure, the fir and poplar wood samples reached a deflection of 5 mm at a load level of around 1000 N. However, this deflection was achieved at a higher load level (~1500 N–1600 N) for hardwoods such as alder and oak. Apart from its impact on the MOE, MOR, and deflection to failure, the degradation affected the failure modes. Figure 5 shows the failure modes after the static bending tests for the oak samples under weathering at three exposure time periods. It was observed that the failure mode was mainly governed by the tensile failure perpendicular to the grain in the radial-longitudinal plane (fiber compression was also observed in parallel). Consequently, increasing the exposure time resulted in having a deeper brittle fracture and splinter in tension. This can be due to the destruction of the middle lamella [65], which contains significant lignin content. Severe checking may develop with longer exposure times in the cell wall components resulting in the loosening and detaching of the fibrils and tracheids from the surface [65]. Feist [64] discussed that the weathering condition can result in the destruction of the middle lamella and cell wall layers. This can impact the cohesive strength of wood tissue [65] that may result in the mechanical degradation of the wood as well.

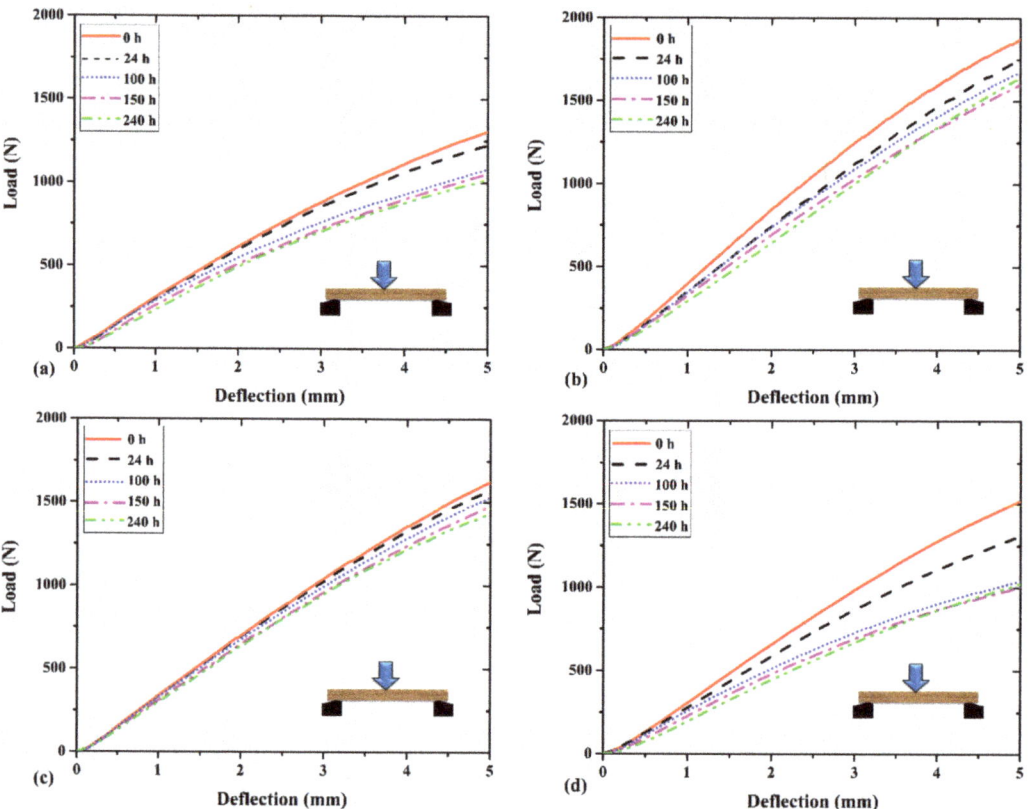

Figure 4. The load–deflection diagram for (**a**) fir, (**b**) alder, (**c**) oak, and (**d**) poplar wood samples exposed to different times of weathering.

3.2. Color Change

Tolvaj and Faix [82] stated that the light reflection decreased after UV radiation, resulting in the wood seeming darker. The reduction in the lightness of the samples with the exposure time can be linked to the negative ΔL, as shown in Figure 6. A similar trend of darkening of the wood appearance after UV radiation has been reported in the literature [35,82]. It can be seen in Figure 6 that after 24 h of exposure, there was a trivial change in the lightness with a minor difference between different wood species. In this time, poplar and oak showed more darkening ($\Delta L = -2.5$). By increasing the exposure time, fir underwent a significant change in ΔL while alder showed a minimal change in ΔL. After 240 h, alder accounted for the minimum change in the lightness ($\Delta L = -4$) followed by oak (ΔL = −11.5) and poplar (ΔL = −13.7), and the maximum change in the lightness occurred in fir wood (ΔL = −25).

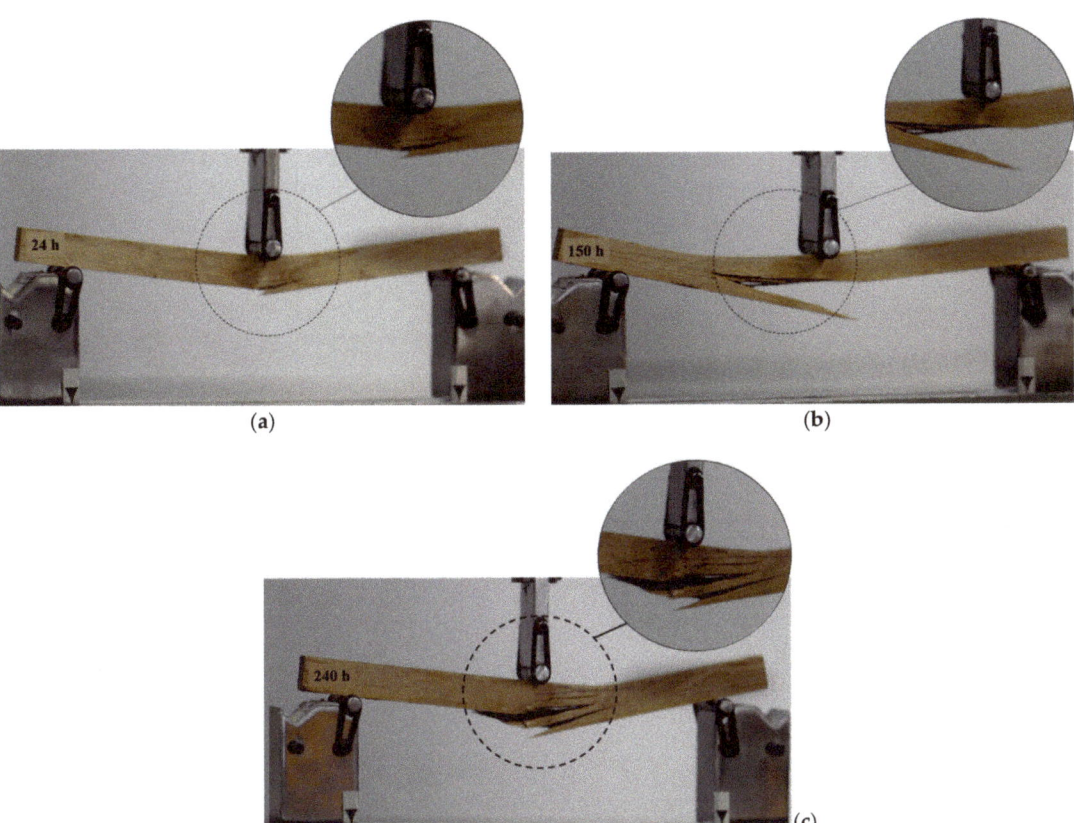

Figure 5. The fracture plane of the weathered oak wood at three exposure times: 24 h (**a**), 150 h (**b**), and 240 h (**c**) under the static bending test.

Variation in the yellowness (Δb) of the wood samples exposed to weathering is shown in Figure 7. It can be seen that the yellowness increased with exposure time and reached its maximum value after 150 h. After this time, fir poplar, and oak experienced a slight reduction in the yellowness, while it stayed almost unchanged for alder. The increase in the yellowness of the samples was discussed to be mainly due to the degradation of lignin [35,36,46]. The continuous increase of the yellowness followed by a slight reduction under artificial weathering was reported by Timar et al. [36] and Pandey [83]. Figure 7 indicates that lignin degradation occurred even after 24 h of exposure. The fir wood accounted for the highest rate of increment in the yellowness following the exposure, with the highest Δb occurring after 150 h of exposure (Δb = 23.8). On the other hand, the smallest change in the yellowness was observed in alder, suggesting a lower lignin degradation of alder wood under ultraviolet conditions compared to other species. The variation in the yellowness was almost similar in the first 100 h of the exposure for poplar and oak wood. For longer exposure times (i.e., 150 h and 240 h), the yellowness of poplar was relatively higher than that for oak.

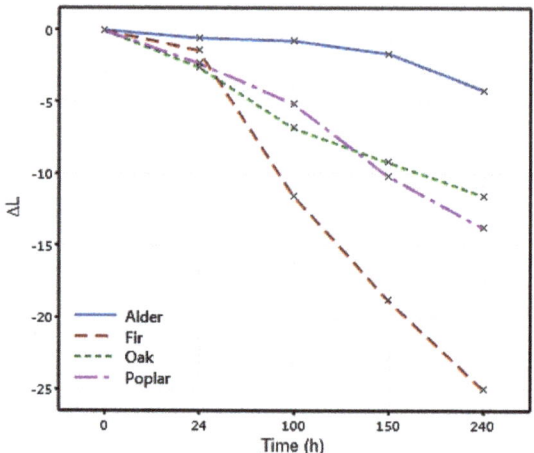

Figure 6. Variation in the lightness (ΔL) of different wood species under artificial weathering.

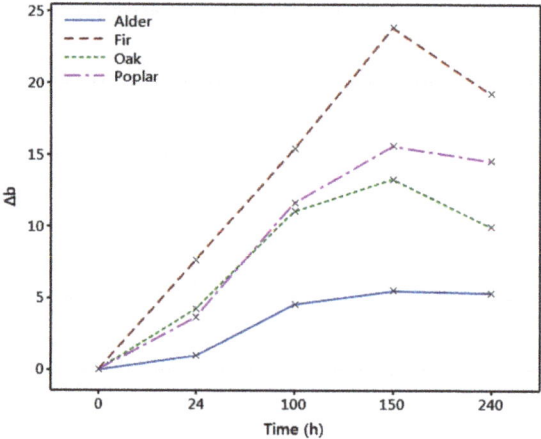

Figure 7. Variation in the yellowness (Δb) of different wood species exposed to artificial weathering.

The variation in the redness (Δa) of the degraded wood samples is shown in Figure 8. It can be seen that there was a reduction in the redness of all wood species during the first 100 h of exposure. However, there was an increasing trend in the redness of the samples after 100 h. While Tolvaj and Faix [82] discussed that it is challenging to explain the wood color change in a definite way, a correlation between the Δa and the extractives content in wood has been reported in the literature [41,84]. For example, Persze and Tolvaj [58] showed that the extractives content has an important role in thermal decomposition during photodegradation. They also reported that the Δa is higher at elevated temperatures, whereas the thermal effect does not alter yellowing. It is also indicated in Figure 8 that after 240 h, fir and oak accounted for the highest increase in the redness (Δa = 1.9), while alder experienced a reduction in it. The color change parameters and the total color change for all wood species (ΔE) are listed in Table 1. It was observed that the total color change increased with the exposure time. A general trend for ΔE of fir > poplar > oak > alder was observed at all exposure times. Interestingly, fir that experienced the maximum color change showed the highest reduction in the mechanical properties, and the alder with the smallest total color change showed the minimum mechanical degradation. This indicates that the color change may be correlated with the mechanical degradation of wood and be used to predict MOE and MOR.

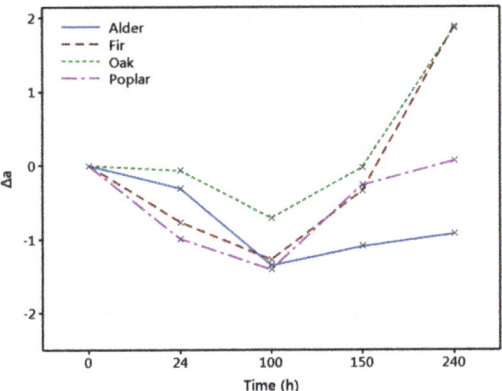

Figure 8. Variation in the redness (Δa) of different wood species exposed to artificial weathering.

Table 1. The mean value and standard deviation of the color change of wood species in different weathering time periods.

Wood	Time (hrs)	ΔL		Δa		Δb		ΔE	
		Mean	Std.	Mean	Std.	Mean	Std.	Mean	Std.
Fir	24	−1.39 [a]	0.18	−0.76 [a]	0.13	7.66 [a]	0.63	7.82 [a]	0.66
	100	−11.51 [b]	1.83	−1.27 [b]	0.18	15.42 [b]	1.17	19.35 [b]	0.95
	150	−18.80 [c]	2.74	−0.35 [c]	0.05	23.88 [c]	1.79	30.42 [c]	2.98
	240	−25.01 [d]	2.58	1.89 [d]	0.14	19.26 [d]	1.49	31.69 [c]	1.72
Alder	24	−0.55 [a]	0.12	−0.31 [a]	0.05	1.01 [a]	0.17	1.19 [a]	0.17
	100	−0.75 [a]	0.18	−1.35 [b]	0.04	4.60 [b]	1.00	4.86 [b]	0.92
	150	−1.63 [b]	0.22	−1.09 [c]	0.07	5.52 [b]	1.46	5.89 [bc]	1.32
	240	−4.19 [c]	0.34	−0.92 [c]	0.09	5.39 [b]	1.10	6.92 [c]	0.84
Oak	24	−2.55 [a]	0.39	−0.07 [a]	0.02	4.26 [a]	1.30	5.02 [a]	1.07
	100	−6.81 [b]	1.06	−0.71 [b]	0.11	11.07 [b]	1.69	13.06 [b]	1.50
	150	−9.20 [bc]	1.60	−0.03 [a]	0.01	13.32 [b]	2.16	16.22 [b]	2.43
	240	−11.50 [c]	1.91	1.86 [c]	0.37	9.97 [b]	1.90	15.44 [b]	1.64
Poplar	24	−2.26 [a]	0.39	−0.99 [a]	0.15	3.67 [a]	0.51	4.45 [a]	0.31
	100	−5.11 [a]	0.94	−1.41 [b]	0.19	11.66 [b]	0.98	12.81 [b]	1.26
	150	−10.14 [b]	1.86	−0.27 [c]	0.04	15.62 [c]	1.42	18.72 [c]	0.78
	240	−13.75 [c]	2.11	0.06 [d]	0.01	14.58 [c]	0.94	20.12 [c]	0.95

* Different letters ([a–d]) within a column for each wood species show the significant difference by Tukey's comparison test ($p < 0.05$).

3.3. MOE and MOR Prediction

The wood color parameters and type of the wood species were used as the input of the decision tree model to predict MOE and MOR. The results of machine learning modeling are shown in Table 2. It was observed that when the wood species type was combined with a, b, and L, the developed decision tree model predicted the mechanical properties of wood with an R^2 of 0.84 and 0.77, respectively (test data). Figure 9 illustrates the relative importance of the selected input parameters used in the decision tree model and indicates that all of them were significant in developing the predictive model. It was shown that following the type of wood species, the level of redness and yellowness had the highest relative importance.

Table 2. The coefficient of determination obtained for the prediction of the wood mechanical properties using different input features.

Model Inputs	R²			
	MOE		MOR	
	Train	Test	Train	Test
Wood species, a, b, L	0.92	0.84	0.81	0.77
Wood species, a, b, L, ΔE	0.92	0.87	0.93	0.88
Wood species, a, b, L, ΔE, Δa, Δb, ΔL	0.90	0.88	0.92	0.90

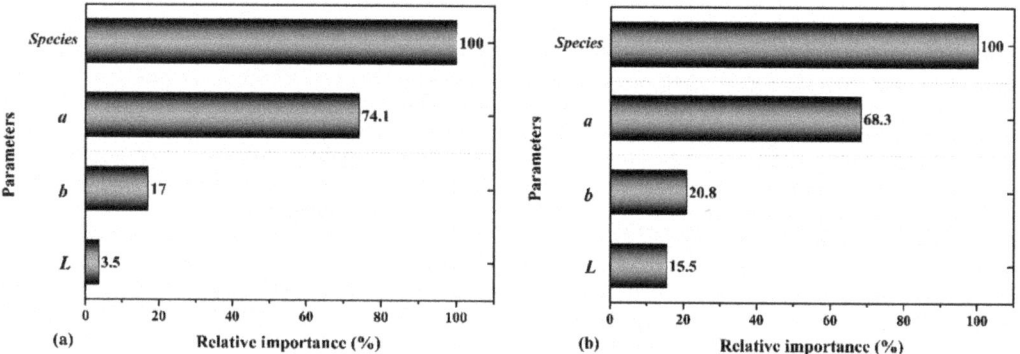

Figure 9. The relative importance of the wood species, a, b, and L for predicting the (a) MOE and (b) MOR.

Table 2 shows that adding the total color change Δ(E) parameter as an input to the model improved the prediction accuracy. In this case, the MOE and MOR were predicted with an R^2 of 0.87 and 0.88, respectively (test data). Figure 10 shows the relative importance of the parameters used in this model. It was indicated that all of the input parameters were important in the performance of the model, while the total color change had relatively a higher importance for the prediction of the MOR than MOE. It was also observed in Figures 9 and 10 that the lightness parameter had higher relative importance when predicting the MOR than MOE.

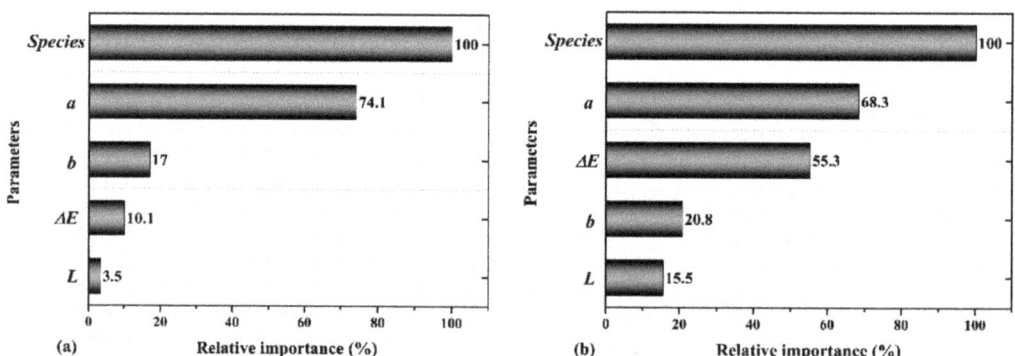

Figure 10. The relative importance of wood species, a, b, L, and ΔE for predicting the (a) MOE and (b) MOR.

The last model, which used the wood species, color parameters (a, b, L), total color change (ΔE), and variation in the color parameters Δ(a, Δb, and ΔL), did not make a noticeable improvement in the prediction accuracy of MOE (R^2 = 0.88 for the test data) and

MOR (R^2 = 0.90 for the test data). Overall, adding the three parameters Δa, Δb, and ΔL had a minor positive impact in the performance of the model. Figures 10 and 11 indicate that the wood species type and redness were still the most important parameters for the prediction of the MOE and MOR. However, in this model, Δb had higher importance than Δa. Markedly, while the level of redness (a), described in the literature to be linked to the extractives content, had relatively a high importance in the performance of the model, its variation (Δa) exhibited less importance in the performance of the predictive model. Overall, these findings indicate that all of the selected input parameters contributed positively to the predictive accuracy of the ML model, albeit with different levels of relative importance. These findings may suggest that there could be a dependency between the mechanical properties of weathered wood and its color parameters. However, further studies with a larger sample size could better clarify the details of these dependencies, especially when combined with the chemical composition analysis.

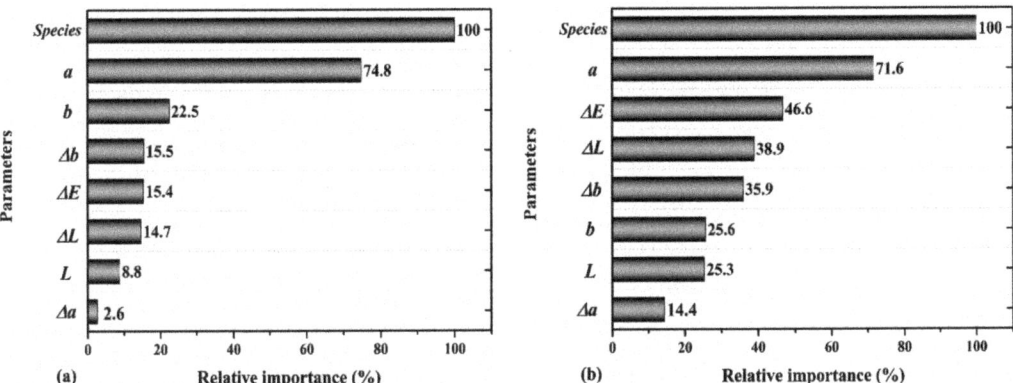

Figure 11. The relative importance of wood species, a, b, L, ΔE, Δa, Δb, and ΔL for predicting the (a) MOE and (b) MOR.

The results of variable clustering analysis were in accordance with those of the decision tree modeling. Figure 12 shows that the MOE and MOR had the highest similarity level with the redness, and form a cluster together. This cluster has the highest similarity with a cluster encompassing the total color change (ΔE), yellowness (b), and variation in the redness (Δa) and yellowness (Δb). The MOE and MOR had the lowest similarity with the lightness (L) and variation in it (ΔL), with a 28.44% similarity level. The general observations of the variable clustering analysis prove that the selected input parameters were correlated with the mechanical properties of degraded wood, which can be used for MOE and MOR prediction. Further study can be performed to outline these dependencies and similarities between the parameters when weathering conditions are changed.

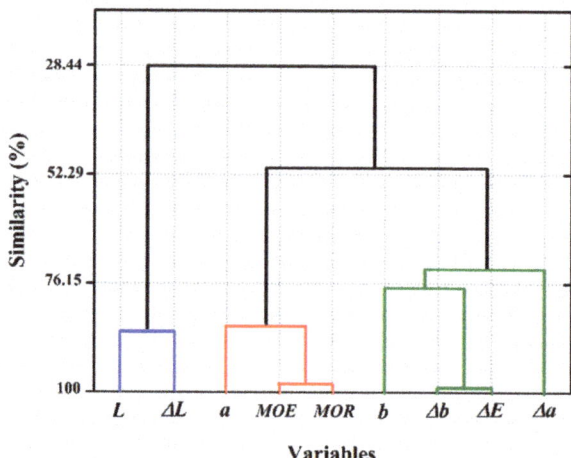

Figure 12. Variable clustering analysis on the color parameters, MOE and MOR (the blue, red and green lines represent the components of the CIELAB three-dimensional color space).

4. Discussion and Remarks

The general trend of the mechanical degradation of the wood samples indicated the better resistance of alder and the higher vulnerability of fir to weathering condition. While it is not within the scope of a monitoring task, further research may be conducted to include the change in the chemical composition of wood to explain the role of extractives and degradation of lignin and how they can be linked to the mechanical properties of wood. The machine learning analysis showed the high importance of change in redness that is linked to the change in the extractives content in the literature. Future research considering the chemical composition of wood could better explain these observed trends. More research could also be performed to assess the relationship between the wood viscoelasticity behavior under artificial weathering and its failure mode and fracture behavior when subjected to loading. This study did not focus on evaluating the UV penetration depth, and aimed only at developing a monitoring model that reflects the combined effects of different parameters accountable for the mechanical degradation of wood under weathering. Future research aiming to explain the mechanism of degradation, rather than in situ monitoring, could be performed to separate the role of ultraviolet radiation, temperature condition, and MC and mass loss on the mechanical behavior of weathered wood.

The performance was obtained on the basis of analyzing a small set of defect-free wood samples. The scope of the experiment could be further expanded while performing the proposed methodology on real-sized timber, including defects such as knots. One of the main challenges associated with color measurement for the characterization of wood properties is non-homogeneous surface color variation [33], especially in real-sized timber, which necessitates multiple color measurements from different locations of the timber. This is an important challenge associated with using color parameters for monitoring the mechanical properties of timber. However, due to the simplicity of the color measurement, the color features can be combined with the data acquisition from other NDE methods (such as Lamb wave propagation) to result in a robust monitoring model. Nasir et al. [59] obtained a similar range for R^2 while using Lamb wave features for predicting the mechanical properties of weathered wood. However, their model did not need to have the type of wood species as an input parameter, as the model was able to understand the difference between the wood species through the extracted Lamb wave features. Thus, if the type of wood species to be monitored is not available, a machine learning model based on the wave propagation features may be chosen. Both the color measurement and wave propagation method should be applied to real-sized timber to have a more reliable comparative study. It should be mentioned that in real applications, other factors related to the environment (rain, decay, etc.) or the timber (grain orientation, defects, annual ring width, etc.) may impact the mechanical properties of degraded wood, and having a single monitoring method may not address both the surface and internal damages to the timber. In this case, combining color parameters that reflect the surface degradation with those of wave propagation that are linked to the internal structure of the wood can better show the mechanical degradation of large-size timber. Additionally, timber monitoring requires dataset of a larger size and perhaps other types of data-driven modeling. The choice of data-driven method depends on both the size and complexity of the data [85,86]. Different machine learning or deep learning models could be studied to choose the one that better fits the dataset of a larger size [87].

5. Conclusions

In this research, a machine learning model was developed based on the color parameters for predicting the MOE and MOR values of weathered wood. Artificial weathering led to a reduction in the wood mechanical properties. Wood species that experienced greater color changes exhibited a more highlighted mechanical degradation. This indicates that the color parameters may be linked to the MOE and MOR of weathered wood. Fir was shown to be more susceptible to artificial weathering, while alder was more resilient and experienced less reduction in its mechanical properties. The mechanical properties of the samples could be predicted by the color parameters, although the developed machine learning model needed the type of wood species as an input parameter for accurate prediction. It was also observed that the deflection to failure of the wood samples increased with the weathering and the failure mode under bending loading also changed. This may be due to the increased viscoelasticity of weathered wood samples. Future studies should be aimed at investigating the relation between the mechanical properties and the color parameters on large-sized timber to evaluate the reliability of the developed model based on the color parameters for NDE of timber structures. Furthermore, the chemical changes caused by weathering and the correlation between the color, mechanical and chemical properties of wood should be studied in future research. This study was performed on small and defect-free wood samples. Additionally, the sample size was small, which is a limitation of the statistical analysis. However, the decision tree modeling was able to successfully handle the dataset. Focusing on large-sized timber containing defects may require expanding the size of dataset in future studies.

Author Contributions: Conceptualization, V.N. and S.K.; methodology, V.N., A.F. and S.K.; experiment and data acquisition, H.F., software, V.N. and H.F.; validation, V.N.; formal analysis, V.N., S.K.; investigation, A.F.; resources, V.N., A.F., S.K., and P.A.; writing—original draft preparation, V.N. and S.K.; writing—review and editing, S.K., F.S., and P.A.; visualization, V.N., S.K. and H.F.; supervision, S.K.; project administration, S.K., F.S., and P.A. All authors have read and agreed to the published version of the manuscript.

Funding: This research received no external funding.

Data Availability Statement: Not applicable.

Conflicts of Interest: The authors declare no conflict of interest.

References

1. Feio, A.; Machado, J.S. In-situ assessment of timber structural members: Combining information from visual strength grading and NDT/SDT methods—A review. *Constr. Build. Mater.* **2015**, *101*, 1157–1165. [CrossRef]
2. Palma, P.; Steiger, R. Structural health monitoring of timber structures–Review of available methods and case studies. *Constr. Build. Mater.* **2020**, *248*, 118528. [CrossRef]
3. Machado, J.S.; Pereira, F.; Quilho, T. Assessment of old timber members: Importance of wood species identification and direct tensile test information. *Constr. Build. Mater.* **2019**, *207*, 651–660. [CrossRef]

4. Piazza, M.; Riggio, M. Visual strength-grading and NDT of timber in traditional structures. *J. Build. Apprais.* **2008**, *3*, 267–296. [CrossRef]
5. Ayanleye, S.; Avramidis, S. Predictive capacity of some wood properties by near-infrared spectroscopy. *Int. Wood Prod. J.* **2021**, *12*, 83–94. [CrossRef]
6. Zhou, Z.; Rahimi, S.; Avramidis, S.; Fang, Y. Species-and moisture-based sorting of green timber mix with near infrared spectroscopy. *BioResources* **2020**, *15*, 317–330.
7. Zhou, Z.; Rahimi, S.; Avramidis, S. On-line species identification of green hem-fir timber mix based on near infrared spectroscopy and chemometrics. *Eur. J. Wood Wood Prod.* **2020**, *78*, 151–160. [CrossRef]
8. Ayanleye, S.; Nasir, V.; Avramidis, S.; Cool, J. Effect of wood surface roughness on prediction of structural timber properties by infrared spectroscopy using ANFIS, ANN and PLS regression. *Eur. J. Wood Wood Prod.* **2021**, *79*, 101–115. [CrossRef]
9. Nasir, V.; Nourian, S.; Zhou, Z.; Rahimi, S.; Avramidis, S.; Cool, J. Classification and characterization of thermally modified timber using visible and near-infrared spectroscopy and artificial neural networks: A comparative study on the performance of different NDE methods and ANNs. *Wood Sci. Technol.* **2019**, *53*, 1093–1109. [CrossRef]
10. Stefansson, P.; Thiis, T.; Gobakken, L.R.; Burud, I. Hyperspectral NIR time series imaging used as a new method for estimating the moisture content dynamics of thermally modified Scots pine. *Wood Mater. Sci. Eng.* **2021**, *16*, 49–57. [CrossRef]
11. Riggio, M.; Anthony, R.W.; Augelli, F.; Kasal, B.; Lechner, T.; Muller, W.; Tannert, T. In situ assessment of structural timber using non-destructive techniques. *Mater. Struct.* **2014**, *47*, 749–766. [CrossRef]
12. Beall, F.C. Overview of the use of ultrasonic technologies in research on wood properties. *Wood Sci. Technol.* **2002**, *36*, 197–212. [CrossRef]
13. Van Blokland, J.; Olsson, A.; Oscarsson, J.; Adamopoulos, S. Prediction of bending strength of thermally modified timber using high-resolution scanning of fibre direction. *Eur. J. Wood Wood Prod.* **2019**, *77*, 327–340. [CrossRef]
14. Van Blokland, J.; Adamopoulos, S.; Ahmed, S.A. Performance of Thermally Modified Spruce Timber in Outdoor Above-Ground Conditions: Checking, Dynamic Stiffness and Static Bending Properties. *Appl. Sci.* **2020**, *10*, 3975. [CrossRef]
15. Van Blokland, J.; Nasir, V.; Cool, J.; Avramidis, S.; Adamopoulos, S. Machine learning-based prediction of internal checks in weathered thermally modified timber. *Constr. Build. Mater.* **2021**, *281*, 122193. [CrossRef]
16. Van Blokland, J.; Olsson, A.; Oscarsson, J.; Daniel, G.; Adamopoulos, S. Crack formation, strain distribution and fracture surfaces around knots in thermally modified timber loaded in static bending. *Wood Sci. Technol.* **2020**, *54*, 1001–1028. [CrossRef]
17. Nasir, V.; Nourian, S.; Avramidis, S.; Cool, J. Stress wave evaluation for predicting the properties of thermally modified wood using neuro-fuzzy and neural network modeling. *Holzforschung* **2019**, *73*, 827–838. [CrossRef]
18. Fathi, H.; Kazemirad, S.; Nasir, V. Lamb wave propagation method for nondestructive characterization of the elastic properties of wood. *Appl. Acoust.* **2021**, *171*, 107565. [CrossRef]
19. Fathi, H.; Kazemirad, S.; Nasir, V. A nondestructive guided wave propagation method for the characterization of moisture-dependent viscoelastic properties of wood materials. *Mater. Struct.* **2020**, *53*, 1–14. [CrossRef]
20. Kubovsky, I.; Kristak, L.; Suja, J.; Gajtanska, M.; Igaz, R.; Ruziak, I.; Reh, R. Optimization of parameters for cutting of wood-based materials by CO_2 laser. *Appl. Sci.* **2020**, *10*, 8113. [CrossRef]
21. Kubovsky, I.; Kacik, F.; Velkova, V. The effects of CO2 laser irradiation on color and major chemical component changes in hardwoods. *BioResources.* **2018**, *13*, 2515–2529. [CrossRef]
22. Torniainen, P.; Jones, D.; Sandberg, D. Colour as a quality indicator for industrially manufactured ThermoWood®. *Wood Mater. Sci. Eng.* **2021**, *16*, 287–289. [CrossRef]
23. Torniainen, P.; Elustondo, D.; Dagbro, O. Industrial validation of the relationship between color parameters in thermally modified spruce and pine. *BioResources.* **2016**, *11*, 1369–1381. [CrossRef]
24. Ockajova, A.; Kucerka, M.; Kminiak, R.; Kristak, L.; Igaz, R.; Reh, R. Occupational exposure to dust produced when milling thermally modified wood. *Int. J. Environ. Res. Public Health* **2020**, *17*, 1478. [CrossRef]
25. González-Peña, M.M.; Hale, M.D. Colour in thermally modified wood of beech, Norway spruce and Scots pine. Part 1: Colour evolution and colour changes. *Holzforschung* **2009**, *63*, 385–393. [CrossRef]
26. González-Peña, M.M.; Hale, M.D. Colour in thermally modified wood of beech, Norway spruce and Scots pine. Part 2: Property predictions from colour changes. *Holzforschung* **2009**, *63*, 394–401. [CrossRef]
27. Kamperidou, V.; Barboutis, I.; Vasileiou, V. Response of colour and hygroscopic properties of Scots pine wood to thermal treatment. *J. For. Res.* **2013**, *24*, 571–575. [CrossRef]
28. Torniainen, P.; Popescu, C.-M.; Jones, D.; Scharf, A.; Sandberg, D. Correlation of Studies between Colour, Structure and Mechanical Properties of Commercially Produced ThermoWood® Treated Norway Spruce and Scots Pine. *Forests* **2021**, *12*, 1165. [CrossRef]
29. Brischke, C.; Welzbacher, C.R.; Brandt, K.; Rapp, A.O. Quality control of thermally modified timber: Interrelationship between heat treatment intensities and CIE L* a* b* color data on homogenized wood samples. *Holzforschung.* **2007**, *61*, 19–22. [CrossRef]
30. Kamperidou, V.; Barmpoutis, P. Correlation between the changes of Colour and Mechanical properties of Thermally-modified Scots Pine (*Pinus sylvestris* L.) Wood. *Pro Ligno* **2015**, *11*, 360–365.
31. Todaro, L.; Zuccaro, L.; Marra, M.; Basso, B.; Scopa, A. Steaming effects on selected wood properties of Turkey oak by spectral analysis. *Wood Sci. Technol.* **2012**, *46*, 89–100. [CrossRef]
32. Nasir, V.; Nourian, S.; Avramidis, S.; Cool, J. Classification of thermally treated wood using machine learning techniques. *Wood Sci. Technol.* **2019**, *53*, 275–288. [CrossRef]

33. Johansson, D.; Morén, T. The potential of colour measurement for strength prediction of thermally treated wood. *Holz Roh. Werkst.* **2006**, *64*, 104–110. [CrossRef]
34. Nasir, V.; Nourian, S.; Avramidis, S.; Cool, J. Prediction of physical and mechanical properties of thermally modified wood based on color change evaluated by means of "group method of data handling" (GMDH) neural network. *Holzforschung* **2019**, *73*, 381–392. [CrossRef]
35. Oberhofnerová, E.; Pánek, M.; García-Cimarras, A. The effect of natural weathering on untreated wood surface. *Maderas Cienc. Tecnol.* **2019**, *19*, 173–184. [CrossRef]
36. Timar, M.C.; Varodi, A.M.; Gurău, L. Comparative study of photodegradation of six wood species after short-time UV exposure. *Wood Sci. Technol.* **2016**, *50*, 135–163. [CrossRef]
37. Tomak, E.D.; Ermeydan, M.A. A natural flavonoid treatment of wood: Artificial weathering and decay resistance. *Eur. J. Wood Wood Prod.* **2020**, *78*, 1221–1231. [CrossRef]
38. Srinivas, K.; Pandey, K.K. Photodegradation of thermally modified wood. *J. Photochem. Photobiol. B* **2012**, *117*, 140–145. [CrossRef] [PubMed]
39. Todaro, L.; D'Auria, M.; Langerame, F.; Salvi, A.M.; Scopa, A. Surface characterization of untreated and hydro-thermally pre-treated Turkey oak woods after UV-C irradiation. *Surf. Interface Anal.* **2015**, *47*, 206–215. [CrossRef]
40. Herrera, R.; Arrese, A.; de Hoyos-Martinez, P.L.; Labidi, J.; Llano-Ponte, R. Evolution of thermally modified wood properties exposed to natural and artificial weathering and its potential as an element for façades systems. *Constr. Build. Mater.* **2018**, *172*, 233–242. [CrossRef]
41. Rüther, P.; Jelle, B.P. Color changes of wood and wood-based materials due to natural and artificial weathering. *Wood Mater. Sci. Eng.* **2013**, *8*, 13–25. [CrossRef]
42. Kržišnik, D.; Lesar, B.; Thaler, N.; Humar, M. Influence of natural and artificial weathering on the colour change of different wood and wood-based materials. *Forests* **2018**, *9*, 488. [CrossRef]
43. D'Auria, M.; Lovaglio, T.; Rita, A.; Cetera, P.; Romani, A.; Hiziroglu, S.; Todaro, L. Integrate measurements allow the surface characterization of thermo-vacuum treated alder differentially coated. *Measurement* **2018**, *114*, 372–381. [CrossRef]
44. Reinprecht, L.; Mamoňová, M.; Pánek, M.; Kačík, F. The impact of natural and artificial weathering on the visual, colour and structural changes of seven tropical woods. *Eur. J. Wood Prod.* **2018**, *76*, 175–190. [CrossRef]
45. Teacă, C.A.; Roşu, D.; Bodîrlău, R.; Roşu, L. Structural changes in wood under artificial UV light irradiation determined by FTIR spectroscopy and color measurements–A brief review. *BioResources* **2013**, *8*, 1478–1507. [CrossRef]
46. Müller, U.; Rätzsch, M.; Schwanninger, M.; Steiner, M.; Zöbl, H. Yellowing and IR-changes of spruce wood as result of UV-irradiation. *J. Photochem. Photobiol. B* **2003**, *69*, 97–105. [CrossRef]
47. Cogulet, A.; Blanchet, P.; Landry, V. Wood degradation under UV irradiation: A lignin characterization. *J. Photochem. Photobiol. B* **2016**, *158*, 184–191. [CrossRef]
48. Dong, Y.; Wang, J.A.; Zhu, J.; Jin, T.; Li, J.; Wang, W.; Xia, C. Surface colour and chemical changes of furfurylated poplar wood and bamboo due to artificial weathering. *Wood Mater. Sci. Eng.* **2020**, 1–8. [CrossRef]
49. Ouadou, Y.; Aliouche, D.; Thevenon, M.F.; Djillali, M. Characterization and photodegradation mechanism of three Algerian wood species. *J. Wood Sci.* **2017**, *63*, 288–294. [CrossRef]
50. Derbyshire, H.; Miller, E.R. The photodegradation of wood during solar irradiation. *Holz Roh. Werkst.* **1981**, *39*, 341–350. [CrossRef]
51. Sharratt, V.; Hill, C.A.; Kint, D.P. A study of early colour change due to simulated accelerated sunlight exposure in Scots pine (Pinus sylvestris). *Polym. Degrad. Stab.* **2009**, *94*, 1589–1594. [CrossRef]
52. Derbyshire, H.; Miller, E.R.; Turkulin, H. Investigations into the photodegradation of wood using microtensile testing. *Holz Roh. Werkst.* **1995**, *53*, 339–345. [CrossRef]
53. De la Caba, K.; Guerrero, P.; Del Río, M.; Mondragon, I. Weathering behaviour of wood-faced construction materials. *Constr. Build. Mater.* **2007**, *21*, 1288–1294. [CrossRef]
54. Del Menezzi, C.H.S.; de Souza, R.Q.; Thompson, R.M.; Teixeira, D.E.; Okino, E.Y.A.; da Costa, A.F. Properties after weathering and decay resistance of a thermally modified wood structural board. *Int. Biodeterior.* **2008**, *62*, 448–454. [CrossRef]
55. Sonderegger, W.; Kránitz, K.; Bues, C.T.; Niemz, P. Aging effects on physical and mechanical properties of spruce, fir and oak wood. *J. Cult. Herit.* **2015**, *16*, 883–889. [CrossRef]
56. Boonstra, M.J.; Van Acker, J.; Kegel, E. Effect of a two-stage heat treatment process on the mechanical properties of full construction timber. *Wood Mater. Sci. Eng.* **2007**, *2*, 138–146. [CrossRef]
57. Tomak, E.D.; Ustaomer, D.; Yildiz, S.; Pesman, E. Changes in surface and mechanical properties of heat treated wood during natural weathering. *Measurement* **2014**, *53*, 30–39. [CrossRef]
58. Persze, L.; Tolvaj, L. Photodegradation of wood at elevated temperature: Colour change. *J. Photochem. Photobiol. B* **2012**, *108*, 44–47. [CrossRef] [PubMed]
59. Nasir, V.; Fathi, H.; Kazemirad, S. Combined machine learning–wave propagation approach for monitoring timber mechanical properties under UV aging. *Struct. Health Monit.* **2021**, *20*, 1475921721995987. [CrossRef]
60. Fathi, H.; Kazemirad, S.; Nasir, V. Mechanical degradation of wood under UV radiation characterized by Lamb wave propagation. *Struct. Control Health Monit.* **2021**, *28*, e2731. [CrossRef]

61. ISO 13061-3. *Physical and Mechanical Properties of Wood—Test Methods for Small Clear Wood Specimens—Part 3: Determination of Ultimate Strength in Static Bending*; International Organization for Standardization: Geneva, Switzerland, 2014.
62. ISO 13061-4. *Physical and Mechanical Properties of Wood—Test Methods for Small Clear Wood Samples—Part 4: Determination of Modulus of Elasticity in Static Bending*; International Organization for Standardization: Geneva, Switzerland, 2014.
63. Tolvaj, L.; Persze, L.; Albert, L. Thermal degradation of wood during photodegradation. *J. Photochem. Photobiol. B* **2011**, *105*, 90–93. [CrossRef]
64. Feist, W.C. Outdoor wood weathering and protection. Archaeological wood, properties, chemistry, and preservation. *Adv. Chem. Ser.* **1990**, *225*, 263–298.
65. Williams, R.S. Weathering of wood. In *Handbook of Wood Chemistry and Wood Composites*; CRC Press: Boca Raton, FL, USA, 2005; Volume 7, pp. 139–185.
66. ASTM D2244-16, *Standard Practice for Calculation of Color Tolerances and Color Differences from Instrumentally Measured Color Coordinates*; ASTM International: West Conshohocken, PA, USA, 2016.
67. ISO 13061-1. *Physical and Mechanical Properties of Wood—Test Methods for Small Clear Wood Samples—Part 1: Determination of Moisture Content for Physical and Mechanical Tests*; International Organization for Standardization: Geneva, Switzerland, 2014.
68. ISO 13061-2. *Physical and Mechanical Properties of Wood—Test Methods for Small Clear Wood Samples—Part 2: Determination of Density for Physical and Mechanical Tests*; International Organization for Standardization: Geneva, Switzerland, 2014.
69. Mardanshahi, A.; Nasir, V.; Kazemirad, S.; Shokrieh, M.M. Detection and classification of matrix cracking in laminated composites using guided wave propagation and artificial neural networks. *Compos. Struct.* **2020**, *246*, 112403. [CrossRef]
70. Yu, Y.; Dackermann, U.; Li, J.; Niederleithinger, E. Wavelet packet energy–based damage identification of wood utility poles using support vector machine multi-classifier and evidence theory. *Struct. Health Monit.* **2019**, *18*, 123–142. [CrossRef]
71. Ružiak, I.; Koštial, P.; Jančíková, Z.; Gajtanska, M.; Krišťák, Ľ.; Kopal, I.; Polakovič, P. Artificial Neural Networks Prediction of Rubber Mechanical Properties in Aged and Nonaged State. In *Improved Performance of Materials*; Springer: Cham, Switzerland, 2018; pp. 27–35.
72. El Kadi, H. Modeling the mechanical behavior of fiber-reinforced polymeric composite materials using artificial neural networks—A review. *Compos. Struct.* **2006**, *73*, 1–23. [CrossRef]
73. Nasir, V.; Cool, J.; Sassani, F. Intelligent machining monitoring using sound signal processed with the wavelet method and a self-organizing neural network. *IEEE Robot. Autom. Lett.* **2019**, *4*, 3449–3456. [CrossRef]
74. Nasir, V.; Cool, J. Characterization, optimization, and acoustic emission monitoring of airborne dust emission during wood sawing. *Int. J. Adv. Manuf. Technol.* **2020**, *109*, 2365–2375. [CrossRef]
75. Nasir, V.; Cool, J.; Sassani, F. Acoustic emission monitoring of sawing process: Artificial intelligence approach for optimal sensory feature selection. *Int. J. Adv. Manuf. Technol.* **2019**, *102*, 4179–4197. [CrossRef]
76. Nasir, V.; Cool, J. Intelligent wood machining monitoring using vibration signals combined with self-organizing maps for automatic feature selection. *Int. J. Adv. Manuf. Technol.* **2020**, *108*, 1811–1825. [CrossRef]
77. Somvanshi, M.; Chavan, P.; Tambade, S.; Shinde, S.V. A review of machine learning techniques using decision tree and support vector machine. In Proceedings of the 2016 International Conference on Computing Communication Control and Automation (ICCUBEA), Pune, India, 12–13 August 2016; pp. 1–7.
78. Van Blokland, J.; Nasir, V.; Cool, J.; Avramidis, S.; Adamopoulos, S. Machine learning-based prediction of surface checks and bending properties in weathered thermally modified timber. *Constr. Build. Mater.* **2021**, *281*, 124996. [CrossRef]
79. Steinberg, D.; Colla, P. CART: Classification and regression trees. In *The Top Ten Algorithms in Data Mining*; CRC Press: Boca Raton, FL, USA, 2009; Volume 9, p. 179.
80. Breiman, L.; Friedman, J.; Stone, C.J.; Olshen, R.A. *Classification and Regression Trees*; CRC Press: Boca Raton, FL, USA, 1984.
81. Fathi, H.; Nasir, V.; Kazemirad, S. Prediction of the mechanical properties of wood using guided wave propagation and machine learning. *Construct. Build. Mater.* **2020**, *262*, 120848. [CrossRef]
82. Tolvaj, L.; Faix, O. Artificial aging of wood monitored by drift spectroscopy and CIE Lab color measurements. *Holzforschung* **1995**, *49*, 397–404. [CrossRef]
83. Pandey, K.K. Study of the effect of photo-irradiation on the surface chemistry of wood. *Polym. Degrad. Stabil.* **2005**, *90*, 9–20. [CrossRef]
84. Nzokou, P.; Kamdem, D.P. Influence of wood extractives on the photo-discoloration of wood surfaces exposed to artificial weathering. Color Research & Application: Endorsed by Inter-Society Color Council, The Colour Group (Great Britain), Canadian Society for Color, Color Science Association of Japan, Dutch Society for the Study of Color, The Swedish Colour Centre Foundation, Colour Society of Australia, Centre Français de la Couleur. *Color Res. Appl.* **2006**, *31*, 425–434.
85. Nasir, V.; Sassani, F. A review on deep learning in machining and tool monitoring: Methods, opportunities, and challenges. *Int. J. Adv. Manuf. Technol.* **2021**, *115*, 2683–2709. [CrossRef]
86. Serin, G.; Sener, B.; Ozbayoglu, A.M.; Unver, H.O. Review of tool condition monitoring in machining and opportunities for deep learning. *Int. J. Adv. Manuf. Technol.* **2020**, *109*, 953–974. [CrossRef]
87. Yang, J.; Li, S.; Wang, Z.; Dong, H.; Wang, J.; Tang, S. Using Deep Learning to Detect Defects in Manufacturing: A Comprehensive Survey and Current Challenges. *Materials* **2020**, *13*, 5755. [CrossRef]

Article

Influence of the Structure of Lattice Beams on Their Strength Properties

Radosław Mirski [1], Łukasz Matwiej [2], Dorota Dziurka [1,*], Monika Chuda-Kowalska [3], Maciej Marecki [1], Bartosz Pałubicki [4] and Tomasz Rogoziński [2]

1. Department of Wood-Based Materials, Poznań University of Life Sciences, 60-627 Poznań, Poland; radoslaw.mirski@up.poznan.pl (R.M.); macius.marecki@gmail.com (M.M.)
2. Department of Furniture Design, Poznań University of Life Sciences, 60-627 Poznań, Poland; lukasz.matwiej@up.poznan.pl (Ł.M.); tomasz.rogozinski@up.poznan.pl (T.R.)
3. Institute of Structural Analysis, Faculty of Civil and Transport Engineering, Poznań University of Technology, pl. Skłodowskiej-Curie 5, 60-965 Poznań, Poland; monika.chuda-kowalska@put.poznan.pl
4. Department of Woodworking and Fundamentals of Machine Design, Poznań University of Life Sciences, 60-627 Poznań, Poland; bartosz.palubicki@up.poznan.pl
* Correspondence: dorota.dziurka@up.poznan.pl; Tel.: +48-061-848-7619

Abstract: This paper presents the strength properties of wooden trusses. The proposed solutions may constitute an alternative to currently produced trusses, in cases when posts and cross braces are joined with flanges using punched metal plate fasteners. Glued carpentry joints, although requiring a more complicated manufacturing process, on the one hand promote a more rational utilisation of available structural timber resources, while on the other hand they restrict the use of metal fasteners. The results of the conducted analyses show that the proposed solutions at the current stage of research are characterised by an approx. 30% lower static bending strength compared to trusses manufactured using punched metal plate fasteners. However, these solutions make it possible to produce trusses with load-bearing capacities comparable to that of structural timber of grade C24 and stiffness slightly higher than that of lattice beams manufactured using punched metal plate fasteners. The strength of wooden trusses manufactured in the laboratory ranged from nearly 20 N/mm^2 to over 32 N/mm^2. Thus, satisfactory primary values for further work were obtained.

Keywords: eco-friendly wood; lattice beams; mechanical properties; bending strength

1. Introduction

Wooden trusses are an example of structural elements used in the construction industry—both in wooden and brick structures [1–5]. Trusses are members manufactured from several up to around a dozen planks joined to form triangular elements. They are often called truss girders or lattice beams. The chords may be parallel (called flat trusses), whereas in roof structures they are non-parallel. Chords are joined using diagonal and vertical members. An important feature of trusses, also referred to as lattice beams, is related to the elements (members) in truss nodes being connected so that only axial forces are present in those elements. In order to ensure such a distribution, the external load needs to be applied directly onto the truss nodes. Cross-sections of wooden truss members are typically 38 mm × 89 mm or 38 mm × 140 mm. Since the truss member cross-section is much smaller than its length, the effect of the structure dead weight on the level of internal forces is negligible. In such a situation, it may be assumed that only axial forces are present in truss members, whereas the effect of shear forces and bending moments from dead weight is negligible. It is the generally adopted convention that axial forces causing tension have the "+" sign (they are positive axial forces), while forces causing compression are assigned the "−" sign (they are negative axial forces) [6,7].

Relatively simple assumptions used to determine stresses found in truss members make it possible to apply various techniques and algorithms to optimise their shape [8–15].

The popularisation of numerical methods, as well as the availability of CAD software have facilitated the design process for systems based on lattice beams [16–20]. When punched metal plate fasteners appeared on the market in the 1950s, the process of wooden truss manufacture and design was considerably simplified. On the one hand, punched metal plate fasteners provide easy and fast connections of individual truss members, while on the other hand, dedicated computer programmes have been developed, defining plate size and timber cross-sections. The advantages of timber structures are related not only to the simplicity of their manufacture, but also to their durability, light weight, easy modification of shapes and other positive properties of timber itself [21–24]. Another advantageous feature of timber structures is connected to the reduction in the carbon footprint, thus having an essential environmental impact, particularly in view of the sustainable development concept [25,26]. A high ratio of stiffness and load-bearing capacity to the amount of used material is also stressed when talking about wooden trusses. Trusses may be manufactured both as elements of small dimensions and members of huge spans. They may be manufactured both from solid wood and glued laminated timber (glulam) [5,27,28]. Positive aspects of wooden trusses or timber itself may be eliminated as a result of an inappropriate connection of all structure components, which was presented in a specific and comprehensive manner in the AWC document [3]. The connection in the form of an articulated joint in individual truss members needs to be designed so as to carry the assumed load with no loss of system stiffness [27]. As it was mentioned above, it may be relatively easily attained with punched metal plate fasteners or previously used carpentry joints, but steel fasteners assembled using nails or screws may also be used for that purpose [29]. Nevertheless, punched metal plate fasteners seem to serve this purpose most effectively. Metal plate fasteners punched into timber provide greater strength and durability of truss members, since the arrangement of spikes in the plates, their number and their quality after an appropriate assembly provide an adequate fastening area, while the spikes themselves are anchored in timber and exhibit pull-out resistance [3,30–33]. A drawback of such a joint may result from the fact that truss cross-sections are increased so that a punched metal plate fastener with an adequate load-bearing capacity may be assembled (with an adequate cross-section area required). However, increasing the cross-section area of timber has some advantages, since it improves the fire resistance of a given element. A significant advantage of timber is connected with its strength, dependent only on the size of the destroyed cross-section and the load level, rather than the reduction in load-bearing capacity related to the increase in temperature [34–36]. In this approach, all steel elements in the wooden structure are typically considered to be more sensitive than timber itself. The results from extensive studies by Sultan [37] show that flat wooden lattice beams constituting the ceiling structure are characterised by comparable fire resistance, irrespective of the method, with which diagonals and chords are joined. This researcher investigated two types of trusses, one with punched metal plate fasteners and the other being finger-jointed trusses (Figure 1).

Advantages of finger-jointed trusses include the considerable stiffness of such structures, potential application of cross diagonals and posts differing in cross-section areas from chords and easy manufacture, as well as the potential to adjust their length to specific needs on site (although the latter is only to a limited extent).

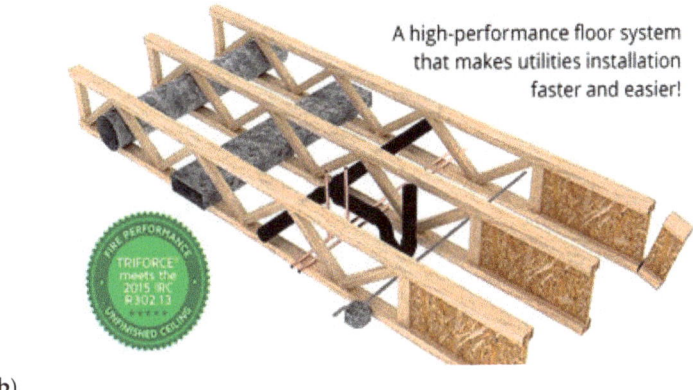

Figure 1. Examples of finger-jointed trusses: (**a**) roof constructions; (**b**) floor structures [38].

In order to be able to vary the length more easily, the end of the truss often includes a chipboard strip. They can be OSB, fibreboards or other modern boards [39]. This is in line with the current trend of the use of eco-friendly wood-based panels in construction [40,41].

For all these reasons, it was decided to propose a similar solution, i.e., manufacturing flat trusses without the use of punched metal plate fasteners as elements at a greater risk of fire damage, requiring the use of timber of equal width for all members. The proposed solution, after being verified in further analysis, should facilitate the rational management of available timber resources.

2. Materials and Methods

It was decided to conduct the tests in two stages. In the first stage, the quality of the model lattice beam was evaluated, while in the second stage, the mechanical properties of actual model lattice beams were tested.

The following assumptions for the structure and loading of lattice beams were adopted:
Height: 240 mm,
Span—the distance between supports: 3240 mm,
Member cross-section: 38 mm × 60 mm,
External force (loading) (P) applied at (truss node): 2.78 kN.

It was assumed that the quality of the designed lattice beams would not be lower than the strength of I-beams made with softwood lumber flanges. I-beams with a height of

240 mm are commonly used in single-family housing. Therefore, it was assumed that the designed lattice beams should be a substitute for these beams. Based on available materials (Steico, Czarnków, Poland; SWISS KRONO Żary, Poland), and taking into account the cross-sections of timber used for chords, it was calculated that the designed beams should carry a load, in a 4-point bending scheme, of min. 22.24 kN.

Six bars of each type were tested (18 bars in total). All of the bars were burdened till destruction. To prevent any possible bar bucklings during bending in the structure of the machine, special vertical runners shifted 5 mm from the side surface of a bar were used. The bars were tested in two burdening schemes: scheme A (Figure 2) and scheme B (Figure 3). In scheme A, a burden of 22.24kN placed evenly and symmetrically in relation to the vertical axis on all 8 lattice knots (2.78kN each) was used. In scheme B, a burden of 22.24kN placed evenly and symmetrically in relation to the vertical axis on 2 selected lattice knots (11.12kN each) was used. During the defining of the terminal bar strength, scheme A was used (burden on all 8 lattice knots). The quality of the analysed lattice beams was tested in two loading schemes, i.e., applying force at each truss node and in two specified truss nodes. Loading points are presented in Figures 2 and 3.

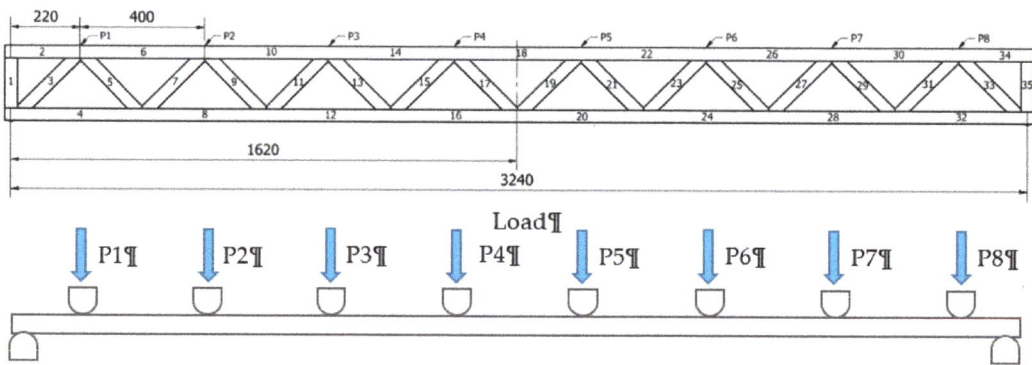

Figure 2. Truss loading scheme at each node ($P_i = P = 2.78$ kN)—scheme A.

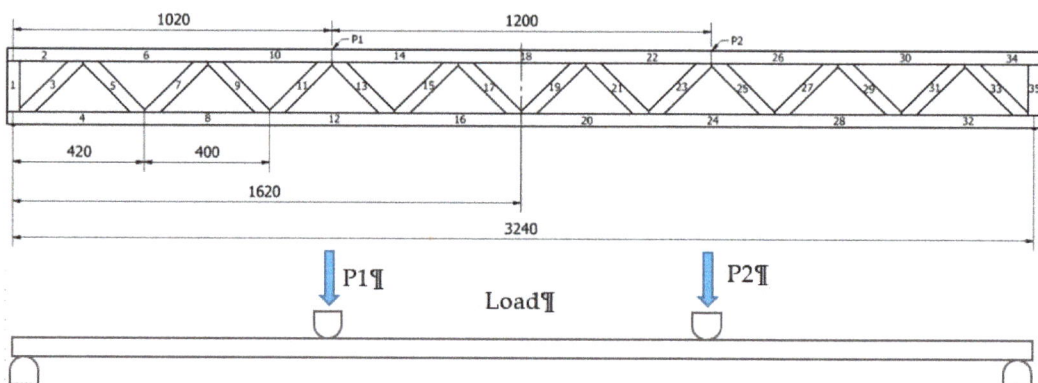

Figure 3. Truss loading diagram at two distinguished nodes ($P_1 = P_2 = 11.12$ kN)—scheme B.

As shown in Figure 3, the second loading scheme may be considered a loading system at 4-point bending. The testing station for this loading system is presented in Figure 4. In addition to the force required to destroy the tested truss, the amount of deflection at a given force was also determined. Deflections were measured using dial indicators installed in 5 points below the deflection zone.

Figure 4. Picture of the test stand.

Deflection was determined using Sylva brand electronic dial sensors. The sensors were connected to a computer to discretely determine the deflection at a given load. Five sensors on laboratory racks were placed under the nodes (components shown in Figure 5). The middle sensor (no. 3) measured deflection at the centre of the beam span. In the tests of commercial lattice beams, two dial indicators were not placed directly under the truss nodes. Based on initial data on deflections provided by the linear regression function, deflections were determined for individual measurement points for pre-determined force values. Next, for the value equal to 1/300 length of the lattice beam, the value of force was established at which the deflection takes place (deflection at the centre of the span). Tests were carried out using testing machine type SAM75 (UPP, Poznań, Poland), which allows the movement of the crosshead with specified speed as well as by specified distance. The speed of crosshead movement was always constant and equal to 10 mm/min. To prevent loss of stability (buckling) of the beam during testing, special stops made of wood were used.

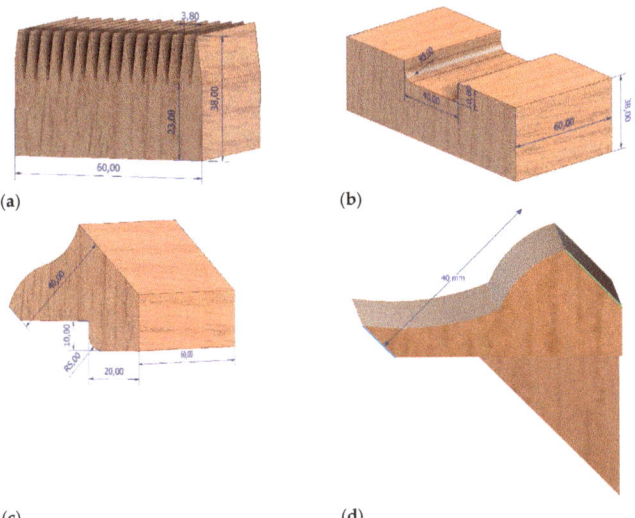

Figure 5. From the left: finger joint—chord of truss, UPP2 (**a**); tenon joint—chord of truss, tenon joint fragment of the lower belt UPP1 (**b**); tenon joint—diagonal member, tenon joint fragment of the crossbone UPP1 (**c**), tenon joint—diagonal member, tenon joint fragment of the crossbone UPP3 (**d**).

For the needs of laboratory analyses, three types of lattice beams were manufactured, with three specimens for each type. Both diagonals and posts were joined with chords using permanent carpentry joints prepared specifically for this project (mortise and tenon joints) by bonding the joined members. The polyurethane (PUR) adhesive was used in the joints. It is a single-component D4 adhesive, cross-linking when it comes into contact with moisture in the air. The two types of connection are presented in Figure 5. The third one may not be currently publicised, as it is patent pending. The presented lattice beams are denoted as UPP_x: UPP_1—mortise and tenon joint, UPP_2—finger joint and UPP_3—classified mortise and tenon design. The mortise and tenon joint was manufactured using CENTATEQ P-110 (HOMAG Group AG, Germany). In turn, the finger joint was produced using a spindle moulder by Felder with the FZK18NS180-002 cutter (Faba, Poland). The fourth type is lattice beams manufactured by the Witkowscy plant (Wieluń, Poland). They were conventional lattice beams with dimensions comparable to those of laboratory specimens. In this case, the elements were joined using punched metal plate fasteners.

3. Results and Discussion

3.1. Calculation of Internal Axial Forces in Truss Members

Table 1 presents values of internal axial forces found in individual lattice beam members. As shown by the data in Table 1, the greatest tensile forces are recorded in members no. 16 and 20, whereas the greatest compression force is found in member no. 18. In both cases, these values are 44.8 kN. In turn, in diagonals the greatest compression forces are recorded in members no. 3 and 33, while the greatest tensile forces are found in members no. 5 and 31. Compression forces in these diagonals amount to 15.84 kN, while tensile forces are 11.88 kN. Based on these values, Table 2 presents the calculated requirement for a specific timber grade needed to manufacture individual members of the designed lattice beam. For the analysed system, the critical strength value refers to the tensile strength of timber. At the assumed chord cross-section of 60 × 38 mm, to meet the assumed load-bearing conditions (strength equivalent to C24-grade timber or GL24h glulam), C35-grade timber is needed, according to EN 338 [42]. In turn, compressed chords may be manufactured from C22-grade timber, while diagonals may be made from out-of-grade C14 construction timber. In turn, when using C24 timber, the tensile chords need to be min. 78 mm × 40 mm in cross-section.

Table 1. Values of forces in bars—scheme A (Figure 2).

No. Member	Force (kN)	Type of Member	No. Member	Force (kN)	Type of Member
1	0.00	s	10	−33.6	tc
2	0.00	pg	11	−7.92	dm
3	−15.84	k	12	39.2	bc
4	11.20	pd	13	3.96	dm
5	11.88	s	14	−42.00	tc
6	−19.60	pg	15	−3.96	dm
7	−11.88	k	16	44.80	bc
8	28.00	pd	17	0.00	dm
9	7.92	s	18	−44.8	tc

s—bar; pg—lower belt; pg—upper belt; k—crossbone.

Table 3 presents the distribution of internal axial forces in individual members in a situation, when the total loading of the lattice beam is identical, but this time the loads were applied only at two truss nodes (Figure 4—system B). In this case, each of the loads are equal to 11.12 kN. The greatest tensile force in the bottom chord is 55.6 kN, while in the diagonals it is 15.7 kN. As a result, irrespective of the scheme of loading and the assumed factors of safety, the required tensile strength of diagonals is approx. 3.5-fold lower than the required tensile strength of chords. As a consequence, this type of lattice beam may be optimised in terms of timber utilisation through the selection of timber quality/strength grades or appropriately adjusting the cross-section of these members.

Table 2. The minimum strength.

No. Member	Force (kN)	Type of Member	$f_{c/t}$ * (N/mm^2)	Min. Class of Timber	$f_{c/t}$ by Standard ** (N/mm^2)
16	44.80	pd	19.65	C35	21
18	−44.80	pg	19.65	C22	20
3	−15.84	s	6.95	C14	16
5	11.88	s	5.21	C14	8

* compression strength/tensile strength, ** EN 338 [42].

Table 3. Values of forces in bars—scheme B (Figure 3).

No. Member	Force (kN)	Type of Member	No. Member	Force (kN)	Type of Member
1	0.00	s	10	−44.45	pg
2	0.00	pg	11	−15.72	k
3	11.12	k	12	55.57	pg
4	−15.72	pd	13	0.00	k
5	15.72	k	14	−55.57	pg
6	−22.23	pg	15	0.00	k
7	−15.72	k	16	55.57	pg
8	33.35	pg	17	0.00	k
9	15.72	k	18	−55.57	pg

s—bar; pg—lower belt; pg—upper belt; k—crossbone.

Another important observation from these analyses is connected to the value of the force, with which the diagonals will be pulled out from the chord, when it works as the tensile member. The preliminary assumption that the diagonals are assembled at a 45° angle to chords simplifies the calculations, since they are reduced to one value. Thus, the force causing detachment of the diagonal from the chord amounts to 0.707 of the tensile force acting on the diagonal. As a result, for the first system (with each truss node being loaded) it is 8.4 kN and for the second system (two forces) it is 11.1 kN. This analysis shows that the strength of the glue line for the analysed PUR adhesive is min. 6 N/mm^2 (mean 6.51 N/mm^2). Thus, the surface area of the proposed joint should be min. 1400 mm^2 or 1850 mm^2 in order to transfer the loads in the first scheme (onto each truss node—scheme A) or in the second scheme (onto two specified nodes—scheme B). In mortise and tenon joints, the glue line area is min. 3300 mm^2, while the glue line area for the finger joint is min. 21,000 mm^2.

3.2. Lattice Beam as a Solid Beam

Since relatively detailed static calculations may be made by replacing the spandrel beam/truss with a solid model, it was decided to introduce such transformations [43]. However, in this case, different stiffness values are used for such a model depending on the load acting on this system. In the analysed case, i.e., for loads acting on the lattice beam, the equivalent moment of inertia takes the form (1):

$$J_{sec.} = \frac{J_o}{m_1} \quad (1)$$

where:

J_o—moment of inertia of lattice beam chords at the centre of the span (2):

$$J_o = e^2 \frac{F_g \times F_d}{F_g + F_d} \quad (2)$$

where:

e—theoretical height of the spandrel beam at the span centre, i.e., the dimension measured at the axes of gravity of chords;

F_d—cross-section area of the bottom chords at the centre of its span;

F_g—cross-section area of the top chords at the centre of its span;

m_1—coefficients being the function of $h_{śr}/h_0$ ($h_{śr}$—height of the spandrel beam at the centre of the span, h_0—height of the spandrel beam at the support). For flat lattice beams, $m_1 = 1$.

For the designed lattice beams, the following assumptions were made:

e = 202 mm;
$F_d = 2280$ mm^2;
$F_g = 2280$ mm^2;
$h_{śr}/h_0 = 1$;
$m_1 = 1$.

Thus, the equivalent moment of inertia for the designed lattice beams (UPP) is 4651.2 cm^2, while for commercial lattice beams it is higher by almost 8%, i.e., 5017.6 cm^4.

The index describing the bending strength of the cross-section is defined as the quotient of the second moment of area in relation to the principal axis of gravity of the cross-section and the distance to the most extreme fibres of the section (3):

$$W_z = \frac{J}{\varepsilon_{max}} \qquad (3)$$

where:

J—the principal moment of inertia of the cross-section in relation to axis z overlapping with the neutral axis of the cross-section;

ε_{max}—maximum distance of the most extreme fibres from the neutral axis.

Assuming that we deal with symmetric systems, the distance of the most extreme fibres is equal to a half of the lattice beam height. Table 4 presents the values of moments of inertia and cross-section strength indexes for the analysed lattice beam systems.

Table 4. Results for trusses.

Type of Truss	Height (mm)	$J_{sec.}$ (cm^4)	W_z (cm^3)	U_{max} (mm) *
UPP	240	4651.2	387.64	10.8
Witkowski	240	5017.6	418.13	10.8

* allowable deflection/maximum permissible deflection—L/300; L—length of the truss.

From the conducted analyses it can be seen that the bending strength of the model lattice beam for the 8-point loading scheme (A) is $\sigma_{8-p} = 23.07$ N/mm^2, while for 2-point loading (4-point bending—scheme B) it is $\sigma_{2-p} = 28.67$ N/mm^2.

Since 4-point bending is the basic system used to evaluate the quality of structural elements, it needs to be stated that properties obtained in this test are more reliable when comparing structural elements differing in their structure. Nevertheless, we are aware that the first analysed scheme (A) is closer to the real situation, and is thus more reliable for the use of trusses as structural elements. Yet, this may not be readily referred to the numerous publications discussing beams as structural elements. Moreover, the authors also intended to relate the results to members manufactured from solid wood. For this reason, when referring the designed lattice beams to construction timber of grade C24 (most commonly used in practice), it needs to be stated that the adopted preliminary values are comparable to those of this timber grade. When analysing the obtained results in more detail, the designed lattice beams should thus have the bending moment in a 4-point bending test greater than 9.31 kN m, or the force required for the failure of the lattice beam should exceed 18.62 kN. In such a case, the designed systems will exhibit strength comparable to that of a solid timber beam.

3.3. Analysis of Laboratory Testing Results

Table 5 presents the mechanical properties of lattice beam models evaluated in the 4-point bending test. The preliminary assessment of the potential applicability of the manufactured lattice beams as an alternative to lattice beams using punched metal plate fasteners indicates that only those produced with finger joints of chords with diagonals are characterised by comparable mechanical properties. Lattice beams manufactured by joining diagonals with chords using mortise and tenon joints exhibited an approx. 30% lower strength. In the case of mortise and tenon joints, UPP_1 failure was observed typically in timber within the truss joint.

Table 5. Evaluation of the mechanical properties of the manufactured truss models.

Joint Type	$M_{g\,max}$ (kN·m)	f_m (N/mm^2)
UPP_1	7.67 (0.82) *	19.79
UPP_2	12.48 (1.39)	32.18
UPP_3	8.32 (0.33)	21.42
Steel truss plates	14.06 (1.34)	33.62

* SD (standard deviation).

Depending on the annual growth increment, systems cracking was observed at the transition zone from early wood to late wood (Figure 6a). In turn, in the case of the finger joint, failure was recorded in the most extreme truss joint, and it consisted of a shearing of the finger joint (Figure 6b). In UPP_3 lattice beams, failure was observed in the tensile chord. However, it was at the finger joint connecting the chord member lengthwise. In the case of commercial lattice beams, in the strength testing, failure was observed either in the timber of the tensile chord, or the punched metal plate fasteners were destroyed or detached. This failure type is shown in Figure 6c. Due to the too-low image-recording frequency of the camera used to record the tests, in many cases it was difficult to assess (or estimate) which failure mechanism initiated the destruction of the beam. Lattice beams manufactured at the laboratory are characterised by a slightly lower moisture content than commercial lattice beams (7–8%). To a certain extent, this may be reflected in the stiffness of the analysed specimens, which will decrease with an increase in wood moisture content.

(a)

(b)

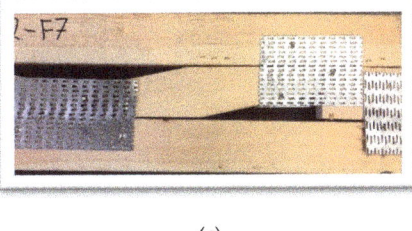

(c)

Figure 6. Failure images of tested lattice beams: (**a**) UPP1; (**b**) UPP2; (**c**) industrial.

Another significant criterion for frame systems is connected with the volume of deflections appearing at loading. Admissible deflections in the span and the bracket are determined following the PN-EN 1995-1-1:2010 standard [44], while for floor beams admissible deflections amount to 1/300 of their length (L/300). The values of deflections and bending moments at such deflections are given in Table 6. As shown by the presented data, the most advantageous results are obtained for the lattice beam, in which chords were joined with diagonals using a carpentry joint designed specifically for the purpose of this project, i.e., the longitudinal mortise and tenon joint.

Table 6. Evaluation of the mechanical properties of manufactured models of lattice beams.

Connection Type	Sensor Number					M_{g003} * (kN·m)	$M_{g\ max}$ (kN·m)
	1	2	3	4	5		
	Deflection Value (mm)						
Tenons UPP$_1$	−8.02	−10.14	−10.68	−10.42	−8.45	5.11 (0.30) **	7.67
Finger joints UPP$_2$	−4.85	−8.42	−10.73	−9.04	−8.86	4.16 (0.26)	12.48
Tenons UPP$_3$	−7.69	−9.57	−10.76	−9.76	−7.80	5.65 (0.32)	8.32
Barbed plates	−6.46	−9.19	−11.00	−9.38	−6.97	5.18 (0.16)	14.06

* moment at deflection to equal 1/300; ** SD (standard deviation).

While this is an improvement by only 10% compared to the lattice beam manufactured with members joined using punched metal plate fasteners, it needs to be remembered that this novel solution is still being modified. What is surprising is that the lowest stiffness was recorded for lattice beams, in which chords and diagonals were connected using finger joints. Thus, these lattice beams are characterised by a considerable strength, while at the same time they are highly susceptible to load-induced strain, which probably results from the moulding of finger joints along the entire chord length.

The results obtained are similar to those obtained by other research centres [45,46]. The comparison itself is relatively difficult. Very often, for simplicity, other studies do not maintain the length/height relationship and the trusses are tested in 3-point bending. However, the authors mentioned above indicate that their trusses fail under loads of 6–8 kN to 11–14 kN. Importantly, these authors indicate that glued trusses exhibit superior properties to nailed trusses.

Table 7 presents testing results from a lattice beam UPP$_3$ loaded at each truss node (scheme A). As shown by the data, deflection amounting to 1/300 length occurs at the bending moment of 3.77 kN·m. In turn, the ultimate bending moment is as high as 9.53 kN·m. The strength of the analysed lattice beam referred to as a solid beam amounts to more than 24 N/mm^2. Strength, even at a relatively low moisture content of used timber, is relatively high. Moreover, the mean value of the load applied at each truss node is 2.91 kN and, thus, it is only approx. 5% higher than the one assumed for calculations for the model lattice beam.

Table 7. Physical and mechanical properties of UPP3 lattice beam loaded at 8 points.

Typ	Mg003 * (kN·m)	Mgmax (kN·m)	F (kN)	fm (N/mm^2)	MC ** (%)
UPP3	3.77 (0.14)	9.53 (0.65)	2.91	24.58	6.7

* bending moment at 1/300; ** wood moisture content.

4. Conclusions

It results from these analyses that it is feasible to manufacture lattice beams with parallel chords to serve as truss beams, when punched metal plate fasteners joining chords with diagonals are replaced by carpentry joints. Preliminary theoretical assumptions were correctly selected and thus provide grounds for design work on such lattice beams. Static analysis showed that cross-sections of elements used as chord fasteners (diagonals, posts) may be considerably reduced or made from inferior-grade timber. Of the two proposed solutions, i.e., perpendicular and longitudinal, more advantageous results were obtained for the longitudinal joint. Moreover, it results from the conducted tests that:

The greatest load-bearing capacity, determined in the bending test, is found for lattice beams manufactured using punched metal plate fasteners;

Only a slightly lower stiffness, amounting to 32.2 N/mm^2, was recorded for lattice beams, elements of which were connected using finger joints; however, they show deflection of 1/300 length even under the smallest load;

Lattice beams, in which chords were connected with diagonals using glued carpentry joints (UPP3), exhibit an approx. 35% lower bending strength, although they show much smaller deflections under the same loading compared to lattice beams manufactured using finger joints.

Although neither of the joints met the requirements in terms of the assumed load-bearing capacity, in view of the observed failure mechanism, it was decided that in further stages of study the mortise and tenon joints UPP$_3$ will be developed. In further studies, it is intended to conduct tests on solid timber chords and to a certain extent modify the shape of the joint itself.

The results of the study provide a valuable basis for further design work. Developers especially are looking for cheaper alternatives to both solid wood and glued laminated timber. Additionally, single-family houses, often in Central Europe, realized in an economic way, are looking for cheaper and at the same time easy-to-install construction materials.

Author Contributions: Conceptualization, R.M. and D.D.; methodology, R.M., M.C.-K. and D.D.; validation, M.M., B.P., T.R. and Ł.M.; formal analysis, M.M., B.P. and R.M.; investigation, M.M., B.P. and M.C.-K.; resources, M.M., B.P., T.R. and Ł.M.; data curation, Ł.M. and T.R; writing—original draft preparation, R.M. and D.D.; writing—review and editing, D.D. and M.C.-K.; visualization, D.D., M.C.-K. and R.M.; supervision, R.M. and M.C.-K.; project administration, R.M.; funding acquisition, R.M. and D.D. All authors have read and agreed to the published version of the manuscript.

Funding: This research was funded by the National Centre for Research and Development, BIOSTRATEG3/344303/14/NCBR/2018. The study was also supported by the funding for statutory R&D activities as the research task No. 506.227.02.00 and 506.224.02.00 of the Faculty of Forestry and Wood Technology, Poznań University of Life Sciences.

Institutional Review Board Statement: Not applicable.

Informed Consent Statement: Not applicable.

Data Availability Statement: The data presented in this study are available on request from the corresponding author.

Acknowledgments: We would like to thank inż. Janusz Dębiński for his help in preparing the research methodology.

Conflicts of Interest: The authors declare no conflict of interest.

References

1. Togan, V.; Durmaz, M.; Daloglu, A. Optimization of roof trusses under snow loads given in Turkish Codes. *Eng. Struct.* **2006**, *28*, 1019–1027.
2. Rinaldin, G.; Amadio, C.; Fragiacomo, M. A component approach for the hysteretic behaviour of connections in cross-laminate wooden structures. *J. Inter. Ass. Earthq. Eng.* **2013**, *1*, 1–21. [CrossRef]
3. American Wood Council. *A Builders Guide to Trusses. Connection Solution for Wood Frame Structures*; Alpine Systems Corporation: Leesburg, VA, USA, 2014. Available online: https://www.awc.org/pdf/education/des/AWC-DES310-Connections-1hr-141128.pdf (accessed on 2 July 2021).
4. Macchioni, N.; Mannucci, M. The assessment of Italian trusses: Survey methodology and typical pathologies. *Int. J. Archit. Herit.* **2018**, *12*, 533–535. [CrossRef]
5. Woods, B.; Hill, I.; Friswell, M.I. Ultra-efficient wound composite truss structures. *Compos. Part A* **2016**, *90*, 111–124. [CrossRef]
6. Lengvarský, P.; Bocko, J. The Static Analysis of the Truss. *Am. J. Mech. Eng.* **2016**, *4*, 440–444.
7. Frans, R.; Arfiadi, Y. Sizing, shape and topology optimizations of roof trusses using hybrid genetic algorithms. *Procedia Eng.* **2014**, *95*, 185–195. [CrossRef]
8. Šešok, D.; Belevičius, R. Use of genetic algorithms in topology optimization of truss structure. *Mechanika* **2007**, *2*, 34–39.
9. Deb, K.; Gulati, S. Design of truss-structures for minimum weight using genetic algorithms. *Finite Elem. Anal. Des.* **2001**, *37*, 447–465. [CrossRef]
10. Rajan, S.D. Sizing, shape and topology optimization of trusses using genetic algorithm. *J. Struct. Eng.* **1995**, *121*, 1480–1487. [CrossRef]
11. Delyová, I.; Frankovský, P.; Bocko, J.; Trebuňa, P.; Živčák, J.; Schürger, B.; Janigová, S. Sizing and Topology Optimization of Trusses Using Genetic Algorithm. *Materials* **2021**, *14*, 715. [CrossRef]
12. Tiachacht, S.; Bouazzouni, A.; Khatir, S.; Wahab, M.A.; Behtani, A.; Capozucca, R. Damage assessment in structures using combination of a modified Cornwell indicator and genetic algorithm. *Eng. Struct.* **2018**, *177*, 421–430. [CrossRef]
13. Sivakumar, P.; Natarajan, K.; Rajaraman, A.; Samuel Knight, G.M. Artificial intelligence techniques for optimisation of steel lattice towers. In Proceedings of the Structural Engineering Convention, Honolulu, HI, USA, 24–27 January 2001; pp. 435–445.

14. Gero, M.B.P.; García, A.B.; del Coz Díaz, J.J. Design optimization of 3D steel structures: Genetic algorithms vs. classical techniques. *J. Constr. Steel Res.* **2006**, *62*, 1303–1309. [CrossRef]
15. Neeraja, D.; Kamireddy, T.; Kumar, P.S.; Reddy, V.S. Weight optimization of plane truss using genetic algorithm. *IOP Conf. Ser. Mater. Sci. Eng.* **2017**, *263*, 32015. [CrossRef]
16. Olhoff, N.; Bendsøe, M.P.; Rasmussen, J. On CAD-integrated structural topology and design optimization. *Comput. Methods Appl. Mech. Eng.* **1991**, *89*, 259–279. [CrossRef]
17. Kim, N.H.; Sankar, B.V.; Kumar, A.V. *Introduction to Finite Element Analysis and Design*; John Wiley & Sons: Hoboken, NJ, USA, 2018.
18. Rajeev, S.; Krishnamoorthy, C.S. Genetic Algorithms-Based Methodologies for Design Optimization of Trusses. *J. Struct. Eng.* **1997**, *123*, 350–358. [CrossRef]
19. Nan, B.; Bai, Y.; Wu, Y. Multi-Objective Optimization of Spatially Truss Structures Based on Node Movement. *Appl. Sci.* **2020**, *10*, 1964. [CrossRef]
20. Bocko, J.; Delyová, I.; Frankovský, P.; Neumann, V. Lifetime Assessment of the Technological Equipment for a Robotic Workplace. *Int. J. Appl. Mech.* **2020**, *12*, 2050097. [CrossRef]
21. Massafra, A.; Prati, D.; Predari, G.; Gulli, R. Wooden Truss Analysis, Preservation Strategies, and Digital Documentation through Parametric 3D Modeling and HBIM Workflow. *Sustainability* **2020**, *12*, 4975. [CrossRef]
22. Ruggieri, N. In Situ Observations on the Crack Morphology in the Ancient Timber Beams. *Sustainability* **2021**, *13*, 439. [CrossRef]
23. Shanuka, F.; Hansen, E.; Kozak, R.; Sinha, A. Organizational cultural compatibility of engineered wood products manufacturers and building specifiers in the Pacific Northwest. *Archit. Eng. Desig. Manag.* **2018**, *14*, 398–434. [CrossRef]
24. Dodoo, A.; Gustavsson, L.; Sathre, R. Lifecycle primary energy analysis of low-energy timber building systems for multi-storey residential buildings. *Energ. Build.* **2014**, *81*, 84–97. [CrossRef]
25. Kromoser, B.; Ritt, M.; Spitzer, A.; Stangl, R.; Idam, F. Design Concept for a Greened Timber Truss Bridge in City Area. *Sustainability* **2020**, *12*, 3218. [CrossRef]
26. Svajlenka, J.; Kozlovská, M.; Spisáková, M. The benefits of modern method of construction based on wood in the context of sustainability. *Int. J. Environ. Sci. Technol.* **2017**, *14*, 1591–1602. [CrossRef]
27. Jin, M.; Hu, Y.; Wang, B. Compressive and bending behaviours of wood-based two-dimensional lattice truss core sandwich structures. *Compos. Struct.* **2015**, *124*, 3. [CrossRef]
28. Houlihan, W.; Georges, L.; Dokka, T.H.; Haase, M.; Time, B.; Lien, A.; Mellegard, S.; Maltha, M. A net zero emission concept analysis of a single-family house. *Energy Build.* **2014**, *74*, 101–110. [CrossRef]
29. Rivera-Tenorio, M.; Moya, R.; Navarro-Mora, A. Wooden trusses using metal plate connections and fabricated with Gmelina arborea, Tectona grandis and Cupressus lusitanica timber from forest plantations. *J. Ind. Acad. Wood Sci.* **2020**, *17*, 183–194. [CrossRef]
30. Guo, W.; Song, S.; Jiang, Z.; Wang, G.; Sun, Z.; Wang, X.; Yang, F.; Chen, H.; Shi, S.Q.; Fei, B. Effect of metal-plate connector on tension properties of metal-plate connected dahurian larch lumber joints. *J. Mater. Sci. Res.* **2014**, *3*, 40–47. [CrossRef]
31. Rammer, R.D. Wood: Mechanical fasteners. In *Encyclopedia of Materials: Science and Technology*; Saleem, H., Ed.; Elsevier: Amsterdam, The Netherlands, 2016; p. 3.
32. Rivera-Tenorio, M.; Camacho-Cornejo, D.; Moya, R. Perception of Costa Rican market about the use of prefabricated trusses with wood from forest plantations and joined with metal plates. *Rev. For. Mesoam. Kuru-RFMK* **2019**, *16*, 35–46. [CrossRef]
33. Rivera-Tenorio, M.; Moya, R. Stress, displacement joints of gmelina arborea and tectona grandis wood with metal plates, screws and nails for use in timber truss connections. *CERNE* **2019**, *25*, 172–183. [CrossRef]
34. Gravit, M.; Serdjuks, D.; Bardin, A.; Prusakov, V.; Buka-Vaivade, K. Fire Design Methods for Structures with Timber Framework. *Mag. Civ. Eng.* **2019**, *85*, 92–106.
35. Qin, R.; Zhou, A.; Chow, C.L.; Lau, D. Structural performance and charring of loaded wood under fire. *Eng. Struct.* **2021**, *228*, 111491. [CrossRef]
36. Malanga, R. Fire endurance of lightweight wood trusses in building construction. *Fire Technol.* **1995**, *31*, 44–61. [CrossRef]
37. Sultan, M.A. Fire Resistance of Wood Truss Floor Assemblies. *Fire Technol.* **2012**, *51*, 1371–1399. [CrossRef]
38. TRIFORCE®. The Open Joist TRIFORCE®Adapts to All Types of Projects. Available online: https://www.openjoisttriforce.com/wp-content/uploads/2017/01/How_are_the_joints_made_on_OpenJoist_TRIFORCE (accessed on 2 July 2021).
39. Antov, P.; Jivkov, V.; Savov, V.; Simeonova, R.; Yavorov, N. Structural Application of Eco-Friendly Composites from Recycled Wood Fibres Bonded with Magnesium Lignosulfonate. *Appl. Sci.* **2020**, *10*, 7526. [CrossRef]
40. Antov, P.; Savov, V.; Trichkov, N.; Krišťák, Ľ.; Réh, R.; Papadopoulos, A.N.; Taghiyari, H.R.; Pizzi, A.; Kunecová, D.; Pachikova, M. Properties of High-Density Fiberboard Bonded with Urea–Formaldehyde Resin and Ammonium Lignosulfonate as a Bio-Based Additive. *Polymers* **2021**, *13*, 2775. [CrossRef] [PubMed]
41. Bekhta, P.; Noshchenko, G.; Réh, R.; Kristak, L.; Sedliačik, J.; Antov, P.; Mirski, R.; Savov, V. Properties of Eco-Friendly Particleboards Bonded with Lignosulfonate-Urea-Formaldehyde Adhesives and PMDI as a Crosslinker. *Materials* **2021**, *14*, 4875. [CrossRef]
42. EN 338. *Structural Timber. Strength classes*; European Committee for Standardization: Brussels, Belgium, 2011.
43. Bogudzki, W. *Budownictwo Stalowe. Część 2*; Wydawnictwo Arkady: Warszawa, Poland, 1977.

44. EN 1995-1-1. *Eurocode 5. Design of timber Structures. Part 1-1. General—Common Rules and Rules for Buildings*; European Committee for Standardization: Brussels, Belgium, 2010.
45. Sagara, A.; Johannes, A.T.; Husain, A.S. Experimental study on strength and stiffness connection of wooden truss structure. *MATEC Web Conf.* **2017**, *101*, 0101. [CrossRef]
46. Wang, X.; Hu, Y.C. Preparation and Bending Properties of Lattice Sandwich Structure for Wooden Truss. 3rd Annual International Conference on Advanced Material Engineering (AME). *AER-Adv. Eng. Res.* **2017**, *110*, 343–349.

Article

Structural Application of Lightweight Panels Made of Waste Cardboard and Beech Veneer

Vassil Jivkov, Ralitsa Simeonova, Petar Antov, Assia Marinova, Boryana Petrova and Lubos Kristak

1. Department of Interior and Furniture Design, Faculty of Forest Industry, University of Forestry, 1797 Sofia, Bulgaria; r_simeonova@ltu.bg (R.S.); a_marinova@ltu.bg (A.M.); b.petrova@ltu.bg (B.P.)
2. Department of Interior and Architectural Design, Faculty of Architecture, University of Architecture, Civil Engineering and Geodesy, 1046 Sofia, Bulgaria
3. Department of Mechanical Wood Technology, Faculty of Forest Industry, University of Forestry, 1797 Sofia, Bulgaria; p.antov@ltu.bg
4. Faculty of Wood Sciences and Technology, Technical University in Zvolen, T. G. Masaryka 24, 960 01 Zvolen, Slovakia
* Correspondence: v_jivkov@ltu.bg (V.J.); kristak@tuzvo.sk (L.K.)

Citation: Vassil Jivkov, Ralitsa Simeonova, Petar Antov, Assia Marinova, Boryana Petrova and Lubos Kristak Structural Application of Lightweight Panels Made of Waste Cardboard and Beech Veneer. *Materials* **2021**, *14*, 5064. https://doi.org/10.3390/ma14175064

Academic Editor: Tomasz Sadowski

Received: 10 August 2021
Accepted: 1 September 2021
Published: 4 September 2021

Publisher's Note: MDPI stays neutral with regard to jurisdictional claims in published maps and institutional affiliations.

Copyright: © 2021 by the authors. Licensee MDPI, Basel, Switzerland. This article is an open access article distributed under the terms and conditions of the Creative Commons Attribution (CC BY) license (https://creativecommons.org/licenses/by/4.0/).

Abstract: In recent years, the furniture design trends include ensuring ergonomic standards, development of new environmentally friendly materials, optimised use of natural resources, and sustainably increased conversion of waste into value-added products. The circular economy principles require the reuse, recycling or upcycling of materials. The potential of reusing waste corrugated cardboard to produce new lightweight boards suitable for furniture and interior applications was investigated in this work. Two types of multi-layered panels were manufactured in the laboratory from corrugated cardboard and beech veneer, bonded with urea-formaldehyde (UF) resin. Seven types of end corner joints of the created lightweight furniture panels and three conventional honeycomb panels were tested. Bending moments and stiffness coefficients in the compression test were evaluated. The bending strength values of the joints made of waste cardboard and beech veneer exhibited the required strength for application in furniture constructions or as interior elements. The joints made of multi-layer panels with a thickness of 51 mm, joined by dowels, demonstrated the highest bending strength and stiffness values (33.22 N·m). The joints made of 21 mm thick multi-layer panels and connected with Confirmat had satisfactory bending strength values (10.53 N·m) and Minifix had the lowest strength values (6.15 N·m). The highest stiffness values (327 N·m/rad) were determined for the 50 mm thick cardboard honeycomb panels connected by plastic corner connector and special screw Varianta, and the lowest values for the joints made of 21 mm thick multi-layer panels connected by Confirmat (40 N·m/rad) and Minifix (43 N·m/rad), respectively. The application of waste corrugated cardboard as a structural material for furniture and interiors can be improved by further investigations.

Keywords: lightweight panels; waste cardboard; corner joints; bending strength; stiffness

1. Introduction

Limited wood resources worldwide require efficient utilisation of waste and by-products, cascading use of the available lignocellulosic raw materials, and search for alternative production processes and materials [1–7]. The COVID-19 crisis has also led to changes in the market and a shortage of resources, including solid wood and furniture panels. The reuse of materials provides new opportunities and represents a sustainable way to address this shortage. The use of corrugated cardboard, a mainstream packaging material, is an option to minimise the problem.

One of the pioneers in the use of non-traditional solutions to save raw materials and lighten the construction of furniture is IKEA. The furniture giant applied paper as a honeycomb core in the 1980s when they started producing the Lack series [8], which was significantly further developed in the last 10–15 years. The honeycomb core is usually prefabricated from craft paper and delivered at an ordered thickness as compressed harmonica-like layers [9]. The technology for producing such panels has been significantly developed over the last 10 years [10]. Panels weighing less than 500 kg/m^3 are defined as light panels, below 350 kg/m^3 are very light, and below 200 kg/m^3 are ultra-light ones [11]. Some of the first comprehensive studies on the application, manufacturing, and properties of paper honeycomb panels were done by Hänel and Weinert [12] and Poppensieker and Thömen [13]. Many studies were made to evaluate the mechanical properties of honeycomb panels [14–21]. Other studies investigated the possibilities for optimisation of the structure of paper honeycomb panels. Smardzewski and Prekrat [22] concluded that the stiffness and strength of cell panels were affected significantly by the weight of paper used to manufacture their cores and the shape and dimensions of cells. Other results showed that sandwich panels with a wavy core can sustain higher loads than honeycombs [23]. Słonina et al. [24] reported that impregnating solutions can be applied to improve the paper core's stiffness. The possibility of manufacturing lightweight flat pressed wood plastic composites were investigated by Lyutyy et al. [25].

Gößwald et al. [26] investigated the potential of using planer shavings with a length over 4 mm for manufacturing low-density one-layer particleboard with a thickness of 10 mm as an option to reduce the raw material demand for wood-based panels. A team of scientists [27] investigated the effect of relative humidity on the strength properties of lightweight panels and found out that after exposure to 95% relative humidity, facings and sandwiches lost up to half their original strength properties. By establishing a 3D moisture-displacement finite element model, the influence of constant and cycle humidity and varied temperature on the flexural creep of the sandwich panel containing Kraft honeycomb core and wood composite skins was studied by Chen et al. [28]. The influence of core shape was studied by several authors [23,29–31]. In the last few years, the strength and stiffness of furniture panels with auxetic cores were investigated. The auxetic cores of the sandwich cellular wood panels exhibited strong orthotropic properties [32]. During bending, wood-based honeycomb panels with auxetic cores absorb energy more effectively than the same panels subjected to axial compression [20]. The relative density of cells significantly affects their mechanical strength [33]. Using an auxetic core and facings of plywood and cardboard significantly reduces the amount of dissipated energy [21].

The problematic parts of applying paper honeycomb panels as structural elements are the joints [11]. For this reason, there is a growing industrial and academic interest on the investigation and optimization of these types of joints [16,34–38]. Some of the authors studied the bending strength of the joints [36,38,39], but tensile and shear tests were applied, as well [15,34,35,38,39]. Lightweight honeycomb panels with different thicknesses have been tested. Several authors studied panels with a relatively small thickness, ranged between 15 and 19 mm [21,22,34,35,38–41]. Other studies were focused on the evaluation of panels with a thickness of 38 mm [15,34,36]. Only one research was found with a commercially produced honeycomb panel with 50 mm thickness [37]. Both glued [21,34] and non-glued joints [36–38] appeared in the studies. Screw withdrawal resistance in honeycomb panels was investigated by some researchers [15,37,40,41].

Lightweight paper panels and paper tubes were analysed concerning their suitability as furniture materials by Petutschnigg and Ebner [16]. The authors concluded that the bending strength of the lightweight paper panel is too low for use in furniture applications and development of materials and joints with enhanced strength properties is needed.

Corrugated cardboard has a relatively low weight, is easily available, recyclable, and waste cardboard can be reused. With the development of online commerce, the use of cardboard packaging, including corrugated cardboard, has been significantly increased. In 2018, the production of paper and paperboard packaging was estimated to 256,138 thousand metric tons worldwide [42]. Around 80% of all goods sold in Europe and the US are in cartons [43]. At the same time, only 65.7% of waste cardboard in the European

Union is recycled [44]. This enormous amount of waste cardboard is a challenge to absorb and obtain new products based on recycling or reuse.

In the literature, there are publications about using waste cardboard as a material for producing new panels. In part of the research, the cardboard is processed and incorporated into new composites [45,46]. Several investigations were carried out to evaluate the mechanical properties of cardboard made from recycled beverage cartons [47–51]. Other researchers used the waste cardboard in the form in which it was obtained [52]. There are many examples of furniture made of cardboard, such as shelves, chairs, tables, desks, even sofas [53]. Some companies specialised in the production of cardboard furniture [54,55]. A Japanese bedding company has created a cardboard bed for athletes at the Tokyo 2020 Olympics. The bed frames were made from recycled cardboard [56]. There is also furniture made of second-hand cardboard, but these are usually of relatively simple structure. Even building construction components were objects of experimentation using sheets of waste cardboard collected from the waste stream [57]. Corrugated cardboard has a good strength-to-weight ratio, excellent burst strength, and resistance to crushing, thus being an ideal material for furniture manufacture [58]. Furthermore, adhesives applicable in the wood-processing and furniture industry can be used to bond such panels.

There is limited information about the strength characteristics of the corner joints made of corrugated cardboard.

The aim of the study was to investigate the possibilities of joining lightweight panels made from waste corrugated board and beech veneer, bonded with urea-formaldehyde (UF) resin and evaluate their application in furniture and interior constructions. For strength comparison, joints obtained from conventional lightweight cardboard honeycomb panels with a thickness of 50 mm were also tested.

2. Materials and Methods

2.1. Materials

In order to establish the possibility of using waste cardboard as a structural material for furniture and interior applications, experimental multi-layer panels were produced in the laboratory using a three-ply corrugated waste cardboard and rotary cut beech (*Fagus sylvatica* L.) veneer sheets with a thickness of 1.2 mm and a moisture content of approximately 8%, provided by the factory Welde Bulgaria AD (Troyan, Bulgaria). The waste cardboard sheets with a length of 1150 mm and width of 780 mm were taken from packages of goods, shipping in pallets, where this cardboard is used as a divider between individual goods. The cardboard sheets had the following dimensions: Thickness of 3.94 mm, top layer of 0.15 mm, and corrugated paper core layer of 0.10 mm. Density of the cardboard was 114 kg/m^3. Commercially available UF resin with dry solids content of 64%, density of 1.29–1.31 g.cm^{-3} at 20 °C, pH value of 8.5, and a molar ratio of 1.16, provided by the company Kastamonu Bulgaria AD (Gorno Sahrane, Bulgaria), was used for hot pressing. The experimental panels were fabricated under laboratory conditions on a single opening hydraulic press (PMC ST 100, Italy). The press temperature used was t = 100 °C. The specific bonding pressure was 0.2 N/mm^2 for 21 mm thickness and 0.13 N/mm^2 for 51 mm thickness of the panels, respectively.

Commercial, commonly produced lightweight panels with a thickness of 50 mm were also included in the study to compare the obtained experimental data with industrially produced lightweight panels. The top layers were 8 mm laminated particleboards and a hexagon honeycomb for the core layer. The commercial panels were supplied by Egger (Fritz Egger GmbH & Co. OG, Weiberndorf 20, 6380 St. Johann in Tirol, Austria).

The laboratory-fabricated multi-layer panels were designed with thicknesses similar to the thicknesses of the commercially available panels, i.e., 50 and 20 mm. This was achieved by adjusting the bonding pressure. The structure of the lightweight panels used in the study was as follows:

1. Multi-layer panel with a thickness of 21 mm made of 7 layers of veneer and 6 layers of corrugated cardboard (Figure 1).

2. Multi-layer panel with a thickness of 51 mm made of 13 layers of veneer and 12 layers of corrugated cardboard (Figure 2).
3. Cardboard honeycomb panel with a thickness of 50 mm, manufactured by Egger, Austria (Figure 3).

Figure 1. Multi-layer panel with a nominal thickness of 21 mm made of 7 layers of veneer and 6 layers of corrugated cardboard: 1. Rotary cut beech veneer, δ = 1.2 mm; 2. three-layer corrugated cardboard $δ_H$ = 6 mm.

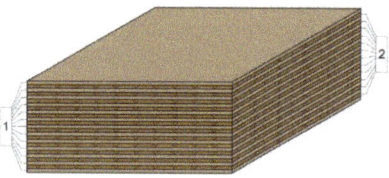

Figure 2. Multi-layer panel with a nominal thickness of 51 mm made of 13 layers of veneer and 12 layers of corrugated cardboard: 1. Rotary cut beech veneer, δ = 1.2 mm; 2. three-layer corrugated cardboard $δ_H$ = 6 mm.

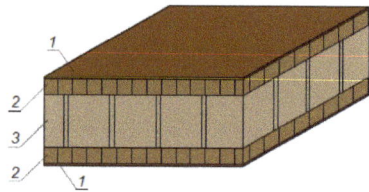

Figure 3. Cardboard honeycomb panel with a thickness of 50 mm, manufactured by Egger, Austria, δ = 50 mm: 1. Decorative foil; 2. PB, δ = 8 mm; 3. honeycomb cardboard core.

After hot pressing, the laboratory-produced panels were conditioned for 7 days at 20 ± 2 °C and 65% relative humidity.

The density of all fabricated panels was below 500 kg/m³ (Table 1). Details about the manufacturing process and the physical and mechanical properties of the developed lightweight panels are given in a previous research [18].

Table 1. Thickness and density of the lightweight panels.

Type of Panel	Thickness, mm	Density kg/m³
1. Multi-layer panel, veneer (7 layers) and corrugated cardboard (6 layers)	21	466
2. Multi-layer panel, veneer (13 layers) and corrugated cardboard (12 layers)	51	349
3. Cardboard honeycomb panel	50	257

2.2. Type of Corner Joints Made of Lightweight Panels

For the purpose of the study, 10 series with a total number of 129 test samples were made. Some well-known connecting solutions were used for the joints, such as a dowel joint, one-element connector Confirmat, an eccentric connector Minifix with a bolt for ø5 mm hole and a plastic corner joint, as well as special connectors and screws designed for lightweight honeycomb panels. Joint types, detailed dimensions, and types of connecting elements are shown in Figure 4. Each joint had only one connecting element.

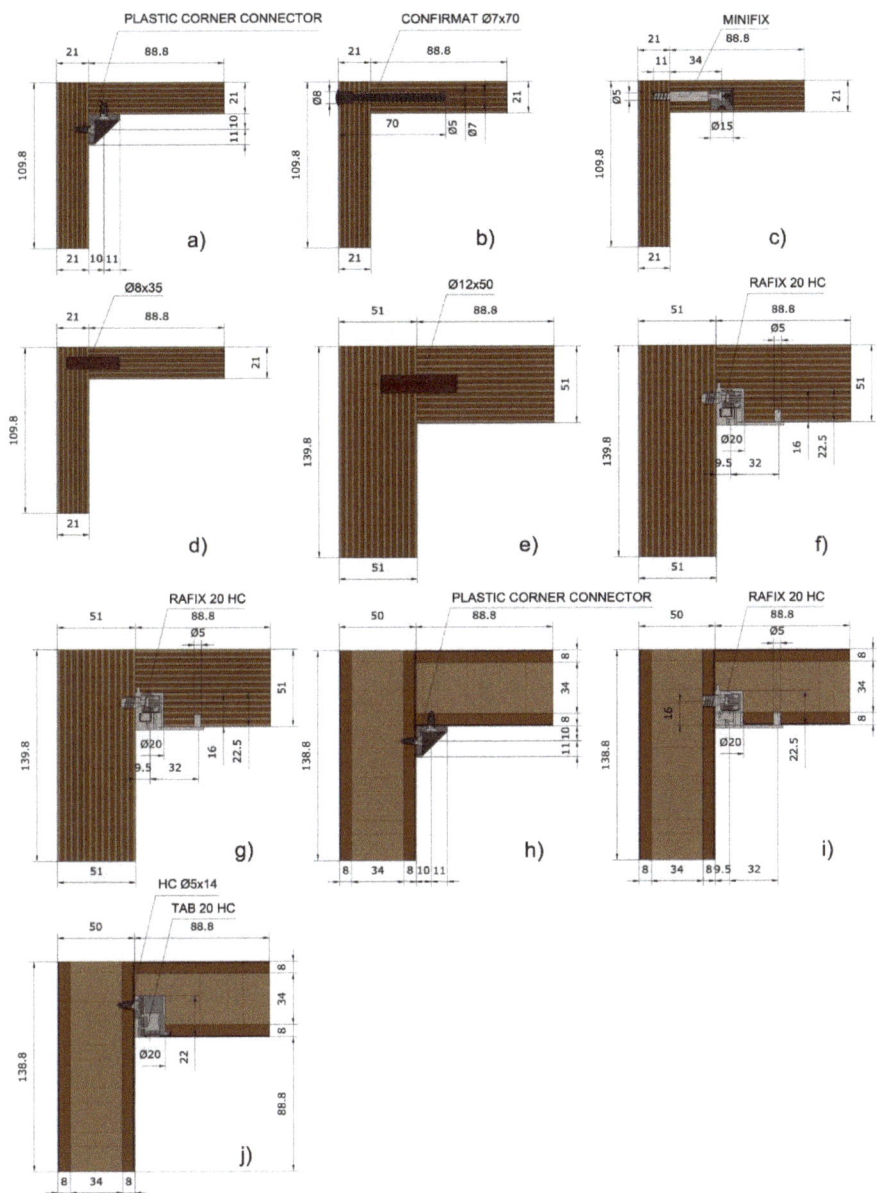

Figure 4. Corner joints made of lightweight panels: (**a**) Multi-layer panel (21 mm) made of veneer and corrugated cardboard,

connected by a plastic corner connector and special screw Varianta; (**b**) multi-layer panel (21 mm) made of veneer and corrugated cardboard, connected by Confirmat ø7 × 70 mm; (**c**) multi-layer panel (21 mm) made of veneer and corrugated cardboard, connected by an eccentric connector Minifix; (**d**) multi-layer panel (21 mm) made of veneer and corrugated cardboard, connected by a dowel ø8 × 35 mm; (**e**) multi-layer panel (51 mm) made of veneer and corrugated cardboard, connected by a dowel ø12 × 50 mm; (**f**) multi-layer panel (51 mm) made of veneer and corrugated cardboard, connected by Rafix 20 HC, inserted longitudinally on the cardboard direction; (**g**) multi-layer panel (51 mm) made of veneer and corrugated cardboard, connected by Rafix 20 HC, inserted across on the cardboard direction; (**h**) cardboard honeycomb panel (50 mm) connected by plastic corner connector and special screw Varianta; (**i**) cardboard honeycomb panel (50 mm) connected by Rafix 20 HC; (**j**) cardboard honeycomb panel (50 mm) connected by TAB 20 HC.

2.3. Test Methods

The type and shape of test samples, made in accordance with the test method, described by Kyuchukov and Jivkov [11], are shown in Figure 5. The dimensions δ_1 and δ_2 are equal and correspond to the thickness of the panels. The dimensions L_1 and L_2 are also equal and depend on the thickness of the panels.

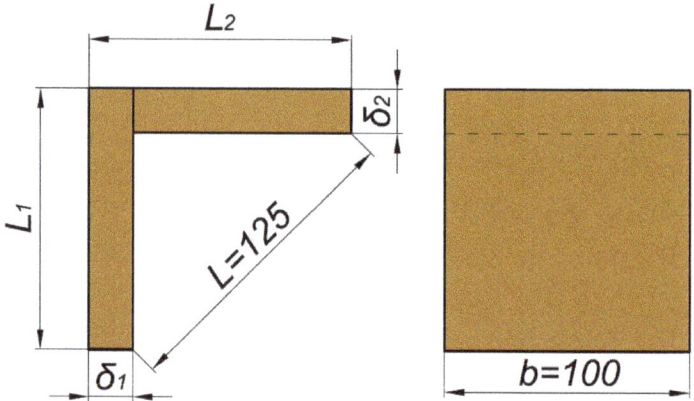

Figure 5. Type and dimensions of the tested samples. Reprinted with permission from Ref. [11]. 2021 BISMAR.

In furniture constructions, the joints are most often loaded in bending with arm compression and arm opening. It has been found from many research studies that the strength and the deformation resistance of the joints are lower under the arm compression loading compared to that under loading with arm opening [11]. Therefore, for the purpose of the present study, it was accepted to investigate end corner joints of lightweight structural elements under bending loading with arm compression. The principal test scheme of the test specimens is given in Figure 6.

The criterion for determining the strength of the tested joints is the maximum bending moment M_{max} calculated according to the formula:

$$M_{max} = F \cdot l, \tag{1}$$

where F is the maximum force under arm compression bending, N, and l is arm, m.

The criterion for determining the deformation characteristic of the corner joints is the stiffness coefficient c [11,59].

The deformation of the joints under the compression bending test gives as a result changes in both the right angle between the joint arms and the bending arms l of the forces (Figure 7).

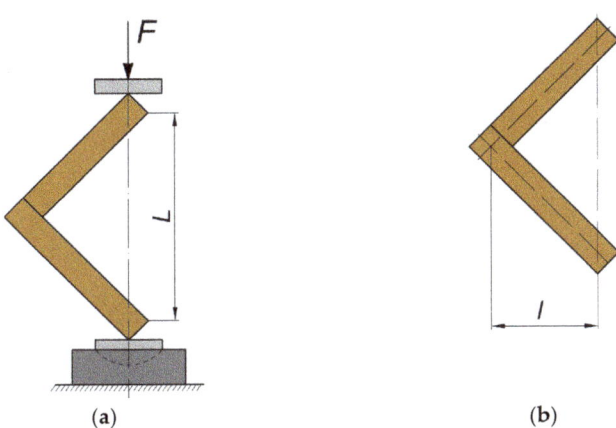

Figure 6. Test procedure. Reprinted with permission from Ref. [11]. 2021 BISMAR. (**a**) Type of loading of the tested samples; (**b**) determination of the bending arm.

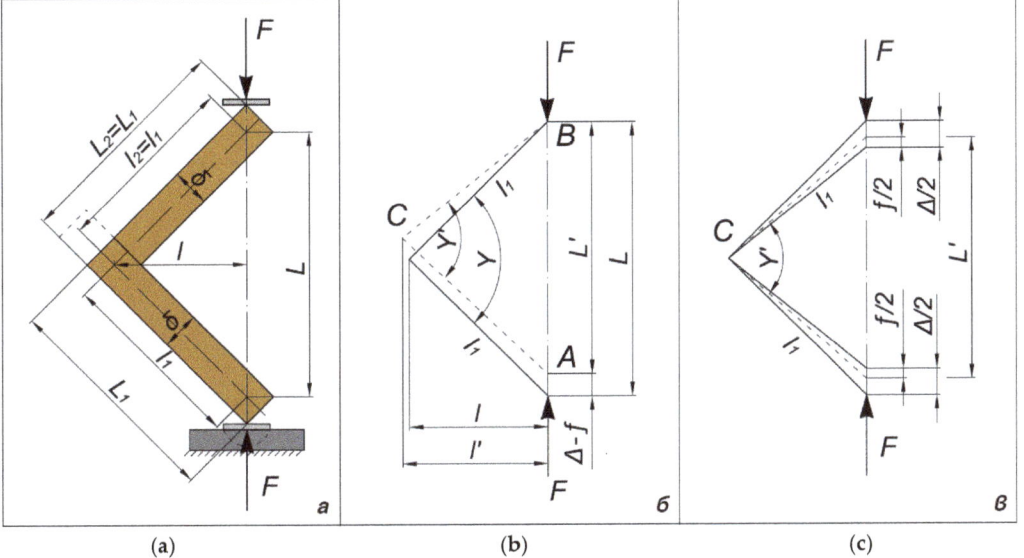

Figure 7. Test scheme and deformation under arm compression bending load of the test samples of end corner joints made of lightweight panels. Reprinted with permission from Ref. [11]. 2021 BISMAR. (**a**) Type of loading and dimensions of the tested samples; (**b**,**c**) scheme of loading and determination of the deformation of the tested samples.

The linear displacement f_i of the application points of the forces F_i is recorded for each test sample at each loading level. It represents a sum of displacement resulting from turning the joint arms and additional displacement Δ_i resulting from bending of the arms. The displacement Δ_i is calculated by the formula:

$$\Delta_i = \frac{F_i a^3}{3EI} \quad (2)$$

where

F_i is the magnitude of the load forces with arm compression, N;

a is the axial length of the joint arms, m;
E is the modulus of elasticity, N/mm^2;
I is the axial moment of inertia of the cross-section of the joint arms, m^4, which is calculated by the formula:

$$I = \frac{\delta b^3}{12}, \tag{3}$$

where
b is the width of the arms, m;
δ is the thickness of the arms, m.

The distance between the force application points at each level of loading is determined by the formula:

$$L_1 = L - f_i + \Delta_i, \tag{4}$$

The angle γ_i [rad] changed under loading between the joint arms is calculated by the formula:

$$\gamma_i = 2\arcsin\frac{L_i}{2a} = 2\arcsin\frac{L - f_i + \Delta_i}{2a}, \tag{5}$$

The changed bending arm l_i is determined by the formula:

$$l_i = a\cos\frac{\gamma_i}{2}, \tag{6}$$

The result from the deformation under the compression bending test is the semi-rigid rotation of the joint arms in [rad]:

$$\alpha_i = \frac{\pi}{2} - \gamma_i, \tag{7}$$

For 10 and 40% of the load force F_i, the bending moment in [N·m] is calculated according to the formula:

$$M_i = F_i l_i, \tag{8}$$

The stiffness coefficient under the compression bending test c_i [N·m/rad] is calculated by the formula:

$$c_i = \frac{\Delta M_i}{\Delta \alpha_i}, \tag{9}$$

In (8), the following designations are used:

$$\Delta M_i = M_i - M_0$$

$$\Delta \alpha_i = \alpha_i - \alpha_0$$

where M_i and α_i are determined according to (8) and (7) for the value of force F_i equal to 40% of F_{max}, and M_0 and α_0 according to (8) and (7) or the value of force F_0, equal to 10% of F_{max}.

The stiffness coefficient c as a deformation characteristic of the corner joint under the compression bending test is defined as the arithmetic means of the result of (9) numbers for each test sample when loaded in the section, which corresponds to the linear section on the curve of the correlation between the bending moment and the corner deformation of the joint.

The test was carried out in the scientific and research laboratory at the Institute of Mechanics and Biomechanics, BAS, Sofia, on a TIRA test 2000 universal type testing machine (TIRA GmbH, Schalkau, Germany). A sensor with a range of 1 kN was used to measure the force. The accuracy was 1%, for force over 10 N. The exact speed of the machine's moving traverse was measured with the aid of the built-in incremental displacement measurement system and an electronic stopwatch. Using a Protek D470 handheld digital multi-meter (Protek Instrument Co., Ltd, Gwangmyeong Gwangmyeong, Korea) and a computer, the force-time relationship was recorded. The traverse displacement was calculated according to time and speed. In determining the deflection of the test sample, the deflection of the loaded parts of the machine was taken into account. The resulting force-displacement diagrams were centred. The maximum force, deflection at the maximum force, force and deflection at a point taken as the failure point, $F_f = 0.8 F_{max}$, were determined. Tests were carried out at the temperature of 20 ± 1 °C and a relative humidity of 55 ± 5%.

Descriptive statistical analysis of the results was done with XLSTAT, version 2020.2.3, Addinsoft, New York, NY, USA (2021). One-way ANOVA was performed on the results for the bending strength and stiffness coefficient of L-type corner joints to analyse variance at a 95% confidence interval ($p < 0.05$). The statistical differences between mean values were evaluated using Tukey's honestly significant difference (HSD) post hoc test.

3. Results

3.1. Bending Moments of Corner Joints Made of Lightweight Panels

The results obtained from the bending strength test were processed statistically and are presented in Table 2 and Figure 8.

Table 2. Bending moments of corner joints made of lightweight panels.

Type of Joints [1]	No. of Test Samples	Mean, N·m	Min., N·m	Max., N·m	Median, N·m	St. Dev., N·m	Var. Coefficient
a	15	7.28	6.10	8.22	7.25	0.64	0.09
b	15	10.53	9.21	12.25	10.29	0.82	0.08
c	15	6.15	5.49	7.52	5.97	0.61	0.10
d	15	15.41	14.20	16.41	15.42	0.62	0.04
e	8	33.22	25.39	39.33	33.46	4.01	0.12
f	10	6.70	6.28	7.58	6.56	0.39	0.06
g	7	7.90	6.70	8.59	7.89	0.56	0.07
h	14	21.66	20.66	23.52	21.33	0.85	0.04
i	15	8.10	6.68	9.91	8.09	0.87	0.11
j	14	9.06	8.05	10.34	8.99	0.59	0.07

[1] The letter index of the type of joints is according to Figure 4.

In analysing the results for the bending strength test of corner L-type joints of lightweight panels, it can be seen that they vary over a fairly wide range, from 33.33 to 6.15 N·m. From the statistical analysis of the one-way ANOVA test and the pairwise comparison performed with the Tukey HSD, a significant difference of α = 0.05 at a confidence level of 95%, was found in seven groups between the obtained bending strength of the L-type end corner joints constructed from lightweight panels. The groups are given in Table 3. The highest strength value of 33.22 N·m was determined for the joints made of multi-layer veneer panels and three-layer cardboard with a nominal thickness of 51 mm and joined by a dowel ø12 × 50 mm. This result can be attributed to the application of adhesive also on the edge of the glued workpiece and the large thickness of the structural elements. The cardboard honeycomb panel (50 mm) connected by a plastic corner connector and special screw Varianta exhibited a bending strength of 21.66 N·m. The multi-layer panel (21 mm) made of veneer and corrugated cardboard, connected by a dowel ø8 × 35 mm, also showed excellent strength characteristics (15.41 N·m).

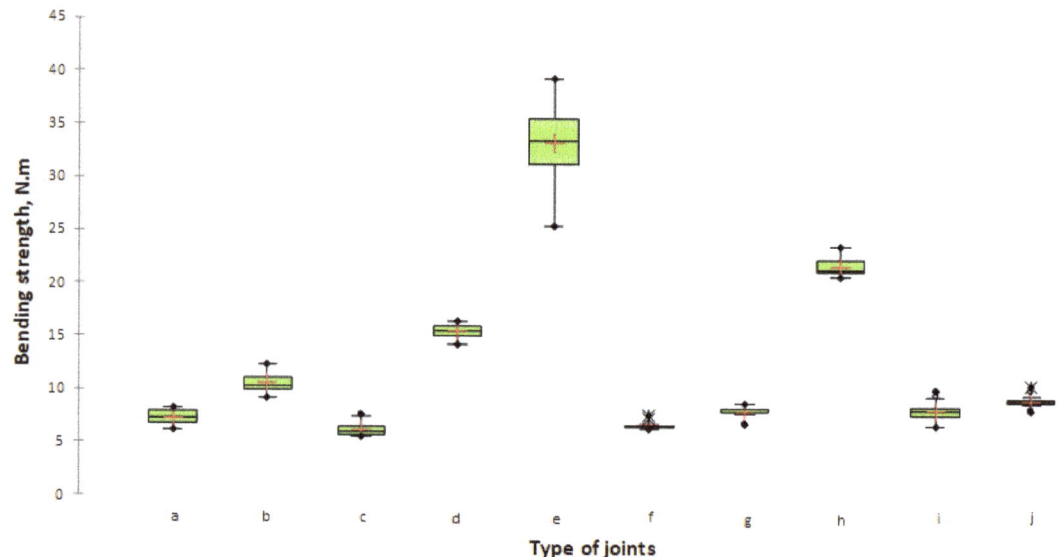

Figure 8. Bending strength under the compression test of L-type corner joints made of lightweight panels. The letter index of the type of joints is according to Figure 4.

Table 3. Tukey HSD analysis of the differences between the groups with a confidence interval of 95% of bending capacity of joints of two materials of all pairwise comparisons.

Index [1]	Type of Joint	Bending Strength N·m	Lower Bound (95%)	Upper Bound (95%)	GH [2]
e	Multi-layer panel (51 mm) made of veneer and corrugated cardboard, connected by a dowel ø12 × 50 mm	33.22	32.34	34.10	A
h	Cardboard honeycomb panel (50 mm) connected by plastic corner connector and special screw Varianta	21.66	21.13	22.18	B
d	Multi-layer panel (21 mm) made of veneer and corrugated cardboard, connected by a dowel ø8 × 35 mm	15.41	14.76	16.05	C
b	Multi-layer panel (21 mm) made of veneer and corrugated cardboard, connected by Confirmat ø7 × 70 mm	10.53	10.03	11.04	D
j	Cardboard honeycomb panel (50 mm) connected by TAB 20 HC	9.06	8.42	9.70	DE
i	Cardboard honeycomb panel (50 mm) connected by Rafix 20 HC	8.10	7.46	8.74	EF
g	Multi-layer panel (51 mm) made of veneer and corrugated cardboard, connected by Rafix 20 HC, inserted across on the cardboard direction	7.90	6.96	8.41	EFG
a	Multi-layer panel (21 mm) made of veneer and corrugated cardboard, connected by a plastic corner connector and special screw Varianta	7.28	6.33	7.91	FG
f	Multi-layer panel (51 mm) made of veneer and corrugated cardboard, connected by Rafix 20 HC, inserted longitudinally on the cardboard direction	6.70	5.90	7.56	FG
c	Multi-layer panel (21 mm) made of veneer and corrugated cardboard, connected by an eccentric connector Minifix	6.15	5.64	6.65	G

[1] The letter index of the type of joints is according to Figure 4. [2] Groups of homogeneities ($\alpha = 0.05$).

The remaining joints demonstrated satisfactory strength characteristics with bending strength values ranging from 6.15 to 10.53 N.m. In this group, the highest strength had series *b*—multi-layer panel (21 mm) made of veneer and corrugated cardboard, connected by Confirmat ø7 × 70 mm (10.53 N·m), followed by series *j*—cardboard honeycomb panel (50 mm) connected by TAB 20 HC (9.06 N·m), and series *i*—cardboard honeycomb panel (50 mm) connected by Rafix 20 HC (8.10 N·m). Finally, the lowest bending strength of 6.15 N·m was recorded in series *c*—multi-layer panel (21 mm) made of veneer and corrugated cardboard, connected by an eccentric connector Minifix.

3.2. Stiffness Characteristics of Corner Joints Made of Lightweight Panels

When analysing the results for the stiffness of the corner L-type joints of lightweight panels, it can be seen that the picture was slightly different. The stiffness values varied from 327 to 40 N·m/rad (Table 4). From the statistical analysis of the one-way ANOVA test and the pairwise comparison performed with the Tukey HSD (Table 5), a significant difference of $\alpha = 0.05$ at a confidence level of 95% was found in six groups between the obtained stiffness coefficients L-type end corner joints constructed from lightweight panels. The greatest stiffness was exhibited by joints made of conventional honeycomb lightweight 50 mm thick panels. This can be explained with an outer layer that is 8 mm thick, which gives more rigidity to the panel and joints. The highest stiffness value of 327 N·m/rad had the joints made of honeycomb panel—series *h*, connected by a plastic corner connector and special screw Varianta, followed by a cardboard honeycomb panel (50 mm) connected by Rafix 20 HC (225 N·m/rad), and a cardboard honeycomb panel (50 mm) connected by TAB 20 HC (195 N·m/rad). A multi-layer panel (51 mm) made of veneer and corrugated cardboard, connected by a dowel ø12 × 50 mm was the joint with the highest stiffness of the corrugated board joints (154 N·m/rad), followed by a multi-layer panel (21 mm) made of veneer and corrugated cardboard, connected by Confirmat ø7 × 70 mm (111 N·m/rad). A very low stiffness was shown by the joints made of a multi-layer panel made of veneer and corrugated cardboard with a thickness of 21 mm. Joints from these panels connected by a plastic corner connector and special screw Varianta had a stiffness coefficient of 49 N·m/rad, followed by an eccentric connector Minifix (43 N·m/rad) and Confirmat ø7 × 70 mm (40 N·m/rad). A graphical representation of the results obtained is shown in Figure 9.

Table 4. Stiffness coefficients of corner joints made of lightweight panels.

Type of Joints [1]	No. of Test Samples	Mean, N·m/Rad	Min., N·m/Rad	Max., N·M/Rad	Median, N·M/Rad	St. Dev., N·m/Rad	Var. Coefficient
a	15	49.19	36.68	61.40	48.77	5.43	0.11
b	15	40.28	27.81	57.14	40.09	7.83	0.19
c	15	42.72	27.46	65.32	41.23	10.56	0.25
d	15	111.02	92.99	134.05	110.47	9.92	0.09
e	8	153.61	117.33	200.91	139.03	31.41	0.20
f	10	81.27	56.86	102.07	81.77	11.62	0.14
g	7	83.76	42.50	134.66	84.69	28.59	0.34
h	14	327.34	267.24	419.56	321.33	39.26	0.12
i	15	224.61	117.49	342.42	215.01	58.87	0.26
j	15	194.71	157.93	245.26	188.06	24.97	0.13

[1] The letter index of the type of joints is according to Figure 4.

Table 5. Tukey HSD analysis of the differences between the groups with a confidence interval of 95% of stiffness coefficients of joints of two materials of all pairwise comparisons.

Index [1]	Type of Joint	Stiffness Coeff. c, N·m/rad	Lower Bound (95%) N·m/rad	Upper Bound (95%) N·m/rad	GH [2]
h	Cardboard honeycomb panel (50 mm) connected by a plastic corner connector and special screw Varianta	327	311.79	342.89	A
i	Cardboard honeycomb panel (50 mm) connected by Rafix 20 HC	225	209.59	239.64	B
j	Cardboard honeycomb panel (50 mm) connected by TAB 20 HC	195	177.58	208.68	BC
e	Multi-layer panel (51 mm) made of veneer and corrugated cardboard, connected by a dowel ø12 × 50 mm	154	133.04	174.19	C
d	Multi-layer panel (21 mm) made of veneer and corrugated cardboard, connected by a dowel ø8 × 35 mm	111	96.00	126.05	D
g	Multi-layer panel (51 mm) made of veneer and corrugated cardboard, connected by Rafix 20 HC, inserted across on the cardboard direction	84	61.77	105.76	DE
f	Multi-layer panel (51 mm) made of veneer and corrugated cardboard, connected by Rafix 20 HC, inserted longitudinally on the cardboard direction	81	62.87	99.67	DE
a	Multi-layer panel (21 mm) made of veneer and corrugated cardboard, connected by a plastic corner connector and special screw Varianta	49	34.17	64.21	EF
c	Multi-layer panel (21 mm) made of veneer and corrugated cardboard, connected by an eccentric connector Minifix	43	27.70	57.74	EF
b	Multi-layer panel (21 mm) made of veneer and corrugated cardboard, connected by Confirmat ø7 × 70 mm	40	25.26	55.31	E

[1] The letter index of the type of joints is according to Figure 4. [2] Groups of homogeneities ($\alpha = 0.05$).

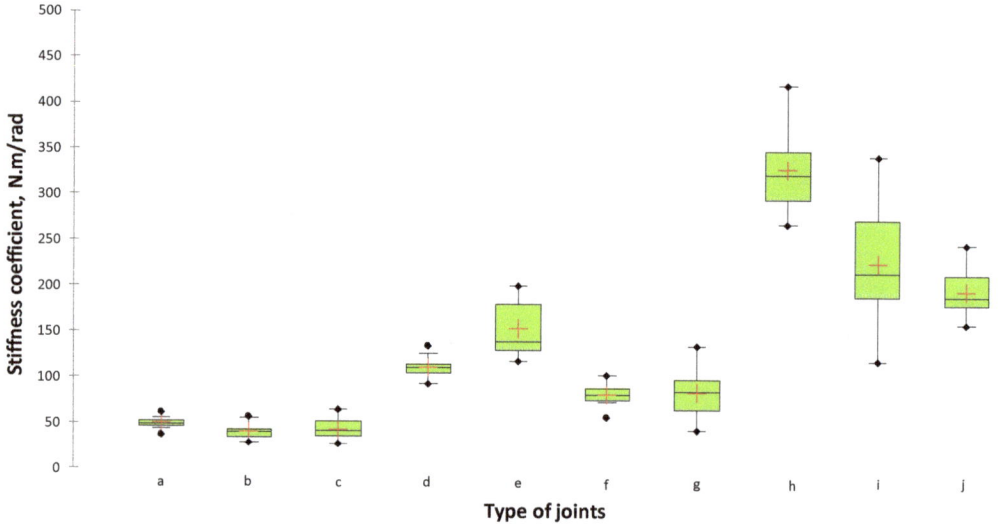

Figure 9. Stiffness coefficients under the compression test of L-type corner joints made of lightweight panels. The letter index of the type of joints is according to Figure 4.

4. Discussion

4.1. Bending Strength of Corner Joints Made of Lightweight Panels

From the results obtained for the bending strength of the corner joints, made of lightweight panels comprising waste cardboard and beech veneer, bonded with UF resin, it can be concluded that they possess the required strength for application in furniture constructions or as interior elements. The joints in both variants of multi-layer corrugated cardboard, with a thickness of 21 and 51 mm, exhibited good bending strength characteristics when connected by gluing with dowels. The results were within the range of bending strength values achieved by [36], where, however, the authors used two connecting elements in each joint. Joints of cardboard honeycomb panel (50 mm) connected by a plastic corner connector and special screw Varianta showed excellent bending strength. More than twice is the strength exhibited by the joints with TAB 20HC in the present study compared to the study conducted by Joscak et al. [60]. To note, multi-layer panels with a thickness of 21 mm made of veneer and corrugated cardboard, connected by an eccentric connector Minifix, showed lesser strength compared to the results of other studies [38,61]. However, it was in the range of similar joints made of laminated particleboards and medium-density fiberboard (MDF) with a thickness of 18 mm [11,59]. There was no significant difference in the joint's strength of multi-layer panel (51 mm) made of veneer and corrugated cardboard, connected by Rafix 20 HC, inserted across and longitudinally on the cardboard direction. Cardboard honeycomb panel (50 mm) and multi-layer panel (51 mm) made of veneer and corrugated cardboard connected by Rafix 20 HC shows a similar joint strength. Although the bending strength of the joints of 21 mm thick multi-layer panels made of waste cardboard and beech veneer connected by Confirmat did not reach the values obtained with particleboards, plywood, and MDF [62–67], still their strength properties were sufficient to be used as an option for joining.

4.2. Stiffness Characteristics of Corner Joints Made of Lightweight Panels

When analysing the results for the stiffness of the joints, it can be seen that the picture is different. The highest stiffness values were obtained for the joints made of 50 mm thick cardboard honeycomb panel coupled by a plastic corner connector with a special screw Varianta, Rafix 20 HC, and TAB 20 HC, and might be explained by the excellent stiffness of the outer layers of laminated chipboard and the good construction of the special connectors and screws developed for honeycomb panels. In the present study, joints connected with TAB 20 HC showed significantly higher stiffness compared to the results obtained by [37].

In contrast to the bending strength of the joints, the stiffness of the joints of cardboard honeycomb panel (50 mm) is more than two times higher compared to the multi-layer panel (51 mm) made of veneer and corrugated cardboard when connected by Rafix 20 HC.

Both multi-layer panels made of veneer and corrugated cardboard with a thickness of 51 and 21 mm, joined with a dowel ø12 × 50 mm and dowel ø8 × 35 mm, respectively, had sufficient rigidity to be used in furniture construction due to the application of an adhesive. Furthermore, the stiffness coefficients were in the range of other commonly used joints [68–73].

Joints exhibited relatively low stiffness with Confirmat and fasteners where screws were involved. The low stiffness of the 21 mm thick multi-layer veneer and corrugated cardboard joints resulted from the material's lack of a rigid and dense structure, which prevented a high screw-holding capacity.

5. Conclusions

The COVID-19 pandemic posed new challenges to the normal business practice and competitiveness of industrial enterprises, connected with significant changes in the market and shortage of resources, including solid wood and furniture panels [74–77]. This study investigated the potential of using waste corrugated cardboard and beech veneer for manufacturing lightweight panels for furniture and interior structural elements application.

For this purpose, various joints, fixed with adhesive and demountable joints were tested. As a result, the following conclusions were drawn:

1. The waste corrugated cardboard in combination with beech veneer is suitable for furniture constructions and interior elements.
2. The use of adhesive significantly increased the bending strength and stiffness of the joints of the veneer and corrugated cardboard panels in both studied panel thicknesses, i.e., 21 and 51 mm.
3. In terms of the stiffness coefficients, the best behaviour was exhibited by the joints, made of cardboard honeycomb panels with a thickness of 50 mm.
4. Except for the cardboard honeycomb panels, all of the tested panels where the demountable joints were used, showed a significantly lower strength compared to bonding.
5. There was no significant difference in the joint strength of the multi-layer 51 mm thick panel made of veneer and corrugated cardboard, connected by Rafix 20 HC, inserted across and longitudinally on the cardboard direction.
6. The cardboard honeycomb panel (50 mm) and multi-layer panel (51 mm) made of veneer and corrugated cardboard connected by Rafix 20 HC show a similar joint strength.
7. Although lightweight corrugated cardboards did not have sufficient density, the conventional joints such as Minifix, Confrimat, and plastic corner joints demonstrated comparable strength characteristics and stiffness to particleboard joints.

Future research should be focused on the development of the multi-layer waste corrugated cardboard and veneer lightweight panels as a prospective and sustainable material for furniture and other applications, in order to improve the strength and other features of the joints.

Author Contributions: Conceptualization, V.J. and R.S.; methodology, V.J., A.M. and R.S.; software, V.J. and B.P.; validation, V.J., P.A., A.M., B.P. and L.K.; formal analysis, V.J. and B.P.; investigation, R.S.; resources, V.J., R.S. and L.K.; data curation, V.J. and B.P.; writing—original draft preparation, V.J. and B.P.; writing—review and editing, P.A. and L.K.; visualisation, B.P.; supervision, V.J.; project administration, V.J. All authors have read and agreed to the published version of the manuscript.

Funding: The research was funded by the Scientific and Research Sector at the University of Forestry, Sofia, Bulgaria (project no. 77/2011) and projects by the Slovak Research and Development Agency under contracts no. APVV-18-0378, APVV-19-0269, and APVV-20-0004.

Data Availability Statement: Not applicable.

Acknowledgments: This publication is the result of the implementation of the following projects: Project no. 77/2011 and project no. НИС-Б-1145/04.2021 "Development, Properties and Application of Eco-Friendly Wood-Based Composites" carried out at the University of Forestry, Sofia, Bulgaria, as well as the National Programme "Young Scientists and Postdoctoral Students" of the Bulgarian Ministry of Education and Science.

Conflicts of Interest: The authors declare no conflict of interest.

References

1. Korol, J.; Hejna, A.; Wypiór, K.; Mijalski, K.; Chmielnicka, E. Wastes from Agricultural Silage Film Recycling Line as a Potential Polymer Materials. *Polymers* **2021**, *13*, 1383. [CrossRef] [PubMed]
2. Arias, A.; Feijoo, G.; Moreira, M.T. Evaluation of Starch as an Environmental-Friendly Bioresource for the Development of Wood Bioadhesives. *Molecules* **2021**, *26*, 4526. [CrossRef] [PubMed]
3. Savov, V.; Valchev, I.; Yavorov, N.; Sabev, K. Influence of press factor and additional thermal treatment on technology for production of eco-friendly MDF based on lignosulfonate adhesives. *Bulg. Chem. Com.* **2020**, *52*, 48–52.
4. Taghiyari, H.R.; Majidi, R.; Esmailpour, A.; Sarvari Samadi, Y.; Jahangiri, A.; Papadopoulos, A.N. Engineering composites made from wood and chicken feather bonded with UF resin fortified with wollastonite: A novel approach. *Polymers* **2020**, *12*, 857. [CrossRef]
5. Rammou, E.; Mitani, A.; Ntalos, G.; Koutsianitis, D.; Taghiyari, H.R.; Papadopoulos, A.N. The Potential Use of Seaweed (*Posidonia oceanica*) as an Alternative Lignocellulosic Raw Material for Wood Composites Manufacture. *Coatings* **2021**, *11*, 69. [CrossRef]

6. Taghiyari, H.R.; Militz, H.; Antov, P.; Papadopoulos, A.N. Effects of Wollastonite on Fire Properties of Particleboard Made from Wood and Chicken Feather Fibers. *Coatings* **2021**, *11*, 518. [CrossRef]
7. Kwidziński, Z.; Bednarz, J.; Pędzik, M.; Sankiewicz, Ł.; Szarowski, P.; Knitowski, B.; Rogoziński, T. Innovative Line for Door Production TechnoPORTA—Technological and Economic Aspects of Application of Wood-Based Materials. *Appl. Sci.* **2021**, *11*, 4502. [CrossRef]
8. IKEA Home Page. Available online: https://www.ikea.com/ms/en_SG/about_ikea/the_ikea_way/history/ (accessed on 18 April 2021).
9. Barbu, M.C. Evolution of lightweight wood composites. *Pro. Ligno.* **2015**, *11*, 21–26.
10. Barbu, M.C.; Lüdtke, J.; Thömen, H.; Welling, J. Innovative production of wood-based lightweight panels. In *Processing Technologies for the Forest and Biobased Products Industries*; Barbu, M.C., Ed.; Salzburg University of Applied Sciences: Kuchl, Australia, 2010; pp. 115–122.
11. Kyuchukov, G.; Jivkov, V. *Furniture Construction. Structural Elements and Furniture Joints*, 1st ed.; Bismar: Sofia, Bulgaria, 2016; p. 452.
12. Hänel, W.; Weinert, M. *Diätkur für Schwergewichte. Vorschläge zum Einsatz leichter Plattenwerkstoffe bei der Konstruktion von Möbeln*; Ihd, Institut für Holztechnologie, Adam Verlag: Dresden, Germany, 2010.
13. Poppensieker, J.; Thömen, H. *Wabenplatten für den Möbelbau*; Arbeitsbericht, Instituts für Holzphysik und Mechanische Technologie des Holzes, Bundesforschungsanstalt für Forst- und Holzwirtschaft: Hamburg, Germany, 2005.
14. Barboutis, I.; Vassiliou, V. Strength properties of lightweight paper honeycomb panels for the furniture. In Proceedings of the 10th International Scientific Conference of Engineering Design (Interior and Furniture Design), Sofia, Bulgaria, 17–18 October 2005; pp. 17–18.
15. Ebner, M.; Petutschnigg, A.J. Lightweight constructions-paper materials as a new option to build furniture. In Proceedings of the International Scientific Conference Interior and Furniture Design, University of Forestry, Sofia, Bulgaria, 1–5 October 2005; pp. 119–131.
16. Petutschnigg, A.J.; Ebner, M. Lightweight paper materials for furniture—A design study to develop and evaluate materials and joints. *Mater. Des.* **2007**, *28*, 408–413. [CrossRef]
17. Sam-Brew, S.; Semple, K.; Smith, G. Preliminary experiments on the manufacture of hollow core composite panels. *For. Prod. J.* **2011**, *61*, 381–389. [CrossRef]
18. Jivkov, V.; Simeonova, R.; Kamenov, P.; Marinova, A. Strength properties of new lightweight panels for furniture and interiors. In Proceedings of the 23rd International Scientific Conference "Wood Is Good–With Knowledge and Technology to a Competitive Forestry and Wood Technology Sector", Zagreb, Croatia, 12 October 2012; pp. 49–58.
19. Chen, Z.; Yan, N.; Sam-Brew, S.; Smith, G.; Deng, J. Investigation of mechanical properties of sandwich panels made of paper honeycomb core and wood composite skins by experimental testing and finite element (FE) modelling methods. *Eur. J. Wood Wood Prod.* **2014**, *72*, 311–319. [CrossRef]
20. Smardzewski, J. Experimental and numerical analysis of wooden sandwich panels with an auxetic core and oval cells. *Mater. Des.* **2019**, *183*, 108159. [CrossRef]
21. Peliński, K.; Smardzewski, J. Bending behavior of lightweight wood-based sandwich beams with auxetic cellular core. *Polymers* **2020**, *12*, 1723. [CrossRef] [PubMed]
22. Smardzewski, J.; Prekrat, S. Modelling of thin paper honeycomb panels for furniture. In Proceedings of the 23rd International Scientific Conference "Wood is Good–With Knowledge and Technology to a Competitive Forestry and Wood Technology Sector", Zagreb, Croatia, 12 October 2012; pp. 179–186.
23. Smardzewski, J. Mechanical properties of wood-based sandwich panels with a wavy core. In Proceedings of the XXVII International Conference Research for Furniture Industry, Ankara, Turkey, 17–18 September 2015; pp. 242–255.
24. Słonina, M.; Dziurka, D.; Smardzewski, J. Experimental research and numerical analysis of the elastic properties of paper cell cores before and after impregnation. *Materials* **2020**, *13*, 2058. [CrossRef]
25. Lyutyy, P.; Bekhta, P.; Ortynska, G. Lightweight flat pressed wood plastic composites: Possibility of manufacture and properties. *Drv. Ind.* **2018**, *69*, 55–62. [CrossRef]
26. Gößwald, J.; Barbu, M.C.; Petutschnigg, A.; Krišťák, Ľ.; Tudor, E.M. Oversized Planer Shavings for the Core Layer of Lightweight Particleboard. *Polymers* **2021**, *13*, 1125. [CrossRef] [PubMed]
27. Semple, K.E.; Sam-Brew, S.; Deng, J.; Cote, F.; Yan, N.; Chen, Z.; Smith, G.D. Properties of commercial kraft paper honeycomb furniture stock panels conditioned under 65 and 95 percent relative humidity. *For. Prod. J.* **2015**, *65*, 106–122. [CrossRef]
28. Chen, Z.; Yan, N.; Deng, J.; Semple, K.E.; Sam-Brew, S.; Smith, G.D. Influence of environmental humidity and temperature on the creep behavior of sandwich panel. *Int. J. Mech. Sci.* **2017**, *134*, 216–223. [CrossRef]
29. Smardzewski, J.; Majewski, A. Mechanical properties of auxetic honeycomb core with triangular cells. In Proceedings of the 25th International Scientific Conference: New Materials and Technologies in the Function of Wooden Products, Zagreb, Croatia, 17 October 2014; pp. 103–112.
30. Sawal, N.; Nazri, A.B.M.; Akil, H.M. Effect of cell size material on the mechanical properties of honeycomb core structure. *Int. J. Sci. Res.* **2013**, *4*, 80–84.
31. Smardzewski, J.; Gajęcki, A.; Wojnowska, M. Investigation of elastic properties of paper honeycomb panels with rectangular cells. *BioResources* **2019**, *14*, 1435–1451.

32. Smardzewski, J. Elastic properties of cellular wood panels with hexagonal and auxetic cores. *Holzforschung* **2013**, *67*, 87–92. [CrossRef]
33. Smardzewski, J.; Prekrat, S. Auxetic structures in layered furniture panels. In Proceedings of the 29th International Conference Wood Science Technology, Zagreb, Croatia, 6–7 December 2018; pp. 163–172.
34. Petutschnigg, A.J.; Koblinger, R.; Pristovnik, M.; Truskaller, M.; Dermouz, H.; Zimmer, B. Leichtbauplatten aus Holzwerkstoffen—Teil I: Eckverbindungen. *Holz Roh Werkst.* **2004**, *62*, 405–410. [CrossRef]
35. Heimbs, S.; Pein, M. Failure behaviour of honeycomb sandwich corner joints and inserts. *Compos. Struct.* **2009**, *89*, 575–588. [CrossRef]
36. Koreny, A.; Simek, M.; Eckelman, C.A.; Haviarova, E. Mechanical properties of knock-down joints in honeycomb panels. *BioResources* **2013**, *8*, 4873–4882. [CrossRef]
37. Joscak, P.; Krasula, P.; Vimpel, P. Strength properties of corner joints and extending joints on honeycomb boards. *Ann. Warsaw Univ. Life Sci.-SGGW. Forest. Wood Technol.* **2014**, *87*, 97–104.
38. Smardzewski, J.; Słonina, M.; Maslej, M. Stiffness and failure behaviour of wood based honeycomb sandwich corner joints in different climates. *Compos. Struct.* **2017**, *168*, 153–163. [CrossRef]
39. Majewski, A.; Krystofiak, T.; Smardzewski, J. Mechanical Properties of Corner Joints Made of Honeycomb Panels with Double Arrow-Shaped Auxetic Cores. *Materials* **2020**, *13*, 4212. [CrossRef]
40. Petutschnigg, A.J.; Koblinger, R.; Pristovnik, M.; Truskaller, M.; Dermouz, H.; Zimmer, B. Leichtbauplatten aus Holzwerkstoffen. Teil II: Untersuchungzum Schraubenausziehwiderstand. *Holz Roh Werkst.* **2005**, *63*, 19–22. [CrossRef]
41. Jivkov, V.; Kyuchukov, B.; Simeonova, R.; Marinova, A. Withdrawal capacity of screws and Confirmat into different wood-based panels. In Proceedings of the International Conference on Research for Furniture Industry, Poznan, Poland, 21–22 September 2017; pp. 68–82.
42. Statista: Global Paper Production Volume from 2008 to 2018 by Type. Available online: https://www.statista.com/statistics/270317/production-volume-of-paper-by-type// (accessed on 9 May 2021).
43. Cardboard Balers: Amazing Facts about Cardboard Waste & Recycling. Available online: https://www.cardboardbalers.org/cost-agricultural-solar-panels/ (accessed on 9 May 2021).
44. Eurostat: Recycling Rate of Packaging Waste by Type of Packaging. Available online: https://ec.europa.eu/eurostat/databrowser/view/cei_wm020/default/table?lang=en/ (accessed on 9 May 2021).
45. Buratti, C.; Belloni, E.; Lascaro, E.; Lopez, G.A.; Ricciardi, P. Sustainable panels with recycled materials for building applications: Environmental and acoustic characterisation. *Energy Procedia* **2016**, *101*, 972–979. [CrossRef]
46. Sair, S.; Mandili, B.; Taqi, M.; El Bouari, A. Development of a new eco-friendly composite material based on gypsum reinforced with a mixture of cork fibre and cardboard waste for building thermal insulation. *Compos. Commun.* **2019**, *16*, 20–24. [CrossRef]
47. Ayrilmis, N.; Candan, Z.; Hiziroglu, S. Physical and mechanical properties of cardboard panels made from used beverage carton with veneer overlay. *Mater. Des.* **2008**, *29*, 1897–1903. [CrossRef]
48. Sen, S.; Ayrilmis, N.; Zeki, C. Fungicide and insecticide properties of cardboard panels made from used beverage carton with veneer overlay. *Afr. J. Agric. Res.* **2010**, *5*, 159–165.
49. Rhamin, H.; Madhoushi, M.; Ebrahimi, A. Faraji, F. Effect of resin content, press time and overlaying on physical and mechanical properties of carton board made from recycled beverage carton and MUF resin. *Life Sci. J.* **2010**, *10*, 613.
50. Ebadi, M.; Farsi, M.; Narchin, P.; Madhoushi, M. The effect of beverage storage packets (Tetra Pak™) waste on mechanical properties of wood–plastic composites. *J. Thermoplast. Compos. Mater.* **2016**, *29*, 1601–1610. [CrossRef]
51. Aranda-García, F.J.; González-Pérez, M.M.; Robledo-Ortíz, J.R. Influence of processing time on physical and mechanical properties of composite boards made of recycled multi-layer containers and HDPE. *J. Mater. Cycles Waste Manag.* **2020**, *22*, 2020–2028. [CrossRef]
52. Asdrubali, F.; Pisello, A.L.; D'Alessandro, F.; Bianchi, F.; Cornicchia, M.; Fabiani, C. Innovative cardboard based panels with recycled materials from the packaging industry: Thermal and acoustic performance analysis. *Energy Procedia* **2015**, *78*, 321–326. [CrossRef]
53. Architecture Art Designs. Available online: https://www.architectureartdesigns.com/30-amazing-cardboard-diy-furniture-ideas// (accessed on 18 April 2021).
54. Stange Design: Cardboard Furniture. Available online: https://www.pappmoebelshop.de/ (accessed on 18 April 2021).
55. Petit & Small: Folding Creativity into Kids Furniture. Available online: https://petitandsmall.com/cardboard-furniture-cardboard-guys/?fbclid=IwAR11YDeR2N-ocpzByv-8aovz0Ck1oqmD3KssF7VM_n5H7xneO3frceHXSD0/ (accessed on 18 April 2021).
56. Dezeen: Airweave Creates Cardboard Beds and Modular Mattresses for Tokyo 2020 Olympics. Available online: https://www.dezeen.com/2021/07/11/cardboard-beds-modular-mattresses-airweave-tokyo-2020-olympics/ (accessed on 9 August 2021).
57. Diarte, J.; Shaffer, M.; Obonyo, E. Developing a Panelised Building System for Low-Cost Housing Using Waste Cardboard and Repurposed Wood. In Proceedings of the 18th International Conference on Non-Conventional Materials and Technologies (NOCMAT) "2019 Construction Materials & Technologies for Sustainability", Nairobi, Kenya, 5–7 November 2019.
58. Suarez, B.; Muneta, M.L.M.; Sanz-Bobi, J.D.; Romero, G. Application of homogenisation approaches to the numerical analysis of seating made of multi-wall corrugated cardboard. *Compos. Struct.* **2021**, *262*, 113642. [CrossRef]

59. Jivkov, V.; Marinova, A. Investigation on ultimate bending strength and stiffness under compression of corner joints from particleboard with connectors for DIY furniture. In Proceedings of the International Scientific Conference Interior and Furniture Design, Sofia, Bulgaria, 17–18 October 2005; pp. 155–165.
60. Joscak, P.; Langova, N.; Tvrdovsky, M. Withdrawal resistance of wood screw in wood-based materials. *Ann. Warsaw Univ. Life Sci.-SGGW. Forest. Wood Technol.* **2014**, *87*, 90–96.
61. Šimek, M.; Haviarová, E.; Eckelman, C. The end distance effect of knockdown furniture fasteners on bending moment resistance of corner joints. *Wood Fiber Sci.* **2008**, *42*, 92–98.
62. Jivkov, V. Influence of edge banding on banding strength of end corner joints from 18 mm particleboard. Symposium "Furniture 2002". Technical University-Zvolen, Zvolen, Slovakia, 24–25 October 2002.
63. Župčić, I.; Grbac, I.; Bogner, A.; Hadžić, D. Research corner joints in corpus furniture. In Proceedings of the International Conference Ambienta "Wood is Good–With Knowledge and Technology to a Competitive Forestry and Wood Technology Sector", Zagreb, Croatia, 12 October 2012; pp. 229–235.
64. Yuksel, M.; Kasal, A.; Erdil, Y.Z.; Acar, M.; Kuşkun, T. Effects of the panel and fastener type on bending moment capacity of L-type joints for furniture cases. *Pro Ligno* **2015**, *11*, 426–443.
65. Langová, N.; Joščák, P. Mechanical properties of Confirmat screws corner joints made of native wood and wood-based composites. *For. Wood Technol.* **2019**, *105*, 76–84. [CrossRef]
66. Sydor, M.; Rogozinski, T.; Stuper-Szablewska, K.; Starczewski, K. The accuracy of holes drilled in the side surface of plywood. *BioResources* **2020**, *15*, 117–129. [CrossRef]
67. Sydor, M. Geometry of Wood Screws: A Patent Review. *Eur. J. Wood. Prod.* **2019**, *77*, 93–103. [CrossRef]
68. Máchová, E.; Langová, N.; Réh, R.; Joščák, P.; Krišťák, L'.; Holouš, Z.; Igaz, R.; Hitka, M. Effect of moisture content on the load carrying capacity and stiffness of corner wood-based and plastic joints. *Bioresources* **2019**, *14*, 8640–8655. [CrossRef]
69. Sydor, M.; Majka, J.; Langova, N. Effective Diameters of Drilled Holes in Pinewwod in Response to Changes in Relative Humidity. *Bioresources* **2021**, *16*, 5407–5421. [CrossRef]
70. Langova, N.; Beno, P.; Luptakova, J.; Fragassa, C. Stress concentration around circular and elliptic holes in wood laminates. *FME Trans.* **2020**, *48*, 102–108. [CrossRef]
71. Machova, E.; Holous, Z.; Langova, N.; Balazova, Z. The effect of humidity on the shear strength of glued wood based and plastic joints. *Acta Fac. Xylologiae* **2018**, *60*, 113–120.
72. Hitka, M.; Joscak, P.; Langova, N.; Kristak, L.; Blaskova, S. Load-carrying capacity and the size of chair joints determined for users with a higher body weight. *Bioresources* **2018**, *13*, 6428–6443. [CrossRef]
73. Branowski, B.; Starczewski, K.; Zablocki, M.; Sydor, M. Design issues of innovative furniture fasteners for wood-based boards. *BioResources* **2020**, *15*, 8472–8495. [CrossRef]
74. Sedliacikova, M.; Moresova, M.; Alac, P.; Drabek, J. How do behavioral aspects affect the financial decisions of managers and the competetiveness of enterprises? *J. Compet.* **2021**, *13*, 99–116.
75. Sedliacikova, M.; Strokova, Z.; Hitka, M.; Nagyova, N. Employees versus implementing controlling to the business practice. *Entrep. Sustain. Issues* **2020**, *7*, 1527–1540.
76. Pirc Barčić, A.; Kitek Kuzman, M.; Vergot, T.; Grošelj, P. Monitoring Consumer Purchasing Behavior for Wood Furniture before and during the COVID-19 Pandemic. *Forests* **2021**, *12*, 873. [CrossRef]
77. Ratnasingam, J.; Khoo, A.; Jegathesan, N.; Wei, L.C.; Latib, H.A.; Thanasegaran, G.; Liat, L.C.; Yi, L.Y.; Othman, K.; Amir, M.A. How are small and medium enterprises in Malaysia's furniture industry coping with COVID–19 Pandemic? Early evidences from a survey and recommendations for policymakers. *Bioresources* **2020**, *15*, 5951–5964. [CrossRef]

Article

Properties of Eco-Friendly Particleboards Bonded with Lignosulfonate-Urea-Formaldehyde Adhesives and pMDI as a Crosslinker

Pavlo Bekhta, Gregory Noshchenko, Roman Réh, Lubos Kristak, Ján Sedliačik and Petar Antov et al.

1. Department of Wood-Based Composites, Cellulose and Paper, Ukrainian National Forestry University, 79057 Lviv, Ukraine
2. Department of Chemistry, Ukrainian National Forestry University, 79057 Lviv, Ukraine; noschenkog@gmail.com
3. Faculty of Wood Sciences and Technology, Technical University in Zvolen, 960 01 Zvolen, Slovakia; reh@tuzvo.sk (R.R.); sedliacik@tuzvo.sk (J.S.)
4. Faculty of Forest Industry, University of Forestry, 1797 Sofia, Bulgaria; p.antov@ltu.bg (P.A.); victor_savov@ltu.bg (V.S.)
5. Department of Wood-Based Materials, Poznań University of Life Sciences, 60-627 Poznań, Poland; rmirski@up.poznan.pl
* Correspondence: bekhta@nltu.edu.ua (P.B.); kristak@tuzvo.sk (L.K.)

Abstract: The purpose of this study was to evaluate the feasibility of using magnesium and sodium lignosulfonates (LS) in the production of particleboards, used pure and in mixtures with urea-formaldehyde (UF) resin. Polymeric 4,4′-diphenylmethane diisocyanate (pMDI) was used as a crosslinker. In order to evaluate the effect of gradual replacement of UF by magnesium lignosulfonate (MgLS) or sodium lignosulfonate (NaLS) on the physical and mechanical properties, boards were manufactured in the laboratory with LS content varying from 0% to 100%. The effect of LS on the pH of lignosulfonate-urea-formaldehyde (LS-UF) adhesive compositions was also investigated. It was found that LS can be effectively used to adjust the pH of uncured and cured LS-UF formulations. Particleboards bonded with LS-UF adhesive formulations, comprising up to 30% LS, exhibited similar properties when compared to boards bonded with UF adhesive. The replacement of UF by both LS types substantially deteriorated the water absorption and thickness swelling of boards. In general, NaLS-UF-bonded boards had a lower formaldehyde content (FC) than MgLS-UF and UF-bonded boards as control. It was observed that in the process of manufacturing boards using LS adhesives, increasing the proportion of pMDI in the adhesive composition can significantly improve the mechanical properties of the boards. Overall, the boards fabricated using pure UF adhesives exhibited much better mechanical properties than boards bonded with LS adhesives. Markedly, the boards based on LS adhesives were characterised by a much lower FC than the UF-bonded boards. In the LS-bonded boards, the FC is lower by 91.1% and 56.9%, respectively, compared to the UF-bonded boards. The boards bonded with LS and pMDI had a close-to-zero FC and reached the super E0 emission class ($\leq$1.5 mg/100 g) that allows for defining the laboratory-manufactured particleboards as eco-friendly composites.

Keywords: magnesium lignosulfonate; sodium lignosulfonate; bio-based adhesives; urea formaldehyde resin; wood-based composites; particleboards; formaldehyde content; physical and mechanical properties; acid-base buffer

Citation: Pavlo Bekhta, Gregory Noshchenko, Roman Réh, Lubos Kristak, Ján Sedliačik and Petar Antov et al. Properties of Eco-Friendly Particleboards Bonded with Lignosulfonate-Urea-Formaldehyde Adhesives and pMDI as a Crosslinker. *Materials* **2021**, *14*, 4875. https://doi.org/10.3390/ma14174875

Academic Editor: Fernão D. Magalhães

Received: 4 August 2021
Accepted: 25 August 2021
Published: 27 August 2021

Publisher's Note: MDPI stays neutral with regard to jurisdictional claims in published maps and institutional affiliations.

Copyright: © 2021 by the authors. Licensee MDPI, Basel, Switzerland. This article is an open access article distributed under the terms and conditions of the Creative Commons Attribution (CC BY) license (https://creativecommons.org/licenses/by/4.0/).

1. Introduction

Particleboards still predominate in the world production of wood-based composites. In 2018, industrial particleboard production reached a record output of 97 million m^3 worldwide [1]. Particleboards remain one of the most important value-added panel products in the wood industry and are widely used in various fields of human activity. With increasing applications of composite materials, the demand for adhesive increases. The cost of resin is about 30–50% of the material costs, whereas the product contains only 2–14% of resin in terms of the amount associated with the dry weight of wood. Even at such low concentrations, the cost of the resin is the main factor significantly affecting the total product price [2].

Today, the main classes of thermosetting adhesives that have been dominating the field of wood composites industry for many decades are amino-based, phenolic, and isocyanate resins [3]. Currently, approximately 95% of the total number of adhesives used for the manufacture of wood composites are formaldehyde-based resins [4], and the most predominant type is urea-formaldehyde (UF) resins, the total consumption of which is estimated to approximately 11 million tons per year [5]. Within these thermosetting adhesives, UF resins are the most widely used adhesives in the manufacture of wood-based composites, including particleboards. The industrial success of these resins is associated with their low-cost raw materials, high reactivity, excellent adhesion to wood, ease of use for a wide range of curing conditions, low temperature of curing, short pressing time, aqueous solubility, and a colourless glue line [6]. However, they have a major drawback, connected to the hazardous emission of volatile organic compounds (VOCs) and free formaldehyde from the finished wood-based composites, which can irritate the eyes, respiratory, and nervous systems and even lead to cancers such as leukaemia [7]. As a result, new formaldehyde emission restrictions have been set for wood-based composites in Europe, the United States, and Japan. In addition, the production of these resins relied on non-renewable oil resources. Today, these adhesives are sufficient for supply, but the shortage of petroleum products may affect the future cost and availability of these petroleum-based adhesives. Therefore, there is a growing interest in the development of environmentally friendly adhesives for wood from renewable resources. This has stimulated the transition from traditional formaldehyde-based synthetic resins to new environmentally friendly adhesives for the production of eco-friendly wood-based panels from renewable resources. Lignin-based products are one of the most promising environmental alternatives to traditional formaldehyde resins [8,9].

The main interest in lignin is due to its phenolic structure with several favourable properties for the manufacture of wood adhesives, such as high hydrophobicity and low polydispersity [10]. However, the low chemical reactivity of lignins requires higher concentrations of catalysts (heat or acid) and longer heating times are required during the production of wood-based composites [9,11–13]. Hence, an additional chemical modification of lignin is required to increase the lignin reactivity to formaldehyde [13–15]. Peng and Riedl [16] proved in their work that the reactivity of lignosulfonate with formaldehyde increases when wheat starch is added as filler. It is estimated that the planet currently contains 3×10^{11} metric tons of lignin with an annual biosynthetic rate of approximately 2×10^{10} tons [13]. There are large quantities of technical lignin (mainly as kraft lignin and lignosulfonate) generated as waste or a by-product from the paper making industry, with an annual global production of approximately 50–75 million tons [17]. Thus, the utilisation of lignin as a renewable component in the production of value-added products, including wood adhesives [18–21], could be an efficient way to achieve sustainable resource management.

Many authors have extensively studied the potential utilisation of different types of lignin including lignosulfonates (LS), Kraft lignin, organosolv lignin, enzymatic hydrolysis lignin, and soda lignin in adhesive applications [12,15,18,20–29]. The drawbacks of using lignins alone as wood adhesives, modifications to enhance the reactivity of lignins, and production of lignin-based copolymer adhesives for composite wood panels are reviewed and discussed in the comprehensive review [30]. In several studies, lignin was used as the binding agent for oriented strand boards [31] and medium density fibreboards (MDF) [27–29], which allowed the production of eco-friendly, low-toxic boards. The main scientific and industrial interest in lignin-based wood adhesives, such as LS, is due to the polyphenolic structure of lignin, allowing a partial replacement of phenol in phenol-formaldehyde (PF) resins. Çetin and Özmen [32] demonstrated that organosolv lignin could be used to replace 20–30% of the phenol in PF resins used to bond particleboards, without adversely affecting bond properties. It was found the phenolated lignin exhibited better mechanical properties than the unmodified lignin. In another study [33], it was also attested the possibility of substitution of up to 30% of phenol by lignin in PF adhesive without significantly affecting the plywood shear strength. Akhtar et al. [34] found that the maximum shear strength and wood failure was obtained by 20% addition of LS to PF resin. da Silva et al. [35], using LS with pure and in mixtures with phenol-formaldehyde (PF) resin to produce particleboards, found that the replacement of PF adhesive by LS in up to 80% was satisfactory in meeting the specific standard for mechanical properties. However, it was not possible to produce boards with minimal strength properties with pure LS-based adhesive. In other works [36,37], particleboards bonded with glyoxalated lignin combined with pMDI showed superior internal bond (IB) strength in dry and boiled conditions. There are also a number of successful attempts to produce MDF in laboratory conditions on the basis of LS [8,21,38–44]. The authors showed that the use of LS as an adhesive is a perspective approach for producing eco-friendly MDF panels without harmful free-formaldehyde emissions. Consequently, various studies have been carried out with a number of different lignin types as substitutes for phenol in PF adhesives. However, to date, the addition of unmodified or modified lignin-based compounds into UF adhesives is limited.

Since recently, due to environmental trends, focus has been drawn to the use of LS in adhesive compositions for wood [38,39,41]; thus, it was of considerable interest to study the effect of LS on the pH of UF compositions. This is due to the fact that the acidity of UF compositions is one of the key factors determining their pot life, cure rate, cure temperature, depth of cure, cohesion and adhesion, moisture resistance and atmospheric influences resistance, and other properties [45–47]. Despite this, to date, the change in pH during the curing of UF resins with natural fillers has not been fully investigated. This information may also be helpful in understanding the effect of pH on the curing of pure UF resins.

Considering the above, the objective of this work was to evaluate the feasibility of using lignosulfonates as an adhesive and the effect of replacing the UF resin by different proportions of lignosulfonate on the physical and mechanical properties of particleboards. The effect of lignosulfonates on the pH of lignosulfonate-urea-formaldehyde adhesive compositions was also investigated.

2. Materials and Methods

2.1. Materials

Factory-produced wood particles comprised of coniferous (75%) and deciduous (25%) species (origin—the Ukrainian Carpathians, Ivano-Frankivsk region) were obtained from the local particleboard plant. The moisture content of wood particles, determined by the drying-weighing method, was approximately 8%. The fractional composition of the particles for the outer and core layers of the boards is presented in Table 1, and the appearance of the particles is shown in Figure 1.

The lignosulfonate-urea-formaldehyde adhesive system (LS-UF) consisted of UF resin grade A (density 1.28 g/cm^3, solid content 66%, Ford cup (4 mm, 20 °C) viscosity 98 s, pH = 8.05, gel time 50 s) (producer Karpatsmoly LLC, Kalush, Ukraine), pMDI resin Ongronat® WO 2750 (NCO content—31.05 wt %, viscosity at 25 °C, 201 MPa·s, acidity as HCl—120 mg/kg), paraffin emulsion, urea, ammonium sulphate, magnesium lignosulfonate (Borregaard, Germany), and sodium lignosulfonate (Domsjö Lignin, Sweden). The water solution with 43% of ammonium sulphate [$(NH_4)_2SO_4$] was used as hardener and mixed with the resin before spraying into wood particles. The water solution with 40% of urea and water solution with 50% of LS were mixed with the resin. Urea and LS were used as water solutions at 40% and 50% working concentrations, respectively. The lignosulfonate addition levels were based on the replacement of 10%, 20%, 30%, 50%, 75%, and 100% of the UF resin in the adhesive system used in outer and core layers. Magnesium lignosulfonate (MgLS) had the following characteristics: total solids content—min 90%;

pH (10%)—4.0 ± 1.0; insoluble matter [%]—max. 0.8; Mg [%]—3; Cl [%]—≤0.1; sucrose [%]—6; density [kg/m³]—450–600. Sodium lignosulfonate (NaLS) had the following characteristics: total solids content > 95%; pH (10% solution)—6 ± 1; sodium Na—9; sulphur, S—8.5; calcium, Ca—0.12; chlorine, Cl—0.01; insoluble substances <0.1; sulphate—7.5 in the form of sulphate ions; sucrose—2.0.

Table 1. Fraction analysis (by % weight) of particles.

Outer Layer		Core Layer	
Screen Hole Size (mm)	Content (%)	Screen Hole Size (mm)	Content (%)
1.25	8.8	5.0	12.0
1.0	1.2	3.15	25.6
0.8	9.4	2.0	31.4
0.63	12.2	1.25	10.6
0.4	26.4	0.63	8.4
0.2	17.6	0.32	1.4
Dust	14.5	Dust	0.6
Total	100	Total	100

(a)   (b)

Figure 1. Appearance of the wood particles used for outer layers (**a**), core layer (**b**).

In order to use lignin of LS as the main component in adhesive formulations, its low reactivity needs to be compensated by using a suitable crosslinker [15]. pMDI was used as an additional component to improve moisture resistance and mechanical properties for LS adhesives due to its high reactivity towards wood surface, LS molecules, and UF resin molecules. In order to find out how the addition of pMDI resin to the adhesive compositions affects the properties of the particleboards, the boards were made using UF and LS adhesives with different content (1%, 3% and 5%) of pMDI resin.

2.2. Measurement of pH

The pH was measured using pH-meter pH-301 and combined electrode with glass membrane (glass/KCl, AgCl/Ag). A two-point calibration procedure was performed with pH buffer solutions of pH 4.01 ± 0.01 and 6.86 ± 0.01 before and after pH measurements. To measure pH, 40 g samples were prepared by weighing the components with an accuracy of 0.001 g. Glass electrode was placed into the sample and then the sample was stirred for at least 5 min. The pH values were measured after stopping stirring. The stirring procedure and measurements were repeated three times, and the average pH value was calculated. When measuring pH, it was observed that to obtain stable pH values for lignosulfonate solutions, stirring for slightly longer periods of time was required compared to standard solutions. Measurement of pH values for each point during titration of lignosulfonates solutions was also performed after 5 min of stirring. To determine the effect of each component on the pH of adhesive composition, the components were added to UF resin one by one, thoroughly mixed, and the pH was measured. The components were mixed in the following sequence—UF resin, LS, urea, ammonium sulphate, paraffin emulsion. The content of UF resin and LS was changed in the compositions and the other components were added in the same amount: paraffin emulsion—7.1%; 33% ammonium sulphate solution—5.4%; 43% urea solution—4.3%.

2.3. Manufacturing of Particleboards

In this study, three-layered particleboards of 290 mm × 290 mm dimensions and a thickness of 16 mm with a target density of 650 kg/m^3 were designed. The amount of UF resin, pMDI resin, urea, hardener, and paraffin emulsion that were needed for the blending process differed between core layer and outer layer. It is due to the temperature difference between surface and core caused by the heat transfer from the surface to the core of particleboard. In addition, different amounts of resin and additives used are due to the difference in surface area of particles used in outer and core of particleboard. In control series, the amount of solid UF resin was 14 wt % and 9 wt % based on the mass of oven dry wood particles for outer and core layers, respectively. During resin mixing, 2.3% and 0.5% of urea solution and 0.2% and 0.6% of ammonium sulphate based on dry particles weights were mixed with 14% and 9% of UF resin for outer and core layers, respectively. On the other hand, 0.8% of paraffin emulsion based on dry particles weight was also incorporated into the resin mixtures. pMDI resin was added to the adhesive system used for core layer. Wood particles were blended with the adhesive by hand. After blending, the resinated particles were hand spread evenly onto a 290 mm × 290 mm wooden box with a caul plate as the base to form the mattress (Figure 2a). The mattress formed was then pre-pressed manually to consolidate the thickness. Next, the mattress was subjected to hot pressing in an automatically controlled hydraulic laboratory press "хоМко"(LLC "ODEK" Ukraine, Ukraine) (Figure 2b) at the pressure of 2.5 MPa, and temperature 200 °C for 600 s (during the last 30 s of the press cycle, the pressure was continuously reduced to 0 MPa). The high pressing temperature and extended pressing time are required for activating lignosulfonates and their binding to wood particles [8,15,30,39,41]. The long pressing time was also to make it possible to evaporate all the water present in the composition of the LS-UF adhesive. The experimental design for this study is summarised in Table 2.

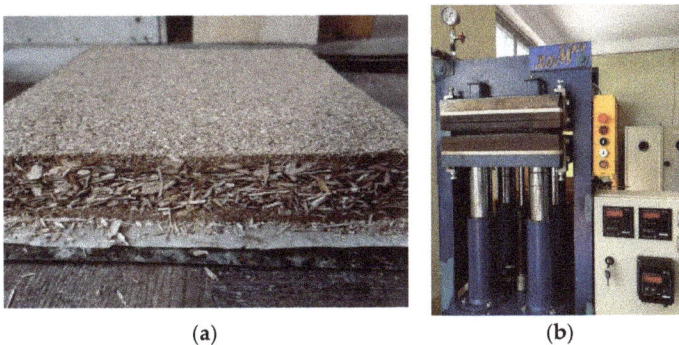

Figure 2. The formed particle mattress (**a**) and hydraulic laboratory press (**b**).

2.4. Particleboards Testing

After pressing, boards were stabilised in air until reaching room temperature. Then, the boards were conditioned for one week in a conditioning room maintained at a relative humidity of 65 ± 5% and 20 ± 2 °C prior to properties evaluation. Three boards were produced for each type of particleboards in the experimental design, i.e., 39 boards in total +1 set of control. The control particleboard was made from adhesive system without LS. The conditioned boards were cut into required testing size according to relevant standards. Three samples of each board were tested according to the European standards for moisture content (EN 322) [48], density (EN 323) [49], bending strength (MOR) (EN 310) [50], modulus of elasticity (MOE) (EN 310) [50], internal bond (IB) (EN 319) [51], and thickness swelling (TS) (EN 317) [52]. On the other hand, for each series, one board was randomly selected for analysis of formaldehyde content (FC) based on EN ISO 12460-5 (perforator method) [53].

Table 2. Manufacturing parameters of particleboards produced in this work.

Board Type	Adhesive Type	UF Resin Content (%)	MgLS Content (%)	NaLS Content (%)
A	MgLS-UF	90	10	0
B	MgLS-UF	80	20	0
C	MgLS-UF	70	30	0
D	MgLS-UF	50	50	0
E	MgLS-UF	25	75	0
F	MgLS	0	100	0
G	NaLS-UF	90	0	10
H	NaLS-UF	80	0	20
I	NaLS-UF	70	0	30
J	NaLS-UF	50	0	50
K	NaLS-UF	25	0	75
L	NaLS	0	0	100
Ref		100	0	0

2.5. Statistical Analysis

The effects of LS content on the properties of the laboratory-fabricated boards were evaluated by analysis of variance (ANOVA) at the 0.05 level of significance. Duncan's Range tests were conducted to determine significant differences between mean values.

3. Results

3.1. The Effect of Lignosulfonates on the pH of the Adhesive Compositions

It is known that the acid value—pH—is an important characteristic of UF adhesive compositions [45–47]. For long-term storage of resin in liquid state, a pH close to 7–8 is maintained in it. To cure it, acidic hardeners are added, which lower the pH. As a result of the decrease in pH, rapid condensation processes of UF oligomers begin, which leads to the curing of the composition [47]. In addition, curing is further accelerated by heating. To better understand the effect of pH, it is worth noting that when the pH of the composition changes by 1, the curing rate of UF resin changes by about 10 times [45]. Hence, it is important to investigate the effect of LS on the pH of adhesive compositions. Since 50% solutions of both LS were acidic, in particular, NaLS is characterised by pH = 5.44 and MgLS has pH = 3.56. At first glance it seems that their addition should accelerate the curing of the adhesive composition. However, the research indicates that the addition of LS can slightly reduce the effectiveness of latent hardeners such as $(NH_4)_2SO_4$.

The results of measuring the pH of individual components and uncured adhesive compositions are shown in Table 3, Table 4, Table 5 and Figure 3. The acidity of pure pMDI was not measured due to its incompatibility with aqueous solutions.

Table 3. pH values of the individual components of the adhesive compositions.

Component	pH
MgLS, 50% aqueous solution	3.565 ± 0.05
NaLS, 50% aqueous solution	5.442 ± 0.05
Paraffin emulsion 7.1%	9.414 ± 0.03
Urea, 1.8% aqueous solution	7.121 ± 0.05
$(NH_4)_2SO_4$, 1.8% aqueous solution	5.682 ± 0.04
UF resin	7.615 ± 0.04

Table 4. pH values of freshly prepared uncured adhesive compositions of NaLS.

Mass Composition of the Mixture NaLS + UF Resin	pH Values of Uncured Adhesive Compositions			
	NaLS + UF	NaLS + UF + Urea	NaLS + UF + Urea + $(NH_4)_2SO_4$	NaLS + UF + Urea + $(NH_4)_2SO_4$ + Paraffin Emulsion
100% NaLS + 0% UF	5.442 ± 0.05	5.454 ± 0.05	5.459 ± 0.05	5.474 ± 0.04
75% NaLS + 25% UF	5.661 ± 0.04	5.667 ± 0.04	5.631 ± 0.05	5.652 ± 0.04
50% NaLS + 50% UF	5.867 ± 0.05	5.864 ± 0.05	5.837 ± 0.04	5.863 ± 0.05
25% NaLS + 75% UF	6.130 ± 0.04	6.146 ± 0.04	6.149 ± 0.04	6.197 ± 0.05
0% NaLS + 100% UF	7.619 ± 0.03	7.508 ± 0.03	7.142 ± 0.03	7.368 ± 0.04

Table 5. pH values of freshly prepared uncured adhesive compositions of MgLS.

Mass Composition of the Mixture MgLS + UF Resin	pH Values of Uncured Adhesive Compositions			
	MgLS + UF	MgLS + UF + Urea	MgLS + UF + Urea + $(NH_4)_2SO_4$	MgLS + UF + Urea + $(NH_4)_2SO_4$ + Paraffin Emulsion
100% MgLS + 0% UF	3.565 ± 0.04	3.569 ± 0.03	3.572 ± 0.03	3.589 ± 0.04
75% MgLS + 25% UF	3.837 ± 0.04	3.869 ± 0.03	3.897 ± 0.03	3.921 ± 0.04
50% MgLS + 50% UF	4.322 ± 0.03	4.334 ± 0.04	4.346 ± 0.03	4.386 ± 0.04
25% MgLS + 75% UF	5.140 ± 0.05	5.161 ± 0.03	5.203 ± 0.03	5.253 ± 0.04
0% MgLS + 100% UF	7.619 ± 0.03	7.508 ± 0.03	7.142 ± 0.04	7.368 ± 0.03

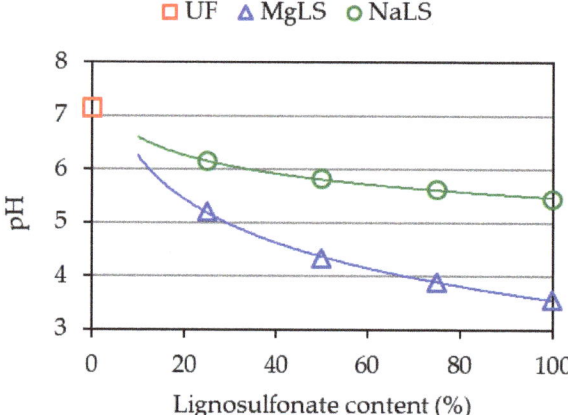

Figure 3. pH of freshly prepared uncured UF compositions as a function of lignosulfonate content.

As is seen from Table 3, the UF resin had a slightly alkaline medium, pH = 7.615. The addition of NaLS significantly reduced the pH of the resin. It stands to mention that despite the acidic nature of LS, when they were added, no thickening and gelation of the UF composition was observed, which would interfere with further work. The addition of urea solution had almost no effect on the pH of the composition. The addition of hardener, $(NH_4)_2SO_4$, also had little effect on the acidity of the compositions. The effect of adding $(NH_4)_2SO_4$ was only noticeable for the pure UF resin not containing NaLS. A similar decrease in the pH of the uncured resin by 0.8 with the addition of NH_4Cl has been noted [47]. The addition of paraffin emulsion slightly moved the pH of the composition towards the alkaline side. Thus, the NaLS had the most significant effect on the pH of adhesive compositions. Figure 3 shows that the effect of NaLS on the acidity of the original composition was proportional to the amount of added NaLS. From the data in Table 4, it is noticeable that the more NaLS is contained in the composition, the less the other components affect its acidity, regardless of whether they are acidic or alkaline in nature. This suggests that NaLS has acid-base buffering properties. Buffering properties mean the ability to resist a change in the pH of a solution when other substances of acidic or alkaline nature are added.

A similar effect was observed for uncured compositions with MgLS (Table 5). The biggest change in the pH of the composition occurred when MgLS was added. $(NH_4)_2SO_4$ and paraffin emulsion had little effect on the pH. $(NH_4)_2SO_4$ markedly changed the pH only in the absence of MgLS. That is, MgLS also exhibits buffering properties with regard to changes in the pH of the composition.

For compositions with a 50 UF/50 MgLS ratio, the effect of pMDI on pH was investigated. First, 3% pMDI was added to the composition containing UF, MgLS, urea, $(NH_4)_2SO_4$, and paraffin emulsion, and thoroughly mixed using an electric stirrer. In this case, the pMDI was dispersed into droplets, which are well distinguished in transmitted light in an optical microscope at 10× magnification. The resulting compositions with pMDI had an average pH of 4.392, which is almost equal to the pH of such compositions without pMDI (4.386). From then on, it was assumed that pMDI does not affect the pH of the compositions, and therefore, it was not added to the compositions intended for the study of pH. In addition, no pMDI was added to avoid adhesion of pMDI to the glass electrode of the pH meter.

It is worth pointing out that pMDI has the potential to increase the pH of compositions when organic co-solvents are used along with water. This is due to the fact that during hydrolysis, the -N=C=O isocyanate groups are converted into -NH$_2$ amino groups, which further neutralise acids [54]. However, pMDI, which remains in the aqueous dispersion in the form of individual droplets, cannot significantly affect pH, since under such conditions, the condensation reaction should prevail in pMDI rather than the hydrolysis reaction. In addition, the hydrolysis product, due to its hydrophobic nature, should remain dissolved in pMDI droplets and not significantly affect the pH of the composition.

It also seemed important to investigate the effect of NaLS and MgLS on the pH of the compositions post curing and to what extent their buffering properties were exhibited in that case. However, pH measurement during the curing process is associated with technical challenges, in particular, with the adhesion of resin onto the electrode of the pH meter. Therefore, it was the pH of the cured compositions that was measured. First, the compositions consisting of UF, NaLS or MgLS, and $(NH_4)_2SO_4$ were prepared. They were cured by heating to 98 °C for 15 min, cooled, left for 24 h, ground to a powdery condition, and the resulting powder was stirred in a double amount of water for 15 min. The pH of the resulting solution was measured. In doing so, a pH value close to that of the cured adhesive composition was obtained. The measurement results are shown in Figure 4. Paraffin emulsion and urea were not added, given that these components affect the pH insignificantly. In addition, this made it possible to see a clearer picture of the competition between lignosulfonates and $(NH_4)_2SO_4$.

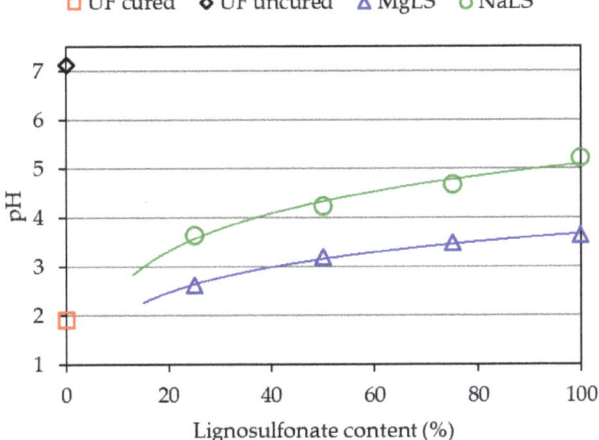

Figure 4. pH of cured UF compositions as a function of lignosulfonate content.

As is seen in Figure 4, the acid value of a composition consisting only of UF resin and $(NH_4)_2SO_4$ post-curing decreased to pH = 1.914 ($pH_{uncured} - pH_{cured} = 5.228$), which is consistent with the data of other authors [55]. On the other hand, the pH of compositions containing NaLS or MgLS changed far less. With an increase in the content of NaLS or MgLS, the pH of the compositions is close to the pH of pure NaLS or MgLS. The effect of NaLS or MgLS is most noticeable if we compare the pH of the compositions with a high content of lignosulfonates. In Figures 3 and 4, it can be seen that in the uncured condition and post-curing, the pH value for them changes little (for example, for 75% NaLS $pH_{uncured} - pH_{cured} = 0.932$, for 75% MgLS $pH_{uncured} - pH_{cured} = 0.407$). Thus, NaLS and MgLS for cured adhesive compositions counteract pH change. Thus, NaLS and MgLS reduce the influence of the $(NH_4)_2SO_4$ latent hardener.

It is known that a decrease in the acidity of UF-based adhesive compositions occurs due to the interaction of latent hardener ($(NH_4)_2SO_4$) with resin formaldehyde according to the equation [45,47]:

$$2(NH_4)_2SO_4 + 6CH_2O = N_4(CH_2)_6 + 6H_2O + 2H_2SO_4$$

In which case, a strong mineral acid H_2SO_4 is released, which reduces the pH of the medium to about 2. To simulate the interaction of the acid released by the latent hardener from lignosulfonates and to compare the effect of LNa and MgLS on pH, we studied the interaction in a system containing acid of known concentration, NaLS or MgLS of known concentration, but not containing UF resin. In parallel, to evaluate how the pH of a solution not containing NaLS or MgLS will change, an experiment adding acid to distilled water was conducted [56].

It should be noted that the effect of acid on the pH of water should be considered quite close to the effect of acid on the UF resin itself, because the resin contains almost no components that could counteract changes in pH. In Figure 5, it can be seen that the addition of 0.74 mL of acid enables reduction of the pH of water to 1.914 (change in pH = 7.000 − 1.914 = 5.086), but the pH of lignosulfonate solutions changes little with the addition of 0.74 mL of acid. For NaLS, the pH decreases only to 4.60 (change in pH = 5.270 − 4.60 = 0.67), and for MgLS, only to 3.20 (change in pH = 3.952 − 3.20 = 0.752). Thus, it is clearly seen from this graph that NaLS and MgLS exhibit buffering properties—that is, they resist pH changes. In this case, the buffer capacity of MgLS is greater than NaLS.

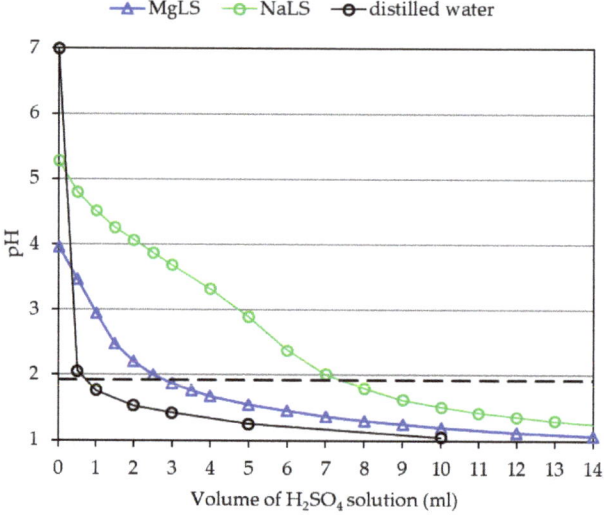

Figure 5. Curves of variation of pH as a function of titration with 0.309 M H_2SO_4 of lignosulfonate solutions or distilled water. Mass of the solution is 40 g. Concentration of lignosulfonate is 6%. The dotted line shows the pH of the cured resin with $(NH_4)_2SO_4$ without lignosulfonates.

The LS neutralisation curve in Figure 5 can be useful for a rough estimation of the amount of acid or acid hardener that needs to be added to the composition to achieve the desired pH. For this, in Figure 5, we draw a horizontal line at pH = 1.914. This pH corresponds to curing the UF resin with $(NH_4)_2SO_4$ without LS. Noticeably, achieving pH = 1.914 requires the addition of 0.74 mL of acid to distilled water. In the presence of 6% MgLS, achieving pH = 1.914 requires addition of 2.84 mL of acid. In the presence of 6% NaLS, achieving pH = 1.914 requires addition of 7.44 mL of acid. Thus, the presence of NaLS in the system requires an increase in the amount of acid by 7.44/2.84 = 2.6 times compared to MgLS. Similarly, an increase in the amount of a latent acid hardener will be required. If the need arises, a recalculation can be made for other contents of lignosulfonates. Of course, it should be remembered that other factors can affect the pH of the composition, too; this calculation is very rough, one might even say semi-quantitative.

The nature of the buffer properties of LS is explained by the presence in the LS molecules of (-COOM where M = Na^+ or Mg^{2+}) neutralised carboxyl groups and (-SO_3M) sulfonic groups, the content of which, according to [57,58], can range within about

0.6–3.0 mmol/g and 1.5–2.0 mmol/g, respectively. These groups can bind H^+ ions generated by the latent hardener according to the scheme [59]:

$$H_2SO_4 = 2H^+ + SO_4^{2-}$$

$$\text{Lignin-COO}^- + H^+ \rightarrow \text{Lignin-COOH}$$

$$\text{Lignin-SO}_3^- + H^+ \rightarrow \text{Lignin-SO}_3H$$

The buffer properties of LS seem important because they can affect the properties of resins in the same way that buffering agents affect the properties of MUF resins [60].

Consequently, the addition of acidic NaLS and MgLS leads to a decrease in the pH of uncured UF resin. In contrast, when curing UF resin, LS counteract the effects of latent hardener (($NH_4)_2SO_4$) due to their acid-base buffering properties. As a result, the cured compositions have higher residual pH values. Thus, LS can be used to adjust the pH of uncured and cured LS-UF compositions.

3.2. Physical Properties of Particleboards

Table 6 presents the average values of density, TS, and water absorption (WA) after 2 and 24 h for the boards produced with the different adhesive compositions. The moisture content of the boards was within the range of 6%.

Table 6. Physical properties of particleboards produced in this work.

Board Type	Density (kg/m^3)	Water Absorption (2 h) (%)	Water Absorption (24 h) (%)	Thickness Swelling (2 h) (%)	Thickness Swelling (24 h) (%)
		Boards bonded with MgLS			
A	625.9 ± 44.1 bc	37.68 ± 3.93 b	47.55 ± 2.63 a	24.92 ± 4.80 b	40.85 ± 6.90 a
B	629.8 ± 44.3 c [1]	37.46 ± 3.22 b	49.35 ± 1.48 ab	23.12 ± 4.73 b	43.85 ± 5.36 a
C	615.0 ± 29.1 abc	42.89 ± 3.43 c	51.50 ± 1.35 c	29.94 ± 5.27 c	52.37 ± 6.85 b
D	623.4 ± 39.9 bc	47.90 ± 1.18 d	58.46 ± 0.76 d	48.46 ± 6.07 d	80.92 ± 7.44 c
E	610.8 ± 37.1 ab	51.37 ± 4.53 e	Samples delaminated	51.69 ± 8.63 d	Samples delaminated
F	606.1 ± 26.8 a	Samples delaminated	Samples delaminated	Samples delaminated	Samples delaminated
Ref	624.5 ± 32.1 bc	31.43 ± 4.68 a	48.43 ± 3.39 a	17.71 ± 2.79 a	40.92 ± 3.56 a
		Boards bonded with NaLS			
G	633.4 ± 38.6 bc	42.51 ± 5.33 b	50.48 ± 1.73 b	26.41 ± 2.98 b	44.73 ± 3.97 a
H	624.2 ± 44.3 abc	44.36 ± 4.65 b	55.55 ± 1.56 c	29.71 ± 4.63 b	49.88 ± 4.81 b
I	644.1 ± 42.9 d	44.36 ± 4.59 b	54.77 ± 2.38 c	32.25 ± 6.02 cb	59.08 ± 8.25 c
J	623.6 ± 35.6 ab	49.11 ± 1.95	56.32 ± 1.56 c	53.17 ± 7.66 d	87.23 ± 7.50 d
K	605.5 ± 52.0 a	50.13 ± 2.07 c	Samples delaminated	72.43 ± 12.61 e	Samples delaminated
L	629.5 ± 43.0 bc	Samples delaminated	Samples delaminated	Samples delaminated	Samples delaminated
Ref	624.5 ± 32.1 abc	31.43 ± 4.68 a	48.43 ± 3.39 a	17.71 ± 2.79 a	40.92 ± 3.56 a

[1] Averages followed by the same letter at the column are statistically equal by the Duncan test at 95% probability.

In general, it was observed that the LS type and its content in the adhesive system had a significant effect on the physical properties of the boards. The average values of the board densities were 618.5 kg/m^3 and 626.6 kg/m^3 for the boards bonded with the addition of MgLS and NaLS, respectively. This difference in density values is small and most likely caused by the conditions of manual mattress forming. No significant differences were observed between the boards bonded by MgLS or NaLS and UF adhesive system. Moreover, the average values obtained were somewhat inferior to the density calculated of 650 kg/m^3 due to the effect of loss of materials during the mattress formation. The average density of reference boards (UF-bonded) was 624.5 kg/m^3.

It was found that the replacement of the UF by both type of LS substantially influenced the TS after 2 h and 24 h, so that the higher the content of LS, the greater the TS. The same behaviour was observed in WA values after 2 h and 24 h. There was no significant difference in TS and WA values after 24 h between boards manufactured with UF adhesive (Ref) and boards manufactured with up to 20% replacement by MgLS (A and B). After 24 h, there was only no difference in TS values between UF adhesive (Ref) and 10% replacement by NaLS (G). That is, the replacement of up to 20% or 10% of UF adhesive by MgLS or NaLS, respectively, did not negatively alter the TS and WA values. The samples manufactured exclusively with 100% LS-based adhesives (F and L) after 2 h and after 24 h of immersion in water were delaminated. The samples manufactured with the replacement of 75% of UF by MgLS (E) and NaLS (K) were also delaminated after 24 h of immersion in water.

The boards produced with the conventional UF adhesive showed that the average thickness swelling values after 2 and 24 h of immersion in water were statistically lower in relation to the boards produced with the lignosulfonate-urea-formaldehyde adhesive. Significant differences were observed between the two types of lignosulfonates for the WA and TS after 2 and 24 h of immersion in water. The boards produced with the MgLS showed means statistically lower in relation to the boards produced with the NaLS; however, the differences in terms of absolute averages were small.

The high WA values in this study could be attributed to the low density of the boards and, accordingly, their high porosity. The low density of boards increased the boards' water absorption for both 2 and 24 h. This result is in good agreement with the concepts mentioned by Maloney [2]. According to this author, the boards with higher density have better closure of its structure by reducing the permeability to water. For the thickness swelling, there was an increase in their average values for boards with higher specific mass. This increase results from the effects of the release of greater compression tensions of the boards produced with higher density. In addition, the higher values for WA and TS recorded in this study have been attributed to the fact that UF resins are characterised by their poor moisture resistance. This has been supported from literature stating that UF-bonded boards always have higher WA and TS than the corresponding PF-bonded board under the same experimental conditions [61].

The mean values of WA and TS after 2 and 24 h of immersion in water obtained in this study were also higher than the results reported by some researchers [62] in particleboards produced with lignin-phenol-formaldehyde resin with density of 750 and 950 kg/m^3. In other works [38,40], it was also indicated that manufactured MDF panels using 15% gluing content of MgLS exhibited a deteriorated dimensional stability (thickness swelling and water absorption (24 h)). da Silva et al. [35] also observed that the replacement of PF resin by calcium and magnesium LS substantially influenced the WA and TS after 2 h and 24 h, so that the higher the proportion of LS, the greater the WA and TS. However, the replacement of up to 40% of PF by LS did not negatively alter the WA and TS values. Gothwal et al. [23] found that the replacement of phenol by lignin in PF adhesive in up to 15% did not alter the particleboard physical properties. On the contrary, Çetin and Özmen [32] observed that the TS and WA were largely unaffected by the presence of lignin in adhesives.

However, it should be noted that the dimensional stability of the boards bonded using LS-UF adhesive could be improved with the application of finishing materials for the board's surface that assist in the waterproofing of the boards. Consequently, MgLS and NaLS, as lignin-based compounds, will require additional chemical modification (such as phenolation and methylolation, among others) in order to increase their chemical reactivity to formaldehyde and bonding efficiency [10,14,15,30].

No clear dependence of formaldehyde content on the LS content in the LS-UF adhesive was found (Figure 6). In general, with the exception of pure adhesives, NaLS-UF-based boards had a lower formaldehyde content than MgLS-UF-based boards. This may be due to the higher pH of the cured compositions containing NaLS (Figure 4), since the influence of the residual acidity of the cured UF compositions on formaldehyde emission is known [55]. Of course, for UF resin compositions with LS, this dependence should be further researched. From the results, it can be also seen that NaLS-UF-based boards had a lower percentage of formaldehyde content than the UF-based boards used as control.

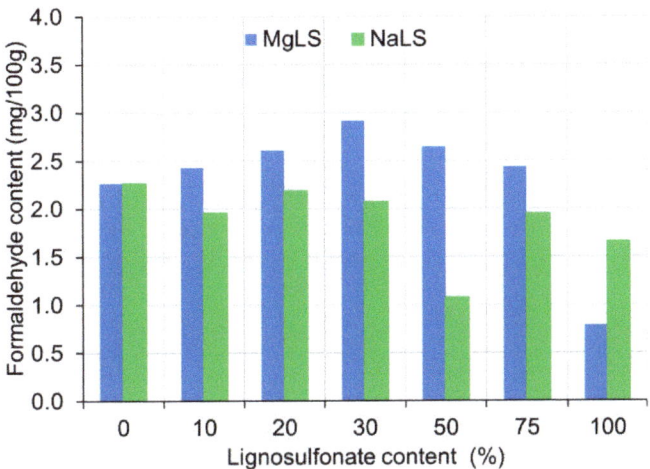

Figure 6. The formaldehyde content of boards made with lignosulfonated-urea-formaldehyde adhesive.

3.3. Mechanical Properties of Particleboards

Graphical representation of the effects of the type and content of LS on the mechanical properties of particleboards are presented in Figures 7–9. In general, it was observed that the LS type and content in the adhesion system had a significant effect on the mechanical properties of the boards.

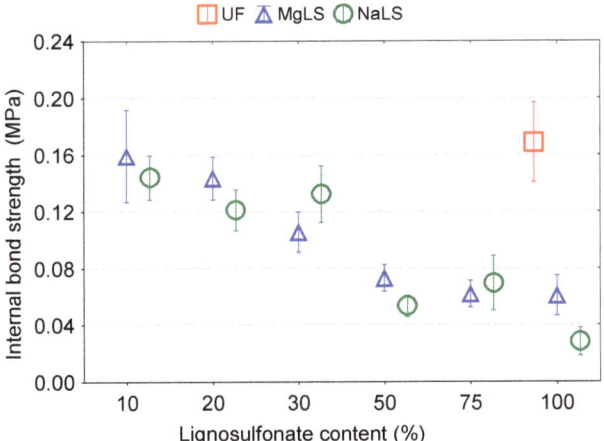

Figure 7. The internal bond (IB) strength of boards made with lignosulfonated-urea-formaldehyde adhesive.

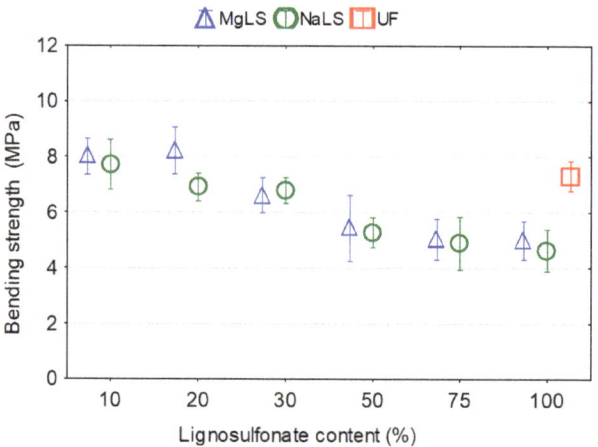

Figure 8. Bending strength (MOR) of boards made with lignosulfonated-urea-formaldehyde adhesive.

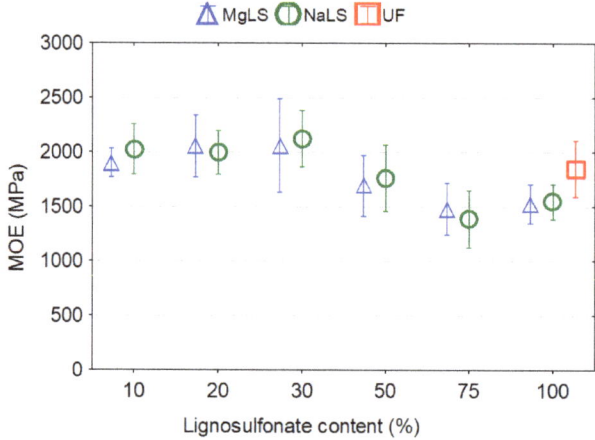

Figure 9. Modulus of elasticity (MOE) of boards made with lignosulfonated-urea-formaldehyde adhesive.

The measured IB values of boards bonded using with LS-UF adhesives are summarised and compared with those of boards bonded UF adhesive in Figure 7. Typically, for both LS used in this work, IB tended to decrease with increasing the LS content. The IB of boards bonded with UF adhesive was significantly ($p < 0.05$) higher than that of boards bonded with LS-UF adhesives. Particleboards bonded with LS-UF adhesives with 10%, 20%, 30%, 50%, 75%, and 100% replacement of UF resin by MgLS showed reductions of 5.9%, 17.6%, 35.3%, 58.8%, 64.7%, and 64.7%, respectively, in the IB values compared to particleboards bonded with UF adhesive. A similar trend was observed in the reductions (17.6%, 29.4%, 23.5%, 70.6%, 58.8%, and 82.3%) in the IB values of boards bonded with 10%, 20%, 30%, 50%, 75%, and 100% replacement of UF resin by NaLS. To note, no significant differences were observed between the IB values for boards bonded with MgLS and NaLS. It can be concluded that there is an apparent correlation between IB and TS; the better a particleboard is bonded, the lower the thickness swelling. This is in a good agreement with similar findings of other researchers [41].

The incorporation of LS to UF resin formulation resulted in a reduction of bonding strength, because unmodified lignin has low reactivity toward formaldehyde [63]. Hence, lignin requires either chemical modification or molecular fractionation to improve its reactivity in the production of UF resins [20]. For example, the chemical reactivity of demethylated lignins was enhanced through diminishing methoxyl group content while increasing hydroxyl group content. With the lignins demethylated under optimum reaction conditions, the bonding strength of lignin-based phenolic resins increased while the formaldehyde emissions decreased [64–66].

Da Silva et al. [35] found that panels with pure PF and 20% replacement by calcium and magnesium LS had the highest average IB values of particleboards. As the PF adhesive was replaced by lignosulfonate in larger proportions, there was a decrease in the values of IB. A similar result was obtained by Akhtar et al. [34], who indicated that the maximum shear strength was obtained by 20% addition of lignosulfonate to PF resin. Çetin and Özmen [32] demonstrated that organosolv lignin could be used to replace 20–30% of the phenol in PF resins used to bond particleboards, without adversely affecting bond properties. Savov et al. [67] reported that MDF panels bonded with different lignosulfonate contents (20%, 30%, 40%) have also met the respective European standard requirements for applications in dry conditions.

The MOR and MOE data of the particleboards made using LS-UF adhesives are illustrated in Figures 8 and 9.

Statistically significant differences were observed in mean values of MOR for the boards produced with UF and LS-UF adhesives. The replacement of 10–20% or 10% of the UF resin by MgLS and NaLS, respectively, caused positive changes in the MOR values by 9.4% and 12.4% for MgLS and by 5.6% for NaLS. From 30% or 20% replacement of UF resin by MgLS and NaLS, respectively, the MOR values decreased. Particleboards bonded with LS-UF adhesives with 30%, 50%, 75%, and 100% replacement of UF resin by MgLS showed reductions of 9.6%, 25.7%, 31.1%, and 31.7%, respectively, in the MOR values compared to particleboards bonded with UF adhesive. This can be explained by the greater fragility of the boards when increasing the content of lignosulfonate and a higher content of the steam–gas mixture in the process of pressing, associated with the increase in moisture content with an increased content of lignosulfonate solution [42,44]. Particleboards bonded with LS-UF adhesives with 20%, 30%, 50%, 75%, and 100% replacement of UF resin by NaLS showed reductions of 5.6%, 7.1%, 27.7%, 33.3% and 36.6%, respectively, in the MOR values compared to particleboards bonded with UF adhesive. There was no significance difference between MOR values for boards bonded with MgLS and NaLS. For both LS, it was evident that as the LS content increased, MOR tended to decrease. Savov and Antov [8] also observed a deterioration in the strength properties of the MDF panels at LS concentrations above 35%. The authors explained such deterioration of the strength properties by increasing the hot-pressing temperature. Antov et al. [42] reported that the increase in LS addition, from 10% to 15%, resulted in lower bending strength of the MDF panels. Quite similar results have been reported by Savov and Mihajlova [44], when they investigated the mechanical properties of MDF bonded with 5% UF resin and calcium LS (0% to 20% addition levels). In another work [35], it was found that the replacement of up to 60% of the PF adhesive by calcium and magnesium LS did not cause a negative change in the MOR values of particleboards. However, from 80% replacement of PF by LS and in boards produced with pure lignosulfonate, the MOR values decrease.

The MOR values found in this work were substantially lower in comparison with the mentioned by other authors [35]. This can be explained by the lower density of the boards and using UF adhesives with worse bonding properties in comparison with PF adhesives. It has been reported that low-density boards have low mechanical properties in general [68]. This might be also attributed to the presence of more sugars in the LS and increased moisture content of the pressed mat material, and higher vapour–gas mixtures at the higher LS addition levels [42].

There was no significance difference between MOE values for boards bonded with UF and LS-UF adhesives. The replacement of up to 30% of the UF resin by LS caused an increase in the MOE values up to 11.44% or 14.82% for MgLS and NaLS, respectively. However, particleboards bonded with LS-UF adhesives with 50%, 75%, and 100% replacement of UF resin by MgLS or NaLS showed reductions of 8.5%, 19.9%, 17.4% and 4.71%, 25.06%, and 16.44%, respectively, in the MOE values compared to particleboards bonded with UF adhesive. Savov and Mihajlova [44] also observed degradation, respectively decreasing, of bending strength and modulus of elasticity in bending of MDF panels but already after passing the calcium lignosulfonate content of 10%. Kouisni et al. [33] attested to the possibility of substitution of up to 30% of phenol by lignin in PF adhesive without significantly affecting the particleboard mechanical properties.

3.4. Effect of pMDI Content on the Properties of Particleboards

It was observed that in the process of manufacturing boards using UF adhesive, increasing the proportion of pMDI resin in the adhesive composition from 1% to 5% increased by 50% the IB of the boards (Figure 10), did not affect the MOR (Figure 11) and at the same time decreased by 25% MOE (Figure 12). The formaldehyde content in the boards increased (Figure 13), although the average values of FC at pMDI content of 1%, 3%, and 5% are lower (1.76, 2.27, and 2.99 mg/100 g, respectively) compared to the boards made with UF adhesive without addition of pMDI resin (3.48 mg/100 g).

In the boards made using MgLS as an adhesive, the addition of pMDI resin increased MOR by 33.2% (Figure 11), MOE by 25.6% (Figure 12), and IB by 366.7% (Figure 10). The high IB strength was probably due to the chemical reaction of LS with pMDI in the cured adhesives, i.e., the formation of urethane linkages [15]. The effect of pMDI resin on the formaldehyde content in MgLS bonded boards is ambiguous (Figure 13). The lowest formaldehyde content was observed in the boards without the addition of pMDI resin (0.31 mg/100 g). Addition of 1% pMDI resin to MgLS adhesive significantly increased the formaldehyde content (2.70 mg/100 g), but a further increase in the pMDI resin content to 3% and 5% leads to a decrease in formaldehyde content (0.79 mg/100 g and 0.57 mg/100 g, respectively).

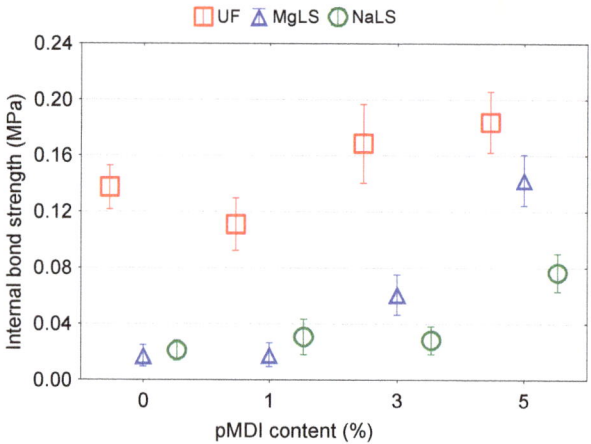

Figure 10. Effect of pMDI content on the IB of particleboards bonded with pure UF and lignosulfonate adhesives.

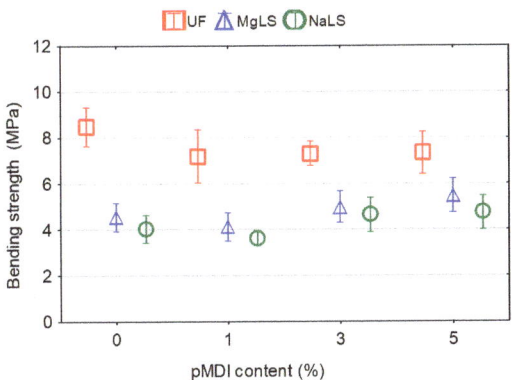

Figure 11. Effect of pMDI content on the bending strength of particleboards bonded with pure UF and lignosulfonate adhesives.

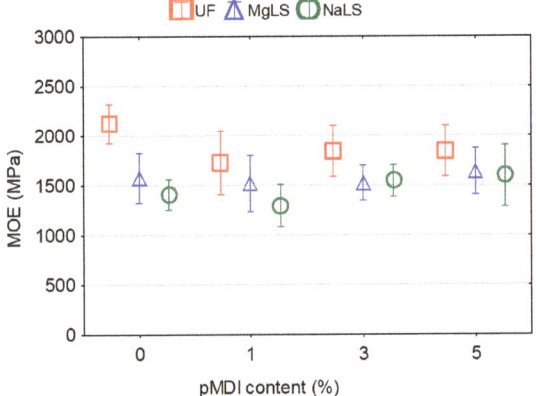

Figure 12. Effect of pMDI content on the MOE of particleboards bonded with pure UF and lignosulfonate adhesives.

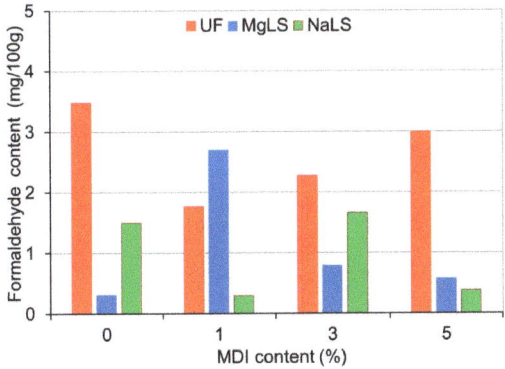

Figure 13. Effect of pMDI content on the formaldehyde content of boards made with pure UF and lignosulfonate adhesives.

In the boards made using NaLS adhesive, the addition of pMDI resin can increase MOR by 35.7% (Figure 11), MOE by 17.9% (Figure 12), and IB by 133.3% (Figure 10). The effect of pMDI resin on the formaldehyde content in the NaLS bonded boards is also ambiguous (Figure 13). The lowest formaldehyde content was observed in the boards with a pMDI resin content of 1% (0.30 mg/100 g). Addition of 3% pMDI resin to NaLS adhesive significantly increases the formaldehyde content (1.66 mg/100 g), but a further increase in the pMDI resin content to 5% leads to a decrease in formaldehyde content (0.38 mg/100 g), reaching the super E0 class ($\leq$1.5 mg/100 g). The determined formaldehyde emission value, which is much lower than 2 mg/100 g, i.e., equivalent to formaldehyde emission of natural wood [14], allows for the defining of manufactured particleboards as eco-friendly composites. The reference board, manufactured bonded with UF adhesive, can be classified under the E1 emission grade ($\leq$8 mg/100 g). The results achieved in this study are in accordance with the results obtained by other several authors, where using different LS types in adhesive formulations for wood-based panels resulted in remarkably low formaldehyde content of the finished composites [9,38,40,42]. This might be attributed to the high amount of reactive groups in LS, which increase its reactivity towards formaldehyde [42,69].

In terms of mechanical properties, the boards based on UF adhesive with the addition of pMDI resin have higher values of MOR, MOE, and IB than boards based on LS adhesives with the addition of pMDI resin. In general, it can be stated that the addition of pMDI resin has a positive effect on the mechanical properties of particleboards. If we compare the properties of boards made using pure UF and LS adhesives (without adding pMDI resin), then here, the boards based on UF adhesive have much better mechanical properties than boards bonded with LS adhesives. However, the boards based on LS adhesives are characterised by a much lower formaldehyde content than the UF-bonded boards. In the boards based on MgLS and NaLS adhesives, the formaldehyde content is lower by 91.1% and 56.9%, respectively, compared to the UF bonded boards.

The previous studies [36,37] on the performance of particleboards bonded with glyoxalated lignin combined with pMDI also showed superior internal bond strength. This can be explained by the fact that the isocyanate groups in pMDI are highly unsaturated and can react with a number of active hydroxyl groups in wood particles' surface and LS molecules, as well as with the moisture contained in the particles and the adhesive system [15]. Younesi-Kordkheili et al. [70] reported that addition of pMDI intensely improved the performance of lignin-urea-formaldehyde (LUF) resins and imparted a positive effect on the formaldehyde emission and water absorption of the panels. An additional small amount of pMDI (2%) was sufficient to yield panels with significantly greater bonding strength. In the work [71], it was shown that the introduction of pMDI to UF resin resulted in an improvement of bending strength and internal bond and in a reduction in formaldehyde content in the boards by as much as 30%.

4. Conclusions

Different adhesive systems were prepared by gradual replacement of UF-resin by magnesium or sodium lignosulfonates. MgLS and NaLS in LS-UF adhesive compositions exhibited acid-base buffer properties, expressed as resistance to pH change with the addition of acids. Compared to ammonium sulphate, NaLS and MgLS decrease the pH of uncured compositions and increase the pH of cured compositions. Lignosulfonates partially neutralise the effect of ammonium sulphate on the pH of cured compositions.

The results obtained demonstrate that when the content of MgLS or NaLS is associated with UF resin with 10–30% replacement by LS, the physical and mechanical properties are comparable with those of the UF-bonded particleboards. There was no significance difference between MOR values for boards bonded with MgLS and NaLS. It was not possible to produce particleboards with satisfactory physical and mechanical properties with pure MgLS and NaLS adhesives. The increased addition level of LS to UF resin formulation showed a reduction in bonding strength. Moreover, all manufactured particleboards bonded with LS-UF adhesives demonstrated significantly deteriorated moisture-related properties, such as WA and TS.

This study demonstrated the potential to combine LS and pMDI in particleboard manufacturing. The formaldehyde content of the particleboards bonded with pure MgLS and NaLS adhesives with the addition of pMDI, tested in accordance with the perforator method, was remarkably low ($\leq$1.5 mg/100 g) and significantly different with the value of the reference UF-bonded boards, which allows for their classification as eco-friendly wood-based composites for indoor applications.

Author Contributions: Conceptualisation, P.B. and G.N.; methodology, P.B. and G.N.; validation, P.B., G.N., R.R., L.K., J.S., P.A., R.M. and V.S.; investigation, P.B., G.N., R.R., L.K., P.A. and V.S.; writing—original draft preparation, P.B. and G.N.; writing—review and editing, P.B. and G.N.; supervision and project administration, P.B.; funding acquisition, P.A. All authors have read and agreed to the published version of the manuscript.

Funding: This research was supported by the project No. НИС-Б-1145/04.2021 "Development, Properties and Application of Eco-Friendly Wood-Based Composites" carried out at the University of Forestry, Sofia, Bulgaria.

Institutional Review Board Statement: Not applicable.

Informed Consent Statement: Not applicable.

Data Availability Statement: The data that support the findings of this study are available upon reasonable request from the authors.

Acknowledgments: This publication was supported by the Slovak Research and Development Agency under contracts No. APVV-18-0378 and APVV-19-0269 and the Polish National Agency for Academic Exchange (NAWA) under contract No. PPN/ULM/2020/1/00188/U/00001. The authors express their sincere thanks to Nataliya Khomyak for her help in performing the research.

Conflicts of Interest: The authors declare no conflict of interest.

References

1. FAO. FAO Yearbook of Forest Products 2018. 2020. Available online: http://www.fao.org/3/cb0513m/CB0513M.pdf (accessed on 11 August 2020).
2. Maloney, T.M. *Modern Particleboard and Dry Process Manufacturing*; Miller Freeman Inc.: San Francisco, CA, USA, 1993.
3. Pizzi, A. Synthetic Adhesives for Wood Panels: Chemistry and Technology—A Critical Review. *Rev. Adhes. Adhes.* **2014**, *2*, 85–126. [CrossRef]
4. Kumar, R.N.; Pizzi, A. Environmental Aspects of Adhesives—Emission of Formaldehyde. In *Adhesives for Wood and Lignocellulosic Materials*; Wiley-Scrivener Publishing: Hoboken, NJ, USA, 2019; pp. 293–312.
5. Pizzi, A.; Papadopoulos, A.N.; Policardi, F. Wood Composites and Their Polymer Binders. *Polymers* **2020**, *12*, 1115. [CrossRef]
6. Dunky, M. Adhesives in the Wood Industry. In *Handbook of Adhesive Technology*, 2nd ed.; Revised and Expanded; Pizzi, A., Mittal, K.L., Eds.; Marcel Dekker, Inc.: New York, NY, USA; Basel, Switzerland, 2003; 71p. [CrossRef]
7. Łebkowska, M.; Załęska–Radziwiłł, M.; Tabernacka, A. Adhesives based on formaldehyde—Environmental problems. *BioTechnologia* **2017**, *98*, 53–65. [CrossRef]
8. Savov, V.; Antov, P. Engineering the Properties of Eco-Friendly Medium Density Fibreboards Bonded with Lignosulfonate Adhesive. *Drv. Ind.* **2020**, *71*, 157–162. [CrossRef]
9. Antov, P.; Savov, V.; Neykov, N. Sustainable bio-based adhesives for eco-friendly wood composites. A review. *Wood Res.* **2020**, *65*, 51–62. [CrossRef]
10. Vázquez, G.; González, J.; Freire, S.; Antorrena, G. Effect of chemical modification of lignin on the gluebond performance of lignin-phenolic resins. *Bioresour. Technol.* **1997**, *60*, 191–198. [CrossRef]
11. Pizzi, A. Natural Phenolic Adhesives II: Lignin. In *Handbook of Adhesive Technology, Revised and Expanded*, 2nd ed.; Pizzi, A., Mittal, K.L., Eds.; CRC Press: Boca Raton, FL, USA, 2003; pp. 589–598.
12. Danielson, B.; Simonson, R. Kraft lignin in phenol formaldehyde resin. Part 1. Partial replacement of phenol by kraft lignin in phenol formaldehyde adhesives for plywood. *J. Adhes. Sci. Technol.* **1998**, *12*, 923–939. [CrossRef]
13. Hu, L.; Pan, H.; Zhou, Y.; Zhang, M. Methods to Improve Lignin's Reactivity as a Phenol Substitute and as Replacement for Other Phenolic Compounds: A Brief Review. *BioResources* **2011**, *6*, 3515–3525. [CrossRef]
14. Pizzi, A. Recent developments in eco-efficient bio-based adhesives for wood bonding: Opportunities and issues. *J. Adhes. Sci. Technol.* **2006**, *20*, 829–846. [CrossRef]
15. Hemmilä, H.; Adamopolus, S.; Hosseinpourpia, R.; Ahmed, S.A. Ammonium Lignosulfonate Adhesives for Particleboards with pMDI and Furfuryl Alcohol as Crosslinkers. *Polymers* **2019**, *11*, 1633. [CrossRef]

16. Peng, W.; Riedl, B. The chemorheology of phenol–formaldehyde thermoset resin and mixtures of the resin with lignin fillers. *Polymers* **1994**, *35*, 1280–1286. [CrossRef]
17. Mandlekar, N.; Cayla, A.; Rault, F.; Giraud, S.; Salaün, F.; Malucelli, G.; Guan, J.-P. An Overview on the Use of Lignin and Its Derivatives in Fire Retardant Polymer Systems. In *Lignin—Trends and Applications*; Poletto, M., Ed.; IntechOpen: London, UK, 2018. [CrossRef]
18. Hemmilä, V.; Adamopoulos, S.; Karlsson, O.; Kumar, A. Development of sustainable bio-adhesives for engineered wood panels—A Review. *RSC Adv.* **2017**, *7*, 38604–38630. [CrossRef]
19. Mantanis, G.I.; Athanassiadou, E.T.; Barbu, M.C.; Wijnendaele, K. Adhesive systems used in the European particleboard, MDF and OSB industries. *Wood Mater. Sci. Eng.* **2018**, *13*, 104–116. [CrossRef]
20. Ferdosian, F.; Pan, Z.; Gao, G.; Zhao, B. Bio-Based Adhesives and Evaluation for Wood Composites Application. *Polymers* **2017**, *9*, 70. [CrossRef]
21. Yotov, N.; Valchev, I.; Petrin, S.; Savov, V. Lignosulphonate and waste technical hydrolysis lignin as adhesives for eco-friendly fibreboard. *Bulg. Chem. Commun.* **2017**, *49*, 92–97.
22. Ghaffar, S.H.; Fan, M. Lignin in straw and its applications as an adhesive. *Int. J. Adhes. Adhes.* **2014**, *48*, 92–101. [CrossRef]
23. Gothwal, R.K.; Mohan, M.K.; Ghosh, P. Synthesis of low cost adhesives from pulp and paper industry waste. *J. Sci. Ind. Res.* **2010**, *69*, 390–395.
24. Olivares, M.; Aceituno, H.; Neiman, G.; Rivera, E.; Sellers, T.J. Lignin-modified phenolic adhesives for bonding radiata pine plywood. *For. Prod. J.* **1995**, *45*, 63–67.
25. Pizzi, A. Wood products and green chemistry. *Ann. For. Sci.* **2016**, *73*, 185–203. [CrossRef]
26. Yang, S.; Zhang, Y.; Yuan, T.Q.; Sun, R.C. Lignin–phenol–formaldehyde resin adhesives prepared with biorefinery technical lignins. *J. Appl. Polym. Sci.* **2015**, *132*, 42493. [CrossRef]
27. Zouh, X.; Tan, L.; Zhang, W.; Chenlong, L.; Zheng, F.; Zhang, R.; Du, G.; Tang, B.; Liu, X. Enzymatic hydrolysis lignin derived from corn stoves as an instant binder from bio-composites: Effect of fiber moisture content and pressing temperature on board's properties. *BioResources* **2011**, *6*, 253–264. [CrossRef]
28. Nasir, M.; Gupta, A.; Beg, M.D.H.; Chua, G.K.; Kumar, A. Physical and mechanical properties of medium density fiberboard using soy-lignin adhesives. *J. Trop. For. Sci.* **2014**, *46*, 41–49.
29. Yuan, Y.; Guo, M.; Liu, F. Preparation and Evaluation of Green Composites Using Modified Ammonium Lignosulfonate and Polyethylenimine as a Binder. *BioResources* **2014**, *9*, 836–848. [CrossRef]
30. Ang, A.F.; Ashaari, Z.; Lee, S.H.; Tahir, P.M.D.; Halis, R. Lignin-based copolymer adhesives for composite wood panels—A review. *Int. J. Adhes. Adhes.* **2019**, *95*, 102408. [CrossRef]
31. Donmez Cavdar, A.; Kalaycioglu, H.; Hiziroglu, S. Some of the properties of oriented strandboard manufactured using kraft lignin phenolic resin. *J. Mater. Process. Technol.* **2008**, *202*, 559–563. [CrossRef]
32. Çetin, N.S.; Özmen, N. Use of organosolv lignin in phenol-formaldehyde resins for particleboard production: II. Particleboard production and properties. *Int. J. Adhes. Adhes.* **2002**, *22*, 481–486. [CrossRef]
33. Kouisni, L.; Fang, Y.; Paleologou, M.; Ahvazi, B.; Hawari, J.; Zhang, Y.; Wang, X.-M. Kraft lignin recovery and its use in the preparation of lignin-based phenol formaldehyde resins for plywood. *Cellul. Chem. Technol.* **2011**, *45*, 515–520.
34. Akhtar, T.; Lutfullah, G.; Ullah, Z. Lignonsulfonate-phenolformaldehyrde adhesive: A potential binder for wood panel industries. *J. Chem. Soc. Pak.* **2011**, *33*, 535–538.
35. Da Silva, M.A.; dos Santos, P.V.; Silva, G.C.; Lelis, R.C.C.; do Nascimento, A.M.; Brito, E.O. Using lignosulfonate and Phenol-Formaldehyde adhesive in particleboard manufacturing. *Sci. For.* **2017**, *45*, 423–433. [CrossRef]
36. El Mansouri, N.E.; Pizzi, A.; Salvado, J. Lignin-based wood panel adhesives without formaldehyde. *Holzals Roh Werkst.* **2007**, *65*, 65–70. [CrossRef]
37. Lei, H.; Pizzi, A.; Du, G. Environmentally friendly mixed tannin/lignin wood resins. *J. Appl. Polym. Sci.* **2008**, *107*, 203–209. [CrossRef]
38. Antov, P.; Jivkov, V.; Savov, V.; Simeonova, R.; Yavorov, N. Structural Application of Eco-Friendly Composites from Recycled Wood Fibres Bonded with Magnesium Lignosulfonate. *Appl. Sci.* **2020**, *10*, 7526. [CrossRef]
39. Antov, P.; Krišťák, Ľ.; Réh, R.; Savov, V.; Papadopoulos, A.N. Eco-Friendly Fiberboard Panels from Recycled Fibers Bonded with Calcium Lignosulfonate. *Polymers* **2021**, *13*, 639. [CrossRef]
40. Antov, P.; Mantanis, G.I.; Savov, V. Development of Wood Composites from Recycled Fibres Bonded with Magnesium Lignosulfonate. *Forests* **2020**, *11*, 613. [CrossRef]
41. Antov, P.; Savov, V.; Krišťák, Ľ.; Réh, R.; Mantanis, G.I. Eco-Friendly, High-Density Fiberboards Bonded with Urea-Formaldehyde and Ammonium Lignosulfonate. *Polymers* **2021**, *13*, 220. [CrossRef]
42. Antov, P.; Savov, V.; Mantanis, G.I.; Neykov, N. Medium-density fibreboards bonded with phenol-formaldehyde resin and calcium lignosulfonate as an eco-friendly additive. *Wood Mater. Sci. Eng.* **2021**, *16*, 42–48. [CrossRef]
43. Savov, V.; Mihajlova, J. Influence of the content of lignosulfonate on physical properties of medium density fiberboard. In Proceedings of the International Conference "Wood Science and Engineering in the Third Millennium"—ICWSE, Brasov, Romania, 2–4 November 2017; pp. 348–352.

44. Savov, V.; Mihajlova, J. Influence of the content of lignosulfonate on mechanical properties of medium density fiberboard. In Proceedings of the International Conference "Wood Science and Engineering in the Third Millennium"—ICWSE, Brasov, Romania, 2–4 November 2017; pp. 353–357.
45. Pizzi, A.; Mittal, K.L. *Handbook of Adhesive Technology*, 2nd ed.; Marcel Dekker: New York, NY, USA, 2003; pp. 628–645.
46. Akyüz, K.C.; Nemli, G.; Baharoğlu, M.; Zekoviç, E. Effects of acidity of the particles and amount of hardener on the physical and mechanical properties of particleboard composite bonded with urea formaldehyde. *Int. J. Adhes. Adhes.* **2010**, *30*, 166–169. [CrossRef]
47. Xing, C.; Zhang, S.Y.; Deng, J.; Wang, S. Urea–formaldehyde-resin gel time as affected by the pH value, solid content, and catalyst. *J. Appl. Polym. Sci.* **2007**, *103*, 1566–1569. [CrossRef]
48. European Committee for Standardization. *Wood-Based Panels—Determination of Moisture Content*; EN 322; European Committee for Standardization: Brussels, Belgium, 1998.
49. European Committee for Standardization. *Wood-Based Panels—Determination of Density*; EN 323; European Committee for Standardization: Brussels, Belgium, 2001.
50. European Committee for Standardization. *Wood-Based Panels—Determination of Modulus of Elasticity in Bending and of Bending Strength*; EN 310; European Committee for Standardization: Brussels, Belgium, 1999.
51. European Committee for Standardization. *Particleboards and Fibreboards—Determination of Tensile Strength Perpendicular to the Plane of the Board*; EN 319; European Committee for Standardization: Brussels, Belgium, 1993.
52. European Committee for Standardization. *Particleboards and Fibreboards—Determination of Swelling in Thickness after Immersion in Water*; EN 317; European Committee for Standardization: Brussels, Belgium, 1998.
53. European Committee for Standardization. *Wood-Based Panels-Determination of Formaldehyde Release—Part 5. Extraction Method (Called the Perforator Method)*; EN ISO 12460-5; European Committee for Standardization: Brussels, Belgium, 2015.
54. Abushammala, H. A Simple Method for the Quantification of Free Isocyanates on the Surface of Cellulose Nanocrystals upon Carbamation using Toluene Diisocyanate. *Surfaces* **2019**, *2*, 444–454. [CrossRef]
55. Myers, G.E.; Koutsky, J.A. Formaldehyde Liberation and Cure Behavior of Urea-Formaldehyde Resins. *Holzforschung* **1990**, *44*, 117–126. [CrossRef]
56. Mitra, R.P.; Atreyi, M. Titration curve of urea-formaldehyde resin. *Naturwissenschaften* **1958**, *45*, 286–287. [CrossRef]
57. Korntner, P.; Schedl, A.; Sumerskii, I.; Zweckmair, T.; Mahler, A.K.; Rosenau, T.; Potthast, A. Sulfonic Acid Group Determination in Lignosulfonates by Headspace Gas Chromatography. *ACS Sustain. Chem. Eng.* **2018**, *6*, 6240–6246. [CrossRef]
58. Petit-Conil, M. Determination of sulfonic and carboxyl acids contents of industrial lignosulfonates. In Proceedings of the 7th Nordic Wood Biorefinery Conference, Stockholm, Sweden, 28–30 March 2017.
59. Yan, M.; Yang, D.; Deng, Y.; Chen, P.; Zhou, H.; Qiu, X. Influence of pH on the behavior of lignosulfonate macromolecules in aqueous solution. *Colloids Surf. A Physicochem. Eng. Asp.* **2010**, *371*, 50–58. [CrossRef]
60. Zanetti, M.; Pizzi, A.; Kamoun, C. Upgrading of MUF particleboard adhesives and decrease of melamine content by buffer and additives. *Holzals Roh Werkst.* **2003**, *61*, 55–65. [CrossRef]
61. Bhaduri, S.K.; Mojumder, P. Medium density particle board from Khimp plant. *Nat. Prod. Radiance* **2007**, *7*, 106–110.
62. Iwakiri, V.T.; Trianoski, R.; Razera, D.L.; Iwakiri, S.; da Rosa, T.S. Production of Structural Particleboard of *Mimosa Scabrella* Benth with Lignin Phenol-formaldehyde Resin. *Floresta Ambiente* **2019**, *26*, e20171006. [CrossRef]
63. Newman, W.H.; Glasser, W.G. Engineering plastics from lignin. XII. Synthesis and Performance of Lignin Adhesives with Isocyanate and Melamine. *Holzforschung* **1985**, *39*, 345–353. [CrossRef]
64. Ferhan, M.; Yan, N.; Sain, M. A new method for demethylation of lignin from woody biomass using biophysical methods. *J. Chem. Eng. Process. Technol.* **2013**, *4*, 160. [CrossRef]
65. Li, J.J.; Wang, W.; Zhang, S.F.; Gao, Q.; Zhang, W.; Li, J.Z. Preparation and characterization of lignin demethylated at atmospheric pressure and its application in fast curing biobased phenolic resins. *RSC Adv.* **2017**, *6*, 67435–67443. [CrossRef]
66. Song, Y.; Wang, Z.; Yan, N.; Zhang, R.; Li, J. Demethylation of wheat straw alkali lignin for application in phenol formaldehyde adhesives. *Polymers* **2016**, *8*, 209. [CrossRef]
67. Savov, V.; Valchev, I.; Antov, P. Processing factors affecting the exploitation properties of environmentally friendly medium density fibreboards based on lignosulfonate adhesives. In Proceedings of the 2nd International Congress of Biorefinery of Lignocellulosic Materials (IWBLCM2019), Cordoba, Spain, 4–7 June 2019; pp. 165–169.
68. Guler, C.; Buyuksari, U. Effect of production parameters on the physical and mechanical properties of particleboards made from peanut (*Arachis hypogaea* L.) hull. *BioResources* **2011**, *6*, 5027–5036.
69. Klapiszewski, Ł.; Oliwa, R.; Oleksy, M.; Jesionowski, T. Calcium lignosulfonate as eco-friendly additive of crosslinking fibrous composites with phenol-formaldehyde resin matrix. *Polymery* **2018**, *63*, 102–108. [CrossRef]
70. Younesi-Kordkheili, H.; Pizzi, A.; Mohammadghasemipour, A. Improving the properties of ionic liquid-treated lignin-urea-formaldehyde resins by a small addition of isocyanate for wood adhesive. *J. Adhes.* **2016**, *94*, 406–419. [CrossRef]
71. Dziurka, D.; Mirski, R. UF-pMDI Hybrid Resin for Waterproof Particleboards Manufactured at a Shortened Pressing Time. *Drv. Ind.* **2010**, *61*, 245–249.

Article

Waste Wood Particles from Primary Wood Processing as a Filler of Insulation PUR Foams

Radosław Mirski, Dorota Dukarska *, Joanna Walkiewicz * and Adam Derkowski

Department of Wood Based Materials, Faculty of Forestry and Wood Technology, Poznań University of Life Sciences, Wojska Polskiego 38/42, 60-627 Poznań, Poland; rmirski@up.poznan.pl (R.M.); adam.derkowski@up.poznan.pl (A.D.)
* Correspondence: dorota.dukarska@up.poznan.pl (D.D.); joanna.siuda@up.poznan.pl (J.W.)

Abstract: A significant part of the work carried out so far in the field of production of biocomposite polyurethane foams (PUR) with the use of various types of lignocellulosic fillers mainly concerns rigid PUR foams with a closed-cell structure. In this work, the possibility of using waste wood particles (WP) from primary wood processing as a filler for PUR foams with open-cell structure was investigated. For this purpose, a wood particle fraction of 0.315–1.25 mm was added to the foam in concentrations of 0, 5, 10, 15 and 20%. The foaming course of the modified PUR foams (PUR-WP) was characterized on the basis of the duration of the process' successive stages at the maximum foaming temperature. In order to explain the observed phenomena, a cellular structure was characterized using microscopic analysis such as SEM and light microscope. Computed tomography was also applied to determine the distribution of wood particles in PUR-WP materials. It was observed that the addition of WP to the open-cell PUR foam influences the kinetics of the foaming process of the PUR-WP composition and their morphology, density, compressive strength and thermal properties. The performed tests showed that the addition of WP at an the amount of 10% leads to the increase in the PUR foam's compressive strength by 30% (parallel to foam's growth direction) and reduce the thermal conductivity coefficient by 10%.

Keywords: polyurethane foams; filler; wood particles; structure; properties

1. Introduction

In recent years, people's environmental awareness has been increasing, which has led to the search for solutions that will allow the use of technologically processed by-products. Due to the increasing development in the wood industry, waste generation is a common problem. Two by-products of wood processing are dust and wood particles. Despite the fact that research is carried out with the use of wood dust in the context of various materials, this material is still a nuisance waste. Nowadays, the most popular composite containing wood (of any form) is a wood plastic composite (WPC) [1]. Research concerning the application of WP was also conducted in order to enhance the properties of thermoplastic starch [2]; as a component in adhesive mixtures for 3D printing [3]; in concrete as a partial replacement for sand [4]; and in the production of new polyurethane foams from liquefied wood powder [5]. Wood waste can also be applied as a potential filler for loose-fill building isolation [6].

PUR represents a wide class of polymeric materials [7,8]. Polyurethane foams account for 2/3 of the world's production of polyurethanes, and because of their numerous applications in the form of rigid, semi-rigid and flexible foams, they are continuously highly ranked among all of available foams [9]. PUR foams are the product of the addition polymerization of polyols and polyisocyanates. Catalysts, surfactants and foaming agents are also used during the production of PUR. These foams may differ in composition, density, color and mechanical properties. There are also studies where fillers were used to lower the cost and increase mechanical properties, e.g., the modulus and strength or density [10].

Polyurethane foams, particularly flexible polyurethane foams, are commonly used in the building, automotive, furniture and packaging industries [11–15]. These foams have many advantages, such as a wide range of flexibility, good shock absorption and high durability in use. Moreover, they can be easily formed into various shapes, and they are characterized by a low density. However, they are not cheap. It is also worth adding that there is an opportunity to easily control and manage their parameters by changing their composition or using fillers [16].

Literature data clearly indicate that the organic or inorganic fillers can significantly improve the mechanical and thermal properties of PUR composites [17–20]. The introduction of inorganic fillers such as glass fibers or carbon fibers in polyurethane foams is a well-known process [21]. The fillers strongly affect the physical properties of foam. Not only inorganic fillers are used in foam formulations, but they can also be filled with materials of natural origin, such as plant and agricultural waste. Agricultural waste has the potential to be used in the production of PUR foams as renewable bio-based fillers [17]. Członka et al., Kerche et al., and De Avila Delucis et al. [19,22,23] claimed that the use of agricultural waste in the production of PUR materials can result in a new approach in the application path and provide a new opportunity for creating a new class of green materials. Paciorek-Sadowska et al. [17] used rapeseed cake as an addition for PUR composite production. Authors observed that 30–60% incorporation of rapeseed cake significantly influenced the apparent density and mechanical properties. Modified sunflower cake was added to PUR foam, and it was found that the addition of plant waste had an impact on the improvement of the mechanical properties, while it did not cause a significant deterioration of the insulating properties [24]. Funabashi et al. [21] researched the effect of filler types on mechanical properties of a rigid polyurethane composite. For this purpose, authors used particles from plant wastes (e.g., bamboo powder, coffee grounds, wood meal and cellulose). Zieleniewska et al. [25] obtained the composites of rigid polyurethane foams (RPURF) with the use of waste hazelnut shells, suitable for applications in the cosmetics industry. Moreover, rice plant wastes were used as a reinforcing material in the concentrations of 5%, 10%, 15% and 20% [26]. According to the literature, wood waste can also be a valuable raw material for the production of biocomposite PUR foams. According to Augaitis et al. [6], their application is beneficial due to their high availability, low cost and content of free hydroxyl groups capable of reacting with isocyanate groups. These authors showed that a biocomposite PUR foam with an apparent density of 150 kg/m^3 and a PUR to pinewood sawdust ratio of 0.7 is characterized by very good physical and insulating properties and high strength. In turn, in this paper, an innovative solution has been proposed to use waste wood particles from primary wood processing as a filler for PUR foam. To the best of our knowledge, this type of filler has not been used in the process of producing insulating PUR foams with open-cell structure so far. It seems that such operations are beneficial from an economic and technological perspective. They enable us to increase the rational utilization of wood waste as full-value raw material without the need for their further processing (e.g., fragmentation, gluing, etc.). It should be emphasized that, for many years, there has been a growing interest in the use of wood waste, mainly in the wood, paper, polymer-wood composite, and energy industries. This is due to the necessity of sustainable management of this type of waste, whose resources increase with the development of the wood industry and the demand for wood-based products. This is evidenced, among others, by current technological trends in the production of wood-based materials and numerous studies indicating the possibility of producing full-value products from recycled wood or directly from primary wood processing [27–33].

The aim of this study was to apply the waste wood particles (WP) from primary wood processing as a filler of open-cell PUR foams. The effects of wood waste's introduction on the apparent density, cellular morphology, mechanical and thermal properties of PUR foams were examined and discussed.

This paper is a continuation of the previous research conducted by the authors concerning the possibility of using by-products from wood processing in order to manufacture materials with improved properties, which are used, e.g., in construction and in the production of interior design elements [30–32].

2. Materials and Methods

2.1. Materials

A two-component foam system for the production of open-cell polyurethane thermal insulation PUREX-WG 2017 (Polychem System, Poznań, Poland) was used in the research. One of the components was a polyol (A component). The isocyanate component (B component) was polymeric methylenediphenyl-4,4′-diisocyanate consisting of 31.14% free isocyanate groups (NCO).

Wood particles (WP) representing a dimensional fraction of 0.315–1.25 mm were used as a filler (Figure 1). WP were obtained as a result of sorting sawdust intended for the production of chipboard. The moisture content of WP ranged between 0.2% and 0.5%. Figure 2 presents their fractional composition. The largest shares were observed for the fractions of 0.315 and 0.630 mm. The wood particles were added to the foam at the concentrations of 0, 5, 10, 15 and 20% determined according to a weight ratio. The amount of filler was determined based on preliminary studies and a literature review on the manufacture of biocomposite PUR foams [6,17,34,35].

Figure 1. Wood particles used as a filler for PUR foams.

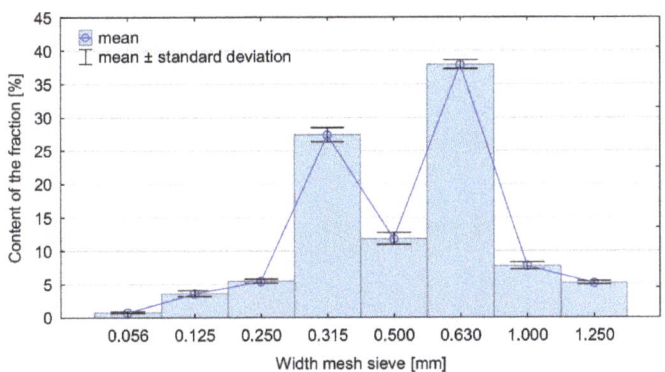

Figure 2. Fractional composition of wood particles added to PUR foams.

2.2. Synthesis of PUR Composite Foams

The course of foaming of the modified PUR foams was characterized on the basis of the times of the successive stages of this process and the maximum foaming temperature. For this purpose, the foam components were mixed in a weight ratio premised in accordance with the manufacturer's recommendations, i.e., A:B = 100:100. The reaction mixture was prepared by mixing the appropriate amounts of wood particles with component B and then adding component A (Figure 3). The reaction mixture was stirred with a low-speed mechanical stirrer at 1200 rpm for 10 s, at a temperature of 23 °C, and then poured into a form with internal dimensions of 250 × 250 × 130 mm^3. After that, PUR-WP composites were allowed to grow and were left at room temperature for 24 h. Each foam variant was prepared in two replicates. The obtained foams were cut with a band saw (Holzstar, Hallstadt, Germany) into specimens of dimensions necessary for testing their properties.

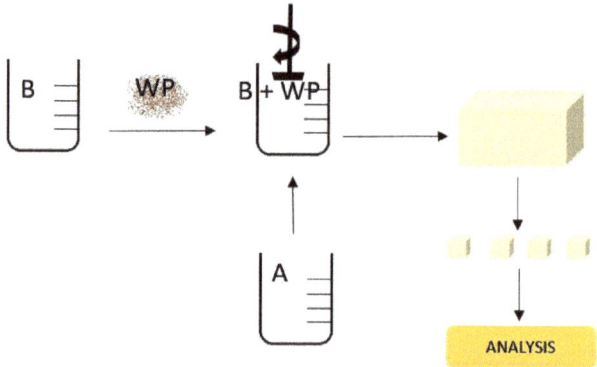

Figure 3. Scheme of PUR manufacturing (A—polyol; B—isocyanate; WP—wood particles).

2.3. Kinetic of PUR Foaming

The influence of wood filler on the foaming process of PUR foams was determined by measuring the following times:

- start of growth—time when the volume of the reaction mixture started to increase;
- gelling—the time after which it was possible to remove the so-called "polyurethane thread";
- growth—the time after which the maximum foam growth was achieved;
- tack-free—the time measured until the foam solidified completely.

The foaming temperatures were measured with a thermocouple immersed in the reaction mixture. The temperature was always read after the foam growth was completed. At the end, the average of 5 individual measurements was evaluated.

2.4. Characterization of PUR Sample

The density of neat PUR foam and PUR-WP composites, defined as the ratio of the sample mass to its volume, was determined in accordance with the PN-EN ISO 845 standard. Samples with the dimensions 50 × 50 × 50 mm^3 were used. The samples were measured with a thickness gauge with an accuracy of 0.01 mm and weighed on an analytical weight with an accuracy of 0.001 g.

The compressive strength in parallel direction to the foam's growth was investigated in accordance with the recommendations of EN 826 standard using a Tinius Olsen H10KT testing machine (Tinius Olsen Ltd., Salfords, UK). The test covered foam samples with dimensions of 50 × 50 × 50 mm^3, which were compressed in the direction of foam growth at a rate of 5 mm/min and the load cell of 250 N. The compressive strength ($\sigma_{10\%}$) was defined as the maximum compressive force achieved when the relative deformation at deflection was less than 10% (based on the initial cross-sectional area of the specimen). The means of the compressive strength were evaluated based on the 7 individual measurements.

The thermal conduction coefficient was determined using a heat flux density sensor type ALMEMO 117 company Ahlborn (Holzkirchen, Germany), with plate dimensions of 100 × 30 × 3 mm^3. The average values of the thermal conduction coefficient were evaluated on the basis of 5 individual measurements. A detailed description of the research method was presented in the previous works by the authors [36,37].

In order to determine the cellular structure, a scanning electron microscope (SEM) and a light microscope were used. SEM analyses were carried out with the use of an SU3500 Hitachi microscope. The images of the sample plane were prepared using a computer image analyzer equipped with a stereoscopic optical microscope (Motic SMZ-168, Hongkong, China) and a camera (Moticam 5.0, Barcelona, Spain). The image of the structure was transferred by a camera to the monitor screen, and then pictures were taken using the Motic Images Plus 3.0 program (Hongkong, China). The cell sizes of pure PUR foam and PUR-WP composite foams were determined using the same equipment.

The samples with the dimensions $50 \times 50 \times 50$ mm^3 were collected in order to evaluate the dispersion of WP in PUR foam using computer tomography which is a type of X-ray tomography allowing to obtain cross-sections of the examined object. Scanning was performed with the use of a Hyperion X9Pro tomography, with objects scanned at a resolution of 0.3 mm at a lamp voltage of 90 kV.

The yielded test results of PUR foams with wood particles addition were analyzed statistically using STATISTICA software v.13.1(StatSoft Inc., Tulsa, OK, USA). Mean values of the parameters were compared in a one-factor analysis of variance—post hoc Tukey's test allowed us to distinguish homogeneous groups of mean values for each parameter for $p = 0.05$.

3. Results and discussion

3.1. The Impact of WP Filler on PUR Foams Manufacture

The parameters characterizing the foaming process of the tested foams are presented in Figures 4 and 5. It can be concluded that the addition of wood particles to the open-cell PUR foam influences the kinetics of the foaming process of the PUR-WP composition. It was observed that the addition of this type of filler accelerates the onset of foam growth. This phenomenon is particularly evident in the case of PUR-WP compositions containing 10% and 15% of particles. In the case of these variants, a reduction in the starting time by approx. 29% was observed. These statistically significant differences were confirmed by the post hoc test. Statistical analysis allowed for the identification of four different groups of average foam expansion start times for the tested variants. However, mixing the foam components with wood particles had an adverse effect, and it led to extensions in the foams increases the growth time and their gelling times. With the maximum addition of wood particles, the growth time was extended by 23%, and the gel time by 33%. The extension of these times is the effect of slowing down the exothermic reaction, which is also evidenced by the decrease in the maximum foaming temperature of the foams. As shown in Figure 5, the addition of wood particles to the components of foamed PUR foam caused a decrease in the foaming temperature from 95 to 82 °C. According to the literature, the reduction in the maximum foaming temperature of PUR foams results from the reduced amount of heat generated during the reactions occurring in the latent stages and growth, used at the stage of foam stabilization and maturation. In addition, as shown in previous studies, organic fillers introduced into PUR foams can absorb some of the heat generated during synthesis, thus lowering the temperature of their foaming [19,38]. An insufficient amount of heat could slow down the cross-linking reaction and, consequently, extend the time required to achieve a track-free time, which is also confirmed by the data presented in Figure 4 [9]. Moreover, the presence of the wood particles and their relatively large dimensions undoubtedly limited both the growth of the foam cells and the susceptibility of the composition to the foaming process. According to Strąkowska et al. [24], the presence of fillers also limits the mobility of the polymer and the speed of the polymerization reaction. Moreover, as reported by Członka et al. [39], hydroxyl groups present in lignocellulosic fillers can react with highly reactive isocyanate groups. This affects the stoichiometry of the system and reduces the number of isocyanate groups capable of reacting with water, which results in a reduction in the amount of CO_2 released.

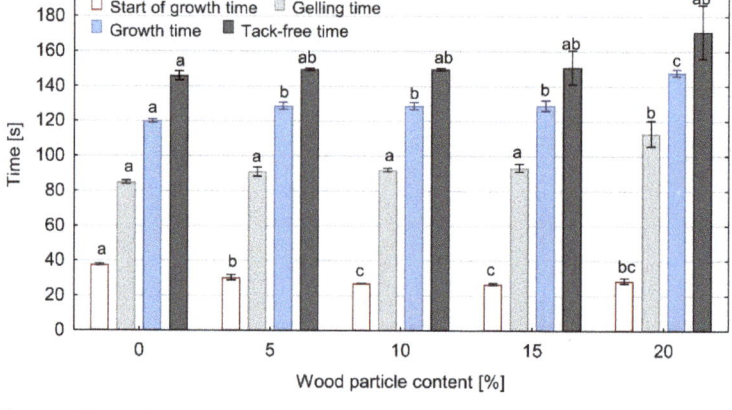

Figure 4. Times characterizing the foaming process of PUR-WP composition depending on wood particle content. Different letters indicate homogeneous groups of mean values determined by one-factor ANOVA with Tukey's test.

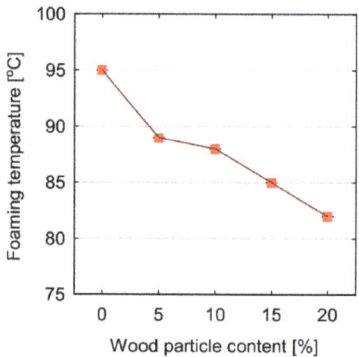

Figure 5. Values of the maximum foaming temperature of PUR-WP composition depending on wood particle content.

3.2. Density, Thermal Conductivity and Microstructure of PUR Foams

As expected, the addition of the wood particles causes a significant increase in the density of the PUR-WP composite. The apparent density of pure PUR foam was 20 kg/m^3 (Figure 6). The introduction of 5% wood filler particles into its structure caused a slight increase in its density; however, the HSD Tukey test did not confirm statistically significant differences. The mean density values obtained for these two variants belong to the same homogeneous groups. A significant increase in the density of the produced foams was noticed only when larger amounts of wood particles were used, i.e., from 10% and more. With their addition in the amount of 20%, the average apparent density of the foam was 34.6 kg/m^3, i.e., it increased in comparison to pure foam by as much as 73%.

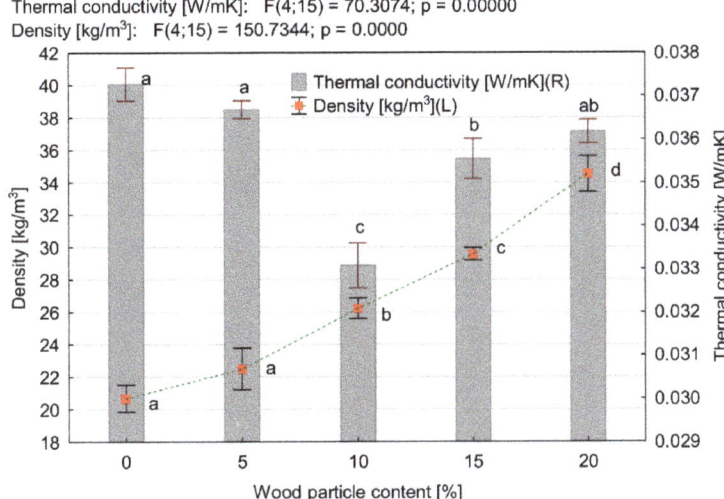

Figure 6. Thermal conductivity and density of PUR-WP composition depending on wood particle content. Different letters indicate homogeneous groups of mean values determined by one-factor ANOVA with Tukey's test.

Along with the increase in the density of the tested foams, the statistically significant changes in their thermal insulation, determined by the thermal conductivity coefficient λ, were also observed. This is confirmed by the results of the post hoc test and the homogeneous groups of mean values λ distinguished on its basis. Moreover, the test probability level was significantly lower than the assumed level of statistical significance, i.e., <0.05. As follows from the data presented in Figure 6, the use of wood particles as a filler for PUR foam at an amount of up to 10% leads to the reduction in the average λ value by approx. 10%, which proves the improvement of thermal insulation of this type of foam.

Unfortunately, a further increase in the wood particle content results in a gradual increase in the value of λ. In the case of the 15% addition of wood filler, the value of λ was lower than the coefficient which the pure foam was characterized by, but higher than that of foams with 10% of wood particles. It should be noted that even with the maximum concentration of wood particles used in the tests (i.e., 20%), the value of the thermal conductivity coefficient was at a level comparable to that of pure PUR foam. Similar observations were made by Tao et al. [34]. The authors noted a decrease in the value of the coefficient λ by as much as 50%, while higher amount of lignocellulosic fibers (up to 20 php) resulted in a gradual increase in thermal conductivity to a level exceeding that of pure PUR foam.

This method of shaping the thermal insulation properties of the tested PUR-WP compositions may result mainly from disturbances in the cell structure of the foams. This might be manifested by changes in the distribution of cell sizes. As shown in the literature, such changes have a significant impact on the insulation properties and mechanical strength of PUR foams. As shown in Figure 7, pure PUR foam is characterized by a uniform cell size distribution, mainly within the range of 150–450 μm. The mean cell size in the range of the highest frequency is 224 μm. Introducing relatively small amount of wood particles (i.e., 5 wt.%) to the system reduces the number of cells in the size range of 200–250 μm. At the same time, it increases the number of cells above this range, i.e., within the range of 300–450 μm. However, the mean cell size with the highest frequency is still in the range of 200–250 μm. A further increase in the proportion of wood shavings in the PUR-WP composition in the amount of up to 10% of the weight results in the shifting of the mean cell size from 250 to 300 μm (average cell size 276 μm). However, taking into account the photos taken with a light microscope and SEM (Figures 8 and 9), it can be concluded that despite the noted changes in the cell size distribution, the structure of foams with the 5% and 10% wood filler added is relatively well-developed, which allows for high thermal insulation parameters. The insulating properties of the wood filler itself are likely to be important as well. Wood itself is an excellent insulator, and also has a heat capacity greater than the PUR foam.

A further increase in the amount of wood filler to 15% and 20% causes a significant disturbance of the foam structure and formation of larger and more irregular pores. In these variants, the cell structure of the composite foams is disturbed by the presence of cells sized above 450 μm. This is particularly visible in the case of variants containing 20% of wood particles (Figures 7 and 8), although a collapse of the cellular structure of the foam is visible even with a filler content of 15%. The presence of larger cells and damage to the structure of the foams result in increased air permeability. As a result, it increases the heat transfer and thus reduces the thermal insulation of this type of foam [6,34]. Similar observations in the case of the modification of closed-cell PUR foam with various types of lignocellulosic fillers were made by Strąkowska et al. and Członka et al. [24,39]. Additionally, as shown in the literature, such disturbances in the morphology of the PUR foam due to the introduction of the organic filler, such as straw particles, may result from poor interfacial adhesion between the polymer matrix and the filler surface, which consequently disrupts the foaming process and, as a result, the structure of modified PUR foams [40]. Additionally, according to Sung et al. [41], during the cell structure formation of PUR foams, the interaction between the filler surface and the polymer matrix can determine the final average cell sizes. The higher the hydrophilicity of the filler surface, as in the case of wood filler, the larger the cell sizes in the microstructure of foams. Moreover, the filler particles may constitute a nucleating agent and cause the nucleation pattern to change from homogeneous to heterogeneous and reduce the nucleation energy. For this reason, smaller cells are formed in the foam structure [42,43]. In the case of our research, the formation of cells of a small size, i.e., about 100 and 150 μm in size, was also noted.

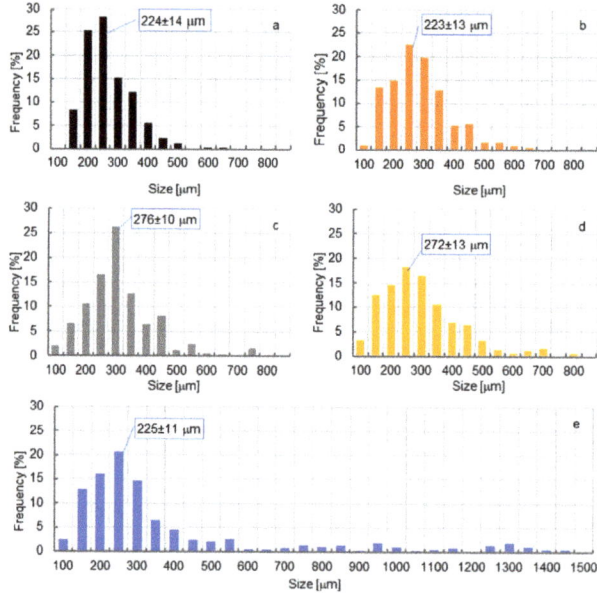

Figure 7. Cell size and cell size distribution of PUR foam depending on wood particle (WP) content: (**a**) control; (**b**) 5%; (**c**) 10%; (**d**) 15%; (**e**) 20%.

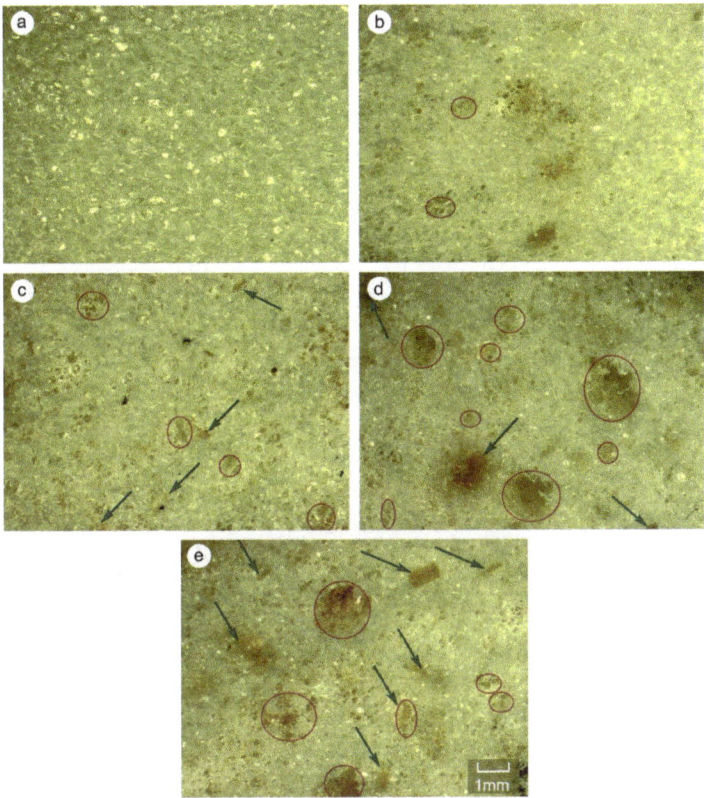

Figure 8. Light microscope photos of PUR samples with addition of different amount of WP: (**a**) control; (**b**) 5%; (**c**) 10%; (**d**) 15%; (**e**) 20% (green arrow—wood particles; red circle—structure disorders, large pores). All scale bars of figures are the same.

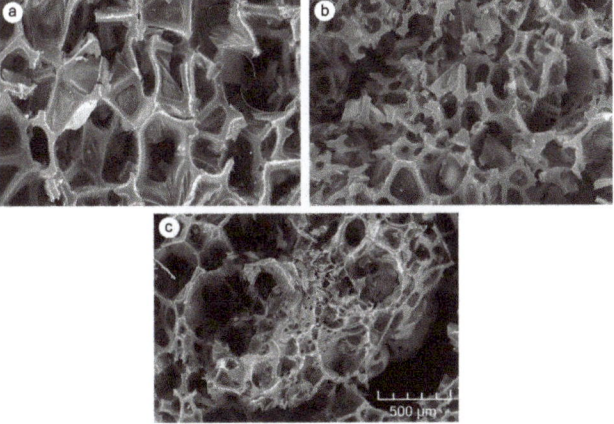

Figure 9. SEM images of PUR samples: (**a**) control sample; (**b**) 10% WP addition; (**c**) 15% WP addition. All scale bars of figures are the same.

This can also be confirmed by the analysis of the cell size distribution and SEM (Figures 7 and 9). It also proves a significant differentiation of the cell size distribution of PUR-WP composite foams. The formation of this type of cell may result from the attachment of filler particles to the foam cell, which leads to damage and weakening of the foam microstructure, and thus lowers its strength.

Analyzing the structure of the produced foams, attention should also be paid to the dispersions of wood particles. As a rule, fillers with smaller particle sizes (e.g., nano-scale) tend to agglomerate, which also interferes with the foaming process and the morphology of PUR foams. The filler of relatively large dimensions and irregular shape used in the research allows for a high degree of dispersion of its particles in the polyurethane matrix. This is evidenced by the 3D photos of foams with 5% and 20% addition of wood particles. The photos were made with the use of the computed tomography technique, presented in Figure 10.

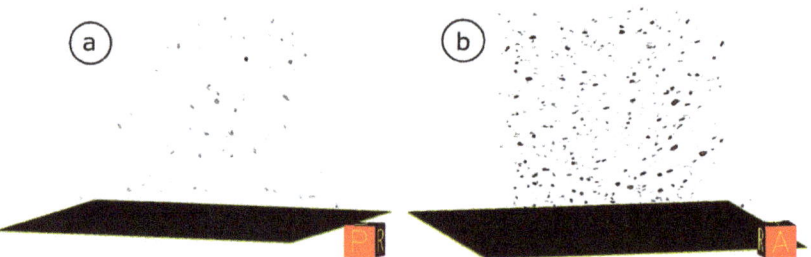

Figure 10. Image of foam with the addition of wood particles made with a computer tomograph: (**a**) 5% WP addition; (**b**) 20 addition WP.

3.3. Compressive Strength

The results of the structure and thermal insulation properties of the produced foams correspond with the results of the analysis of the compressive strength measurements. As shown in Figures 11 and 12, the maximum increase in the mean value of compressive strength was recorded for the composition with a 10% share of foam wood particles—an increase in $\sigma_{10\%}$ by approx. 30% in comparison with control PUR foam. With a 15% addition of wood filler, a tendency towards a reduction in compressive strength can be noticed, but it should be emphasized that the HSD Tukey analysis did not confirm the statistically significant differences noted for the compositions containing 10% and 15% of wood particles (the same homogeneous group b). A statistically significant decrease in compressive strength was noticed only for the composition containing 20% wood particles. It should be emphasized that in this case, the compressive strength was comparable to control samples (the same homogeneous group-a). The increase in the compressive strength of compositions containing up to 10% wood filler can be explained by an increase in the apparent density of the tested foams. At the same time, the foam still has a well-developed cell structure, which is also confirmed by the analysis of the cell size distribution of the tested foams and the photos in Figures 7–9.

Despite a significant increase in the apparent density of foams, a further increase in the amount of wood particles reduced the strength of PUR-WP. Such shaping of the compressive strength of the tested compositions proves that this type of parameter is influenced not only by the apparent density of the foam, but also by its structure. This is confirmed by the research of other authors [38,39,44] and the additionally estimated specific compressive strength of the produced PUR-WP compositions, which is defined as the ratio of the compressive strength $\sigma_{10\%}$ to the density of the tested foams [39]. As demonstrated by the analysis, the use of waste wood particles of such large dimensions as PUR foam filler increases the specific compressive strength by up to 5%. Above this amount, the value of this parameter gradually decreases.

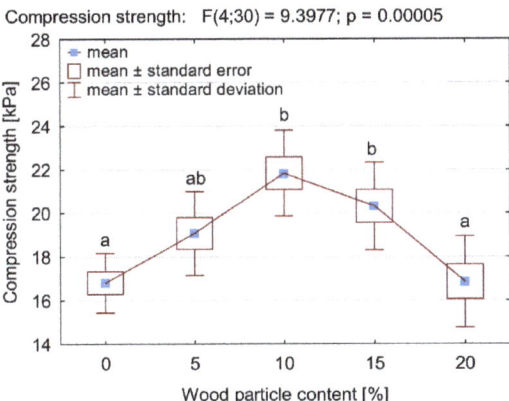

Figure 11. Compression strength of PUR-WP composition depending on wood particle content. Different letters indicate homogeneous groups of mean values determined by one-factor ANOVA with Tukey's test.

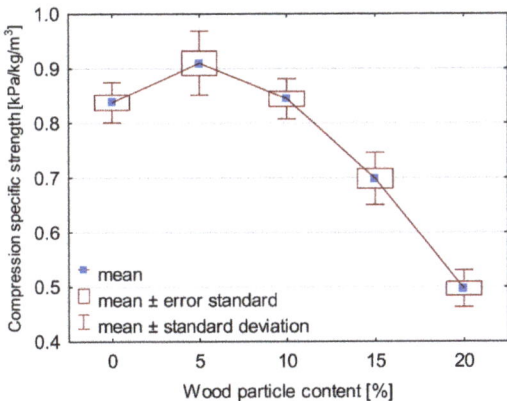

Figure 12. Compression specific strength of PUR-WP composition depending on wood particle content.

This is due to the fact that foams with 5% of wood filler have a homogeneous structure with a narrow range of cell size distribution (like pure foam). The proper filler dispersion with a well-formed foam structure probably facilitates the transfer of strain under compressive load and thus increases its strength. As mentioned earlier, higher filler additions result in a broadening of the range of the cell size distribution and the presence of numerous structural disorders (especially at 15% and 20% of the WP activity). Despite the increase in density, these foams are characterized by lower compressive strength.

4. Conclusions

This article presents the influence of wood particles as a filler in PUR foams. The WP was added in different amounts, i.e., 0, 5, 10, 15 and 20% to PUR and apparent density, cellular morphology, mechanical properties and thermal property were conducted. As expected, the addition of the wood particles causes a significant increase in the density of the PUR-WP composite. A significant increase in the density of the produced foams was noticed only when larger amounts of wood particles were used, i.e., from 10% and more. Along with the increase in the density of the tested foams, statistically significant changes in their thermal insulation were observed. The addition of wood filler in the amount of 10% allows us to improve the insulation properties of PUR foam, which is manifested by a decrease in the value of the thermal conductivity coefficient by 10%. It should be noted that even with the maximum amount of wood particles used in the tests (i.e., 20%), the value of the thermal conductivity coefficient was at a level comparable to that of pure PUR foam. Moreover, the results of the compressive strength in the parallel direction to the foam's growth showed that addition of 10% WP to the foam lead to the increase in $\sigma_{10\%}$ by approx. 30% in comparison with the control PUR foam. The increase in the compressive strength of compositions containing up to 10% can be explained by an increase in the apparent density of the tested foams. Thus, the conducted studies indicated the possibility of using wood waste as fillers for PUR foams with open-cell structure. Such composite foams with 10 wt.% of waste wood particles from primary wood processing can be used as thermal insulation of open diffusivity building partitions in modern prefabricated buildings.

Author Contributions: Conceptualization, R.M.; methodology, R.M., D.D. and J.W.; validation, R.M.; formal analysis, R.M., D.D., J.W. and A.D.; investigation, D.D., J.W. and A.D.; resources, R.M., D.D., J.W.; writing—original draft preparation, D.D., J.W.; writing—review and editing R.M., D.D. and J.W.; visualization, D.D., J.W. and A.D.; project administration, R.M.; funding acquisition, R.M. All authors have read and agreed to the published version of the manuscript.

Funding: This research was funded by the National Center for Research and Development, grant number BIOSTRATEG3/344303/14/NCBR/2018.

Institutional Review Board Statement: Not applicable.

Informed Consent Statement: Not applicable.

Data Availability Statement: The data presented in this study are available on request from the corresponding author.

Conflicts of Interest: The authors declare no conflict of interest.

References

1. Rowell, R.M. *Handbook of Wood Chemistry and Wood Composites*; CRC Press: Boca Raton, FL, USA, 2012.
2. Żelaziński, T.; Ekielski, A.; Tulska, E.; Vladut, V.; Durczak, K. Wood dust application for improvement of selected properties of thermoplastic starch/wykorzystanie pyłu drzewnego do poprawy wybranych właściwości skrobi termoplastycznej. *INMATEH—Agric. Eng.* **2019**, *58*, 37–44. [CrossRef]
3. Kariz, M.; Sernek, M.; Kuzman, M.K. Use of Wood Powder and Adhesive as a Mixture for 3D Printing. *Eur. J. Wood Prod.* **2016**, *74*, 123–126. [CrossRef]
4. Velmurugan, P.; RavindraRraj, B.J. Effect on Strength Properties of M30 Grade of Concrete by Using Waste Wood Powder as Partial Replacement of Sand. *Int. J. Eng. Manag. Res. (IJEMR)* **2017**, *7*, 301–305.
5. Alma, M.H.; Basturk, M.A.; Digrak, M. New Polyurethane-Type Rigid Foams from Liquified Wood Powders. *J. Mater. Sci. Lett.* **2003**, *22*, 1225–1228. [CrossRef]
6. Augaitis, N.; Vaitkus, S.; Członka, S.; Kairytė, A. Research of Wood Waste as a Potential Filler for Loose-Fill Building Insulation: Appropriate Selection and Incorporation into Polyurethane Biocomposite Foams. *Materials* **2020**, *13*, 5336. [CrossRef] [PubMed]
7. Ababsa, H.S.; Safidine, Z.; Mekki, A.; Grohens, Y.; Ouadah, A.; Chabane, H. Fire Behavior of Flame-Retardant Polyurethane Semi-Rigid Foam in Presence of Nickel (II) Oxide and Graphene Nanoplatelets Additives. *J. Polym. Res.* **2021**, *28*, 87. [CrossRef]
8. Gurusamy Thangavelu, S.A.; Mukherjee, M.; Layana, K.; Dinesh Kumar, C.; Sulthana, Y.R.; Rohith Kumar, R.; Ananthan, A.; Muthulakshmi, V.; Mandal, A.B. Biodegradable Polyurethanes Foam and Foam Fullerenes Nanocomposite Strips by One-Shot Moulding: Physicochemical and Mechanical Properties. *Mater. Sci. Semicond. Process.* **2020**, *112*, 105018. [CrossRef]
9. Czech-Polak, J.; Oliwa, R.; Oleksy, M.; Budzik, G. Sztywne Pianki Poliuretanowe o Zwiększonej Odporności Na Płomień. *Polimery* **2018**, *63*, 115–124. [CrossRef]
10. Yuan, J.; Shi, S.Q. Effect of the Addition of Wood Flours on the Properties of Rigid Polyurethane Foam. *J. Appl. Polym. Sci.* **2009**, *113*, 2902–2909. [CrossRef]
11. Kraemer, R.H.; Zammarano, M.; Linteris, G.T.; Gedde, U.W.; Gilman, J.W. Heat Release and Structural Collapse of Flexible Polyurethane Foam. *Polym. Degrad. Stab.* **2010**, *95*, 1115–1122. [CrossRef]
12. Lefebvre, J.; Bastin, B.; Le Bras, M.; Duquesne, S.; Ritter, C.; Paleja, R.; Poutch, F. Flame Spread of Flexible Polyurethane Foam: Comprehensive Study. *Polym. Test.* **2004**, *23*, 281–290. [CrossRef]
13. Lefebvre, J.; Bastin, B.; Le Bras, M.; Duquesne, S.; Paleja, R.; Delobel, R. Thermal Stability and Fire Properties of Conventional Flexible Polyurethane Foam Formulations. *Polym. Degrad. Stab.* **2005**, *88*, 28–34. [CrossRef]

14. de Mello, D.; Pezzin, S.H.; Amico, S.C. The Effect of Post-Consumer PET Particles on the Performance of Flexible Polyurethane Foams. *Polym. Test.* **2009**, *28*, 702–708. [CrossRef]
15. You, K.M.; Park, S.S.; Lee, C.S.; Kim, J.M.; Park, G.P.; Kim, W.N. Preparation and Characterization of Conductive Carbon Nanotube-Polyurethane Foam Composites. *J. Mater. Sci.* **2011**, *46*, 6850–6855. [CrossRef]
16. Wolska, A.; Goździkiewicz, M.; Ryszkowska, J. Influence of Graphite and Wood-Based Fillers on the Flammability of Flexible Polyurethane Foams. *J. Mater. Sci.* **2012**, *47*, 5693–5700. [CrossRef]
17. Paciorek-Sadowska, J.; Borowicz, M.; Isbrandt, M.; Czupryński, B.; Apiecionek, Ł. The Use of Waste from the Production of Rapeseed Oil for Obtaining of New Polyurethane Composites. *Polymers* **2019**, *11*, 1431. [CrossRef]
18. Borowicz, M.; Paciorek-Sadowska, J.; Lubczak, J.; Czupryński, B. Biodegradable, Flame-Retardant, and Bio-Based Rigid Polyurethane/Polyisocyanurate Foams for Thermal Insulation Application. *Polymers* **2019**, *11*, 1816. [CrossRef]
19. Członka, S.; Strąkowska, A.; Kairytė, A.; Kremensas, A. Nutmeg Filler as a Natural Compound for the Production of Polyurethane Composite Foams with Antibacterial and Anti-Aging Properties. *Polym. Test.* **2020**, *86*, 106479. [CrossRef]
20. Hu, X.-M.; Wang, D.-M. Enhanced Fire Behavior of Rigid Polyurethane Foam by Intumescent Flame Retardants. *J. Appl. Polym. Sci.* **2013**, *129*, 238–246. [CrossRef]
21. Funabashi, M.; Hirose, S.; Hatakeyama, T.; Hatakeyama, H. Effect of Filler Shape on Mechanical Properties of Rigid Polyurethane Composites Containing Plant Particles. *Macromol. Symp.* **2003**, *197*, 231–242. [CrossRef]
22. Kerche, E.F.; Silva, V.D.d.; Jankee, G.d.S.; Schrekker, H.S.; Delucis, R.d.A.; Irulappasamy, S.; Amico, S.C. Aramid Pulp Treated with Imidazolium Ionic Liquids as a Filler in Rigid Polyurethane Bio-Foams. *J. Appl. Polym. Sci.* **2021**, *138*, 50492. [CrossRef]
23. de Avila Delucis, R.; Fischer Kerche, E.; Gatto, D.A.; Magalhães Esteves, W.L.; Petzhold, C.L.; Campos Amico, S. Surface Response and Photodegradation Performance of Bio-Based Polyurethane-Forest Derivatives Foam Composites. *Polym. Test.* **2019**, *80*, 106102. [CrossRef]
24. Strąkowska, A.; Członka, S.; Kairytė, A.; Strzelec, K. Effects of Physical and Chemical Modification of Sunflower Cake on Polyurethane Composite Foam Properties. *Materials* **2021**, *14*, 1414. [CrossRef] [PubMed]
25. Zieleniewska, M.; Szczepkowski, L.; Krzyżowska, M.; Leszczyński, M.; Ryszkowska, J. Rigid Polyurethane Foam Composites with Vegetable Filler for Application in the Cosmetics Industry. *Polimery* **2016**, *61*, 807–814. [CrossRef]
26. Olcay, H.; Kocak, E.D. Rice Plant Waste Reinforced Polyurethane Composites for Use as the Acoustic Absorption Material. *Appl. Acoust.* **2021**, *173*, 107733. [CrossRef]
27. Irle, M.; Privat, F.; Couret, L.; Belloncle, C.; Déroubaix, G.; Bonnin, E.; Cathala, B. Advanced Recycling of Post-Consumer Solid Wood and MDF. *Wood Mater. Sci. Eng.* **2019**, *14*, 19–23. [CrossRef]
28. Antov, P.; Mantanis, G.I.; Savov, V. Development of Wood Composites from Recycled Fibres Bonded with Magnesium Lignosulfonate. *Forests* **2020**, *11*, 613. [CrossRef]
29. Antov, P.; Krišťák, L.; Réh, R.; Savov, V.; Papadopoulos, A.N. Eco-Friendly Fiberboard Panels from Recycled Fibers Bonded with Calcium Lignosulfonate. *Polymers* **2021**, *13*, 639. [CrossRef]
30. Mirski, R.; Derkowski, A.; Dziurka, D.; Wieruszewski, M.; Dukarska, D. Effects of Chip Type on the Properties of Chip–Sawdust Boards Glued with Polymeric Diphenyl Methane Diisocyanate. *Materials* **2020**, *13*, 1329. [CrossRef]
31. Mirski, R.; Dukarska, D.; Derkowski, A.; Czarnecki, R.; Dziurka, D. By-Products of Sawmill Industry as Raw Materials for Manufacture of Chip-Sawdust Boards. *J. Build. Eng.* **2020**, *32*, 101460. [CrossRef]
32. Mirski, R.; Derkowski, A.; Dziurka, D.; Dukarska, D.; Czarnecki, R. Effects of a Chipboard Structure on Its Physical and Mechanical Properties. *Materials* **2019**, *12*, 3777. [CrossRef]
33. Iždinský, J.; Vidholdová, Z.; Reinprecht, L. Particleboards from Recycled Wood. *Forests* **2020**, *11*, 1166. [CrossRef]
34. Tao, Y.; Li, P.; Cai, L. Effect of Fiber Content on Sound Absorption, Thermal Conductivity, and Compression Strength of Straw Fiber-Filled Rigid Polyurethane Foams. *BioResources* **2016**, *11*, 4159–4167. [CrossRef]
35. Kuranchie, C.; Yaya, A.; Bensah, Y.D. The Effect of Natural Fibre Reinforcement on Polyurethane Composite Foams—A Review. *Sci. Afr.* **2021**, *11*, e00722. [CrossRef]
36. Dukarska, D.; Czarnecki, R.; Dziurka, D.; Mirski, R. Construction Particleboards Made from Rapeseed Straw Glued with Hybrid PMDI/PF Resin. *Eur. J. Wood Wood Prod.* **2017**, *75*, 175–184. [CrossRef]
37. Mirski, R.; Dziurka, D.; Trociński, A. Insulation Properties of Boards Made from Long Hemp (Cannabis Sativa L.) Fibers. *BioResources* **2018**, *13*, 6591–6599. [CrossRef]
38. Członka, S.; Bertino, M.F.; Strzelec, K. Rigid Polyurethane Foams Reinforced with Industrial Potato Protein. *Polym. Test.* **2018**, *68*, 135–145. [CrossRef]
39. Członka, S.; Strąkowska, A.; Kairytė, A. Coir Fibers Treated with Henna as a Potential Reinforcing Filler in the Synthesis of Polyurethane Composites. *Materials* **2021**, *14*, 1128. [CrossRef] [PubMed]
40. Ruijun, G.U.; Khazabi, M.; Sain, M. Fiber Reinforced Soy-Based Polyurethane Spray Foam Insulation. Part 2: Thermal and Mechanical Properties. *BioResources* **2011**, *6*, 3775–3790. [CrossRef]
41. Sung, G.; Kim, J. Influence of filler surface characteristics on morphological, physical, acoustic properties of polyurethane composite foams filled with inorganic fillers. *Compos. Sci. Technol.* **2017**, *146*, 147–154. [CrossRef]

42. Formela, K.; Hejna, A.; Zedler, Ł.; Przybysz, M.; Ryl, J.; Saeb, M.R.; Piszczyk, Ł. Structural, Thermal and Physico-Mechanical Properties of Polyurethane/Brewers' Spent Grain Composite Foams Modified with Ground Tire Rubber. *Ind. Crop. Prod.* **2017**, *108*, 844–852. [CrossRef]
43. Silva, N.G.; Cortat, L.I.; Orlando, D.; Mulinari, D.R. Evaluation of Rubber Powder Waste as Reinforcement of the Polyurethane Derived from Castor Oil. *Waste Manag.* **2020**, *116*, 131–139. [CrossRef] [PubMed]
44. Paciorek-Sadowska, J.; Borowicz, M.; Czupryński, B.; Liszkowska, J. Kompozyty Sztywnych Pianek Poliuretanowo-Poliizocyjanurowych z Korą Dębu Szypułkowego. *Polimery* **2017**, *62*, 666–672. [CrossRef]

Review

Hemp and Its Derivatives as a Universal Industrial Raw Material (with Particular Emphasis on the Polymer Industry)—A Review

Karol Tutek and Anna Masek *

Faculty of Chemistry, Institute of Polymer and Dye Technology, Lodz University of Technology, Stefanowskiego 16, 90-537 Lodz, Poland; karol.tutek@dokt.p.lodz.pl
* Correspondence: anna.masek@p.lodz.pl

Abstract: This review article provides basic information about cannabis, its structure, and its impact on human development at the turn of the century. It also contains a brief description of the cultivation and application of these plants in the basic branches of the economy. This overview is also a comprehensive collection of information on the chemical composition of individual cannabis derivatives. It contains the characteristics of the chemical composition as well as the physicochemical and mechanical properties of hemp fibers, oil, extracts and wax, which is unique compared to other review articles. As one of the few articles, it approaches the topic in a holistic and evolutionary way, moving through the plant's life cycle. Its important element is examples of the use of hemp derivatives in polymer composites based on thermoplastics, elastomers and duroplasts and the influence of these additives on their properties, which cannot be found in other review articles on this subject. It indicates possible directions for further technological development, with particular emphasis on the pro-ecological aspects of these plants. It indicates the gaps and possible research directions in basic knowledge on the use of hemp in elastomers.

Keywords: hemp; polymer; composites; chemical and physical properties; fiber; extract

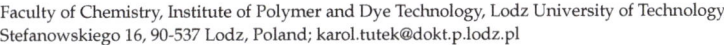

1. Introduction

The polymer industry has grown smoothly and continuously for many decades. Polymers and composites based on them have become one of the basic utility materials, next to wood, concrete, glass or metals, for the production of simple everyday objects, through elements of larger structures, such as vehicles and buildings, to modern and very complex parts, specialized equipment and even spaceships. The wide range of polymers and their properties, which can be further changed by the use of other materials and ingredients, give us as scientists an infinite field for the development of these versatile materials, limited only by our imagination. All this makes polymer composites a versatile product. However, they have their pros and cons, as with any other product. One such ambiguous property is its high durability. As consumers, we most often want polymer materials to have as much as possible, but after use, a problem arises. What to do with such materials? Unfortunately, many of the plastics used so far do not break down too quickly [1]. For some of them, this process can take hundreds of years. For this reason, the growing awareness of researchers, consumers and ecologists put pressure on the development of the field of science dealing with polymer materials and biocomposites, which, in particular, should be characterized by a high biodegradability or compostability potential [2–5]. One of the approaches to the development of environmentally friendly polymer materials is the use of substances derived from fauna and flora in composites. An even better solution is to use materials that are typical production waste from other industrial sectors. In this article, we focus on presenting just such an approach. This review focuses on the material of plant origin, and in particular on hemp and all its derivatives that have or may have potential

Citation: Tutek, K.; Masek, A. Hemp and Its Derivatives as a Universal Industrial Raw Material (with Particular Emphasis on the Polymer Industry)—A Review. *Materials* **2022**, *15*, 2565. https://doi.org/10.3390/ma15072565

Academic Editors: Réh Roman, Ľuboš Krišťák and Petar Antov

Received: 15 February 2022
Accepted: 29 March 2022
Published: 31 March 2022

Publisher's Note: MDPI stays neutral with regard to jurisdictional claims in published maps and institutional affiliations.

Copyright: © 2022 by the authors. Licensee MDPI, Basel, Switzerland. This article is an open access article distributed under the terms and conditions of the Creative Commons Attribution (CC BY) license (https://creativecommons.org/licenses/by/4.0/).

industrial applications, including with particular emphasis on polymers for the creation of biocomposites.

The cannabis plants originally come from Central and East Asia, which later spread to the rest of Asia, and in subsequent periods also to Europe [6]. They are one of the earliest used and cultivated plants in human history. Initially, they were used as plants for the production of food, but also as fiber and medicinal substances. Due to different purposes, man has become interested in two basic types of cannabis, also known as *Cannabis Sativa* fibrous cannabis and *Cannabis Indica*, with a higher content of narcotic compounds. The paleontological records indicated in the research contain information about the discovery of cannabis fragments used by primitive humans at least 10,000 years ago [7]. Based on the collected biological data, it can be concluded that cannabis grows best in temperate climates, where optimal temperatures for growth are between 15 and 27 degrees Celsius. These plants grow best in permeable soils with high fertility and in slightly moist or periodically dry areas. Such conditions for good development can be found mainly on the shores of water reservoirs. Due to their round shape, the seeds are not very well carried by the wind. Rather, cannabis has developed a mechanism that uses birds and other animals to carry their seeds because they are high-value food. Probably the seeds were used as the first parts of the cannabis plant. They were mainly used as food. The next step was to process the stems into strings, fabrics and fibers. It was one of the most important steps in the progress of mankind, which allowed for the production of combining skins, furs and the creation of fabrics or everyday objects, such as baskets, which allowed for faster expansion of people into areas previously inaccessible, cool postglacial. The development of this technology allowed the transition from a nomadic to a more sedentary lifestyle. Further technological advances allowed the production of textile fabrics from fibers and more and more complex elements. Hemp material was widely used in the Columbian era because without hemp fibers, it would not have been possible to create such strong and durable ropes and masts on ships that allowed long and distant sea journeys and the discovery of new lands [8]. Only the development of the cotton industry and then the large-scale plastics has led to the marginalization of the hemp share so far. Currently, these versatile plants are making a comeback thanks to the multitude of possible uses with an ecological approach at the same time. These aspects make it one of the most interesting plants indicated as the future of agriculture in the European Union [9]. This article is a broad overview of cannabis, its structure, composition and properties of individual fractions and, importantly, describes the use of these very important plants in various sectors of the economy, from the food industry, construction and pharmacology, through the automotive industry, with an emphasis on the polymer industry. It is an extensive work containing the most important information on the material and physicochemical properties of this important pro-ecological, multifunctional plant. This review article has an evolutionary structure that resembles the life cycle. This text begins with the characteristics of the historical outline of the use of cannabis for human life, in the following chapters the structure of plants as well as the chemical composition and properties of individual anatomical parts of plants are described. The following chapters describe the ways of using hemp broken down into individual industries, with the final development of the topic of the use of the polymer industry with the division into thermoplastic polymers, elastomers and duroplasts, which are the main topic of our research. It is worth noting that the prepared review is one of the few prepared texts that contain a comprehensive approach to the topic, as it is characterized not only by the composition of individual hemp derivative fractions but also contains a general overview of the use in various industries, taking into account the latest technological innovations of each of them. It is also the only one that describes the use of hemp materials in polymeric materials in more detail, highlighting the knowledge gap regarding the use of fibers and other hemp derivatives in thermosets and, above all, in elastomers as research in this area is insufficient. We believe that this article will be a very good tool to start spreading knowledge about cannabis, improve knowledge about it, and change the attitude in the community, and in particular, for it to be appreciated by researchers, technologists and entrepreneurs.

2. Characteristics of Hemp Plants

2.1. Structure and Composition

2.1.1. Hemp Fibres

Cellulose fibers are the most common biopolymer in the world; their production in 2004 was about 10^{11} tons. They are widely used by man in a variety of technological processes due to their abundance, common occurrence and excellent physicochemical properties [10]. Cellulose belongs to the group of polysaccharides, i.e., polysaccharides such as starch, chitin or dextrins. It is also a polyacetal containing glycosidic bonds linking individual sugar residues, forming long linear polymer chains [11]. This homopolymer is produced indirectly by plants through photosynthesis from the substrates, which include water and carbon dioxide [12]. During its synthesis in plant cells, there is also the necessary energy from sunlight. The cellulose itself is used by plants mainly as a construction material in the construction of conductive tissues in wood. It occurs mainly in stems. It is a solid with a fibrous structure, which consists of crystalline and amorphous areas; thus, it can be characterized as semicrystalline [13]. It has no smell or taste and does not dissolve in cold or warm water and organic solvents [14]. As a polymer of natural origin, cellulose has a number of distinguishing properties among other widely used materials. In short, you can characterize it as:

- Homopolymer that comes from natural sources;
- It has a zero-carbon balance for the environment due to its use in its synthesis, carbon dioxide;
- It is a biopolymer derived from renewable sources, biodegradable, providing good environmental and biological characteristics and high bioorganic compatibility [15–17];
- It is highly pure and non-toxic [18];
- It is characterized by good mechanical strength, which is why it is used as one of the basic natural construction materials [19].

During biodegradation processes, microorganisms decompose biopolymers, and cellulose as an example of a polysaccharide, under aerobic conditions into water, carbon dioxide and biomass, and under anaerobic conditions into CH_4, biomass and water [20].

Cellulose is a linear high molecular weight homopolymer. Its structure includes sections composed of D-glucose, and more specifically β-D-glucopyranose, connected to each other by β-1,4-glycosidic bonds. Native cellulose up to 10 thousand residues β-anhydroxyglucose linked together to form a long chain molecule. This means that the mass of such a molecule is over 1.5 million units. However, the unit length of β-anhydroxyglucose is 0.515 nm, i.e., 5.15 Å. It follows that the total length of the natural cellulose molecule is approximately 5 μm. The cellulose pulp and filter paper used usually contain particles with a degree of polymerization from 500 to 2.1 thousand [21,22]. Each β-anhydroxyglucose unit in the cellulose chain has a chair configuration with hydroxyl groups in equatorial positions and with hydrogen atoms in axial positions. The chair-shaped conformation of the chain (poly-1,4-D-glucosan) is shown in Figure 1. It can be seen that the unit part of the chain is rotated around its main axis by 180°, resulting in an unrestricted rope configuration with minimal steric hindrance. The glycosidic bonds act similarly to the functional group, which, together with the hydroxyl groups, determines the chemical properties of cellulose. All significant chemical reactions take place precisely in the area of the glycosidic bond or the hydroxyl group. Each of the heterocyclic rings has the following groups:

- Primary-CH_2-OH;
- Two secondary hydroxyl groups-OH.

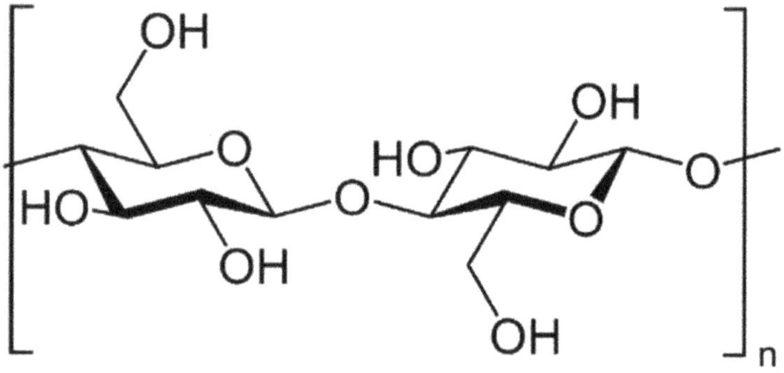

Figure 1. The chair conformation of β-anhydroxyglucose units in the cellulose chain.

The nature of cellulose, and more specifically its chemical, physical and mechanical properties, as well as the fibrous structure, we can relate to its molecular structure. Analogously to other hydrophilic linear polymers, individual cellulose molecules combine together to form a fibril or protofibril about 10 nm in length, 4 nm in width and about 3 nm thick. Unit distribution of cellulose chains is oriented parallel to each other and tightly connected by numerous intermolecular hydrogen bonds. The structure that makes up the cellulose fiber is hierarchical. The smallest basic microfiber building unit and the macrofiber is fibril. Many such structures collectively aggregate into long, thin bundles to form a microfiber. They, on the other hand, form macrofibrils in greater numbers and then fibers [23–28]. The crystalline regions of the linear cellulose chains are laterally linked by hydrogen bonds. They build a kind of mesh that extends across the entire cross-section of the microfiber. The crystalline regions are separated from each other by a layer of cellulose molecules, the arrangement of which is not specifically oriented towards each other, creating spaces characterized by amorphous or otherwise paracrystalline domain. The disordered area allows the degradation of the polymer chain with an aqueous solution of a strong acid. The amorphous portions of the fiber can occur naturally and can also be produced during mechanical degradation. The length of the molecule after acidolysis is variable and depends on the origin of the cellulose. However, the process itself leads to an even degree of polymerization in the obtained micelles or microcrystals [29–34].

The list of the initial stages of cellulose fiber degradation is presented, which can be described as follows:

- An intact fiber-containing crystalline and amorphous regions, with frayed ends at the periphery consisting of a paracrystalline region of cellulose, lignocellulosic or hemicellulose;
- Initial attack on regions with an amorphous structure;
- There will remain residual microcrystallites and decomposition of the remaining free short chain fragments;
- Attack on a crystalline region.

In the literature, there is a division of native cellulose into two crystal structures:

I_α with the structure of a triclinic unit cell;
I_β monoclinic unit cell structure.

Natural cellulose, derived from green plants and wood, contains a mixture of both crystal structures. However, it is primarily the characterization structure of a monoclinic unit cell. For example, cotton or ramie contains as much as 77% of it in their structure. On the other hand, the one derived from algae and bacteria has a higher content of the triclinic form. The more stable thermodynamic form of this biopolymer is I_β because I_α cellulose is transformed by hydrothermal treatment in alkaline solutions or by heat treatment in an inert gas atmosphere at 280 °C into the β form.

It is possible to recognize both structures by means of the ^{13}C-NMR test, while the correlation of this test with the absorption coefficient of the FT-IR spectrum is also performed using the following formula [24,35–39]:

$$f_\alpha = 2.55 \times \left(\frac{A_{750}}{A_{710}}\right) - 0.32$$

where:
- A_{710} absorption intensity at the wavenumber of 710 cm^{-1};
- A_{750} absorption intensity at the wavenumber of 750 cm^{-1}.

Cellulose has several crystalline forms, which may change from one another by simple technological unit operations.

Hemp is one of the main crops grown for its fiber. These fibers very quickly grow up to several mm a day during the intense growth phase. Primaries can reach a length of about 15 mm with a spread from a few to even more than 50 mm, as described in the work of Mussing et al. After the intensive growth phase, the cell walls are lignified and the cellulose content increases. This step leads to a mechanical strengthening of the fibers, in particular their stiffness. The length, chemical composition and properties of the fibers strongly depend on the variety and conditions in which the plant has grown. Compared to other plants, fibrous hemp is characterized by a high content of cellulose from 70 to 74% (including the one with a high degree of crystallinity) and hemicellulose, about 15–20%, but it has a limited amount of lignin, which usually 3.5%, but does not exceed 5.7%. The pectin content is usually around 0.8%, and the fat and wax content is 1.2–6.2% [40–42]. The amount and ratio of these components may vary depending on the degree of purification with NaOH. In the case of mechanical properties, it is difficult to determine individual values for individual fibers due to their short length. However, the length reported in most literature sources ranges from 5 to 55 mm, the crystallinity index is indicated as 55%, the diameter is in the range of 10.9 to 42 micrometers, and the density is about 1.5 g/cm^3. As for Young's modulus, it varies significantly from 14.4 to even 90 GPa, as does the breaking strength 285–1110 MPa and the elongation at break from 0.8 to 3.3% [43]. Large differences in the values in the sources result from irregularities in the diameter and length of the fiber, and the fact that a technical fiber was tested, which is characterized by a lower strength than a single, separated fiber with a shorter length, amounting to a few millimeters [44].

The analysis of the spectrum of infrared spectroscopy with Fourier transformation of hemp fibers shows the presence of several absorption bands characteristic of this material. These data were collected and described by Kaczmar et al. in the form of a table shown below [45]. Table 1 was also confirmed by other references than those mentioned in the above work.

Table 1. Absorption signals of Fourier transform infrared spectroscopy for the spectrum of hemp fibers.

Name of the Function Group	Wavenumber [cm^{-1}]	Bibliographic
C-OH out-of-plane bending vibrations; C-C	557	[46]
Stretch vibrations of the glucose ring; C–H stretching vibrations outside the plane of the aromatic ring	895	[22,23]
-OH; -COO	900–1200	[47]
CO-O-CO	1000–1100	[48]
C-O stretching vibrations; deformation of the C-H aromatic plane	1030–1058	[49,50]
The absorption band of hydroxyl compounds -OH	1100	[51,52]

Table 1. Cont.

Name of the Function Group	Wavenumber [cm^{-1}]	Bibliographic
C-O stretching vibrations; asymmetric bridge C-O-C stretching vibrations	1158	[52,53]
C-O; C=O; C-C-; COOH	1100–1300	[54]
Acyl-oxygen CO-OR stretching vibrations in hemicelluloses; -CH$_3$	1245	[50]
C-H deformation vibrations; -OH bending vibrations	1325	[51]
C-H bending vibrations related to the structure of cellulose and hemicellulose	1369	[53,55]
CH$_2$ stretching vibrations related to the cellulose structure, vibrations of the bonds of the aromatic backbone	1425–1426	[52,53,56,57]
CH deformation vibrations; asymmetric bending vibrations from -CH$_2$ and -CH$_3$ groups	1426–1463	[46]
C=C stretching vibrations in aromatic structures	1508	[51]
C=C stretching of the aromatic ring	1550	[45]
C=C unsaturated bonds;	1592	[51]
COO$^-$ (pectin)	1650	[45]
-OH from absorbed water; C=C	1653	[50,51,56,58]
C=O stretching vibrations in uncoupled ketones and free aldehydes	1736; 1718	[55–57,59,60]
CH stretching vibrations in methyl and methylene groups	2896	[53,55,61]
-OH stretching vibrations (hydrogen bonds)	3331	[53,62]

Based on the knowledge of the structure of hemp fiber, its chemical and physical structure and properties, its possible application in various technical solutions is known. However, some applications require changing these properties. In order to modify them, in basic technological operations, these are mechanical modifications such as cutting or grinding. More advanced techniques that significantly change the properties, however, are based on chemical modifications such as alkalization, acetylation, esterification, silanization, acrylation, or through the use of carboxylic acids, anhydrides or solvent replacement. Each of the mentioned physical and chemical modifications causes a specific change in the properties of the hemp fiber and adjusts it to the most interactive use. This topic is explored more in-depth in the extensive work of Tanasa et al. [28].

2.1.2. Extract

Plants are the main source of natural extracts. Their matrix is used to extract all the natural compounds needed by humans, such as oils, essential oils, compounds with healing properties and others. Hemp, in this case, is also a rich source of these substances. In order to obtain them and collect the appropriate fraction, an appropriate extraction method should be selected. The literature indicates two main ones: the method with the use of organic solvents and the method with the use of supercritical gases. In the case of the first of them, one of the first steps is the comminution of the plant material and then treatment with a suitably selected organic solvent or their mixture at a predetermined temperature. The solvent flushes out specific compounds from the plant matrix as a result of its diffusion through its tissue. Unfortunately, this method suffers from a number of disadvantages, such as low selectivity, contamination of the extract obtained with residues of often toxic solvents, and the effect of high temperature as a factor causing the degradation of unstable natural compounds. The second method, i.e., the use of supercritical gases (most often carbon dioxide), is more innovative and allows avoiding the use of both elevated temperatures and organic solvents that are unfriendly to the natural environment. In this case, the solubility of active natural compounds that depend on such physicochemical factors as gas pressure, the temperature of the extraction process-by controlling these parameters, the gas diffusivity and polarity are controlled, which affects the solubility of the extracted substances. An additional advantage is that this process is carried out under inert gas conditions, which significantly reduces the oxidation of unstable compounds in the air atmosphere. Unfortunately, a small range of substances dissolves very well in supercritical carbon dioxide, which is why small amounts of other solvents are often used as cosolvents to help flush out the expected natural compounds from the matrix [63].

As the research conducted so far shows, over 500 active substances have been discovered in *Cannabis Sativa*, which can be classified into 18 main groups of chemical substances. Among other things, they are rich in 12 fatty acids, about 200 terpenes and 20 heterocyclic compounds containing nitrogen atoms in their ring structure, over 50 hydrocarbons and as many as 100 cannabinoids, of which hemp is the most famous. The main cannabidiols found in this plant include delta-9-tetrahydrocannabinol (THC), cannabidiol (CBD), cannabigerol (CBG), and cannabinol (CBN). The structures of the listed compounds have been collected and presented in Figure 2. These salivae also contain various polyphenols with antioxidant properties, coloring compounds and polysaccharides [64].

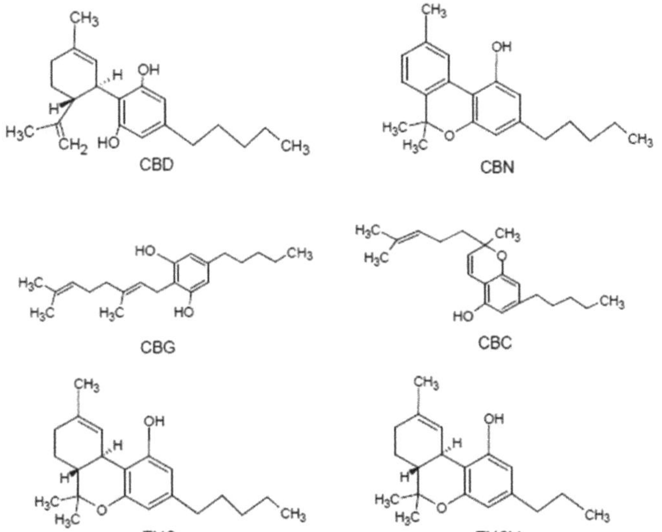

Figure 2. Structural formulas of cannabinoids occurring in hemp (CBD—cannabidiol; CBN—cannabinol; CBG—cannabigerol; CBC—cannabichromene; THC—tetrahydrocannabinol; THCV—tetrahydrocannabivarin).

Previous research in cannabis has discovered CBD was shown to have very strong antioxidant, anti-inflammatory and bactericidal properties and is used in anxiolytic, anticonvulsant and neurological therapies, while CBG also has analgesic properties. All of the mentioned compounds belong to the group of phytocannabinoids occurring depending on the variety and the way of cultivation in various quantities in cannabis. Cannabigerol is a precursor to the formation of compounds such as THC, CBD and cannabichromene (CBC). Unfortunately, THC, due to its psychoactive effects, has made cannabis infamous, as it contains a wide range of active, health-promoting natural compounds. CBD, CBC, and CBG are indicated as one of the main potential medicinal substances that can help people with diseases such as cancer, neurological diseases, bacterial infections and severe inflammation in the body. Strong healing properties are indicated even among drug-resistant bacteria, such as the *Staphylococcus aureus* strain. Hemp also contains various polyphenols with antioxidant properties, coloring compounds and polysaccharides [65–77].

2.1.3. Waxes

Waxes are still a little-known and studied part of the hemp plants. Its source may be hemp dust and waste generated during the processing of entire plants, fibers, seeds and leaves in various technological processes. This dust and waste is usually a waste product, but in the future, it may become a potential source of hemp waxes for use as an ingredient in cosmetics or as a natural polymer plasticizer [69]. They are part of the oil fraction which, according to the data contained in Table 2 given by L. Apostol in his article, has the following composition [78]:

Table 2. Composition of the oily fraction derived from hemp seeds.

Component	Value [%]
The content of the oily fraction in the entire mass of the hemp seed	28.7
Saturated Fatty Acid	
Palmitic acid	6.96
Stearic acid	2.74
Arachidic acid	0.77
Total saturated fatty acid	10.47
Unsaturated Fatty Acid	
Oleic acid	13.64
Linoleic acid	56.35
Gamma-linoleic acid	1.35
Alpha-linoleic acid	17.30
Stearidonic acid	0.50
Eicosenoic acid	0.39
Total unsaturated fatty acid	89.53

Attard et al., as a result of their research, performed an extraction using the supercritical carbon dioxide method and the Soxhlet extraction using heptane. In all the samples tested, they detected the presence of hydrocarbons, fatty acids, alcohols, fatty aldehydes, sterols, cannabinols and wax esters. The last of the mentioned groups of compounds were characterized by chain length from C_{38} to C_{58}. The most common wax ester in the samples was C_{46}, and then C_{44}. Interestingly, almost all wax esters were lost from hemp waste processed during paper production. The largest amounts of waxes were obtained as a result of supercritical extraction carried out at a temperature of about 50 °C and high pressure of 350 bar [69].

Other studies on cannabis samples by Francisco et al. showed the following chemical composition of the waxes obtained from the ethanol suspension [79]. These data are presented in Table 3 below.

Table 3. The content of the fraction in the entire mass of the hemp waxes.

Component	Value [%]
Alkanes 27.02–28.85	
pentacosane	1.92–2.17
heptacosane	6.96–7.55
octacosane	0.75–5.56
nonacosane	9.92–10.51
triacontane	0.44–0.58
dotriacontane	0.49

Table 3. *Cont.*

Component	Value [%]
tritriacontane	1.58–2.06
pentatriacontane	1.13–1.24
heptatriacontane	1.18–1.23
Monoterpens	
sabinene	0.31–0.51
p-cymene	3.32–5.15
Sesquiterpenes	
β-cubebene	0.31–0.40
(−)-trans-caryophyllene	5.90–7.22
β-copaene	0.32–0.40
α-humulene	0.51–0.94
(E,E)-β-farnesene	0.30–0.33
γ-gurjunene	0.27
γ-curcumene	0.59–0.70
valencene	0.51–0.60
germacrene A	0.39–0.44
α-7-epi-selinene	0.42–0.54
α-cadinene	0.20–0.33
α-bisabolene	1.63–2.50
(E,E)-α-farnesene	0.28
Terpenoids 22.92–23.70	
dehydro-1,8-cineole	1.23–1.99
isoborneol	0.38
fenchone	0.26–0.44
cis-thujone	0.27
endo-fenchol	0.26–0.28
cis-nerolidol	2.50–2.84
trans-nerolidol	0.43
caryophyllene oxide	0.49–0.89
humulene epoxide II	0.31–0.37
10-epi-γ-eudesmol	0.61–0.82
1,10-di-epi-cubenol	0.29–0.36
γ-eudesmol	0.29–0.47
α-muurolol	0.25–0.35
β-eudesmol	0.67–1.01
α-bisabolol	0.18
(2Z,6Z)-farnesol	0.49
Cannabinoids 41.67–46.37	
CBD	4.20–9.67
CBC	0.11–0.18
Δ^8-THC	0.12–0.13
Δ^9-THC	0.22–0.37
CBG	0.07–0.22
CBN	1.20–2.40
CBDA	22.91–34.56
THCA	5.78–5.89
Other 1.48–2.09	
heptanal	0.22–0.61
2,4-hexadienal	0.11
nonanal	0.37
vanillin	0.27
tridecanoic acid	0.21–0.31
ethyl tetradecanoate	0.42
hexadecenoic acid	0.25–0.27
ethyl hexadecanoate	0.22–0.31

As the research of the above scientists show, hemp waxes are rich not only in alkanes but also in monoterpenes, sesquiterpenes, terpenoids, but also in cannabinoids. This suggests that apart from the plasticizing and lubricating properties, the cannabis wax esters have strong healing and antioxidant properties. These products play a multi-functional role in their applications. However, these substances still require in-depth study because of the limited knowledge about them.

2.2. Sectors of the Economy Using Cannabis

The subject of the use of cannabis in science, industry or the arts has gained prominence in recent years. As can be seen from the Figure 3 below, the popularity of this keyword in the Scopus database has increased nearly 10-fold over the last 20 years. This indicates a remarkable interest, particularly since 2015, in this plant. The possibility of its use in a wide range of applications and the development of the pro-ecological trend in the world causes newer and more advanced research towards the description of properties and applications of hemp plants in everyday products.

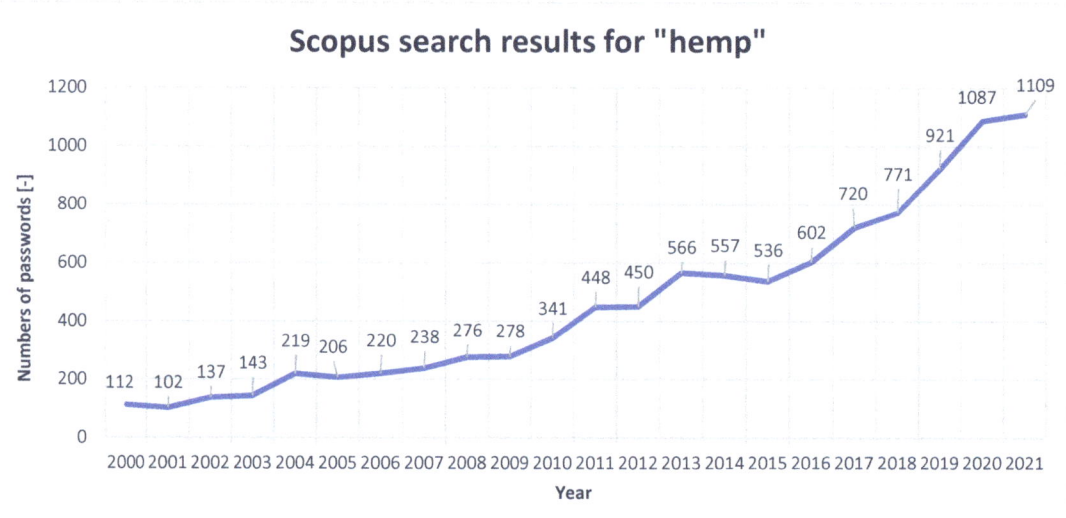

Figure 3. Change in the number of occurrences of the keyword "hemp" in the Scopus database in 2000–2021.

2.2.1. Agriculture and Energetic

Fiber hemp is a species of the annual hemp plant. These plants do not contain psychoactive substances; they are used in many industries, reaching a height of 1.5–3.5 m under favorable conditions. The main direction of the use of hemp in agriculture is the production of straw. As a result of processing the straw of mono-hemp fibrous hemp, we obtain 25–30% of the fiber and about 70% of the shives. Hemp is an interesting plant in terms of ecology and economy. Their cultivation does not require the use of plant protection products or pesticides, and hemp itself inhibits the development of weeds, repels pests, is resistant to diseases and requires only minerals contained in the soil. This has a positive effect on the environment as it contributes to the improvement of soil systems [80]. A suitable example is that these plants have a pile system root, which loosens and ventilates the soil and improves its water conditions, making it more beneficial for all plants that coexist with cannabis. This brings about positive effects, positively influencing the development of the economy, especially everything in agricultural countries. A favorable pro-ecological effect on the environment may, to some extent, reduce the need to increase expenditure on environmental protection and climate change. As indicated in the work of Żuk-Gołaszewska et al., one hectare of hemp plants is capable of absorbing about 2.5 tons of carbon dioxide [80]. Agriculture is the main and basic source of food, but despite this, it is not an economically competitive sector of the economy compared to other industries. Hemp straw in agriculture is used as a source of fodder with very good nutritional parameters for farm animals, mainly cattle. However, apart from that, it is also used as highly efficient biomass in the processes of generating both thermal and electric energy, which was presented in Figure 4 below [81].

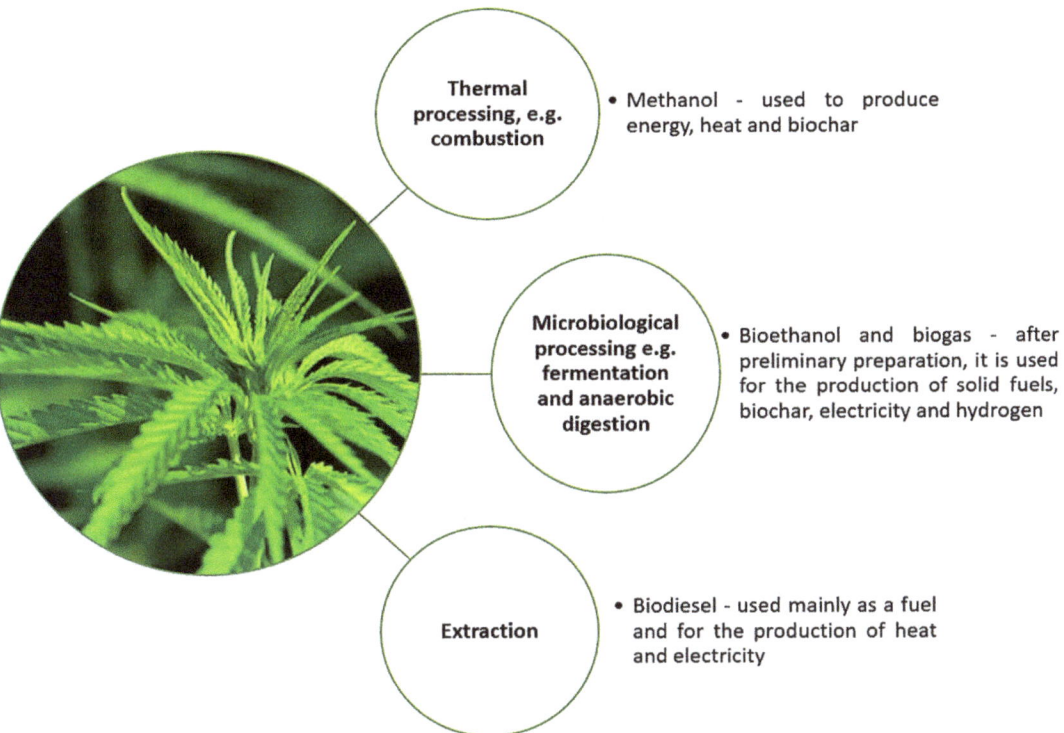

Figure 4. Processing cannabis for energy purposes.

2.2.2. Food Industry

One of the industries that use hemp is the food industry. Hemp food has been known for thousands of years and, at the same time, is a modern, fashionable and healthy food supplement containing valuable ingredients. The hemp seed based on which most hemp foods are prepared contains all of the amino acids and Omega 3, 6 and 9 fatty acids needed for the proper functioning of the human body, especially the brain, in appropriate proportions. About 35% of the seed content is easily digestible high-quality protein, but also contains dietary fiber to support the digestive system and proper digestion, as well as vitamins B and E. On the other hand, about 30% of the seed content is carbohydrates, providing energy for the body. Hemp seeds can be eaten raw, sprouted or powdered as flour. They are used for baking, as well as hemp milk made from them, similar to soy. About 27–38% of the seed weight can be extracted into hemp oil rich in unsaturated fatty acids, as Fike pointed out in his article [9]. The competition for hemp-based food products is the entire food market, especially organic food. Food products made on the basis of hemp have a positive health-promoting effect on our body. They affect cell regeneration, slow down the aging processes, inhibit the development of cancer cells and have a significant effect on immunity [9].

As Kaniewski pointed out in his work, hemp seeds are a rich source of edestin, phytic acid, choline, trigonelline, lecithin, chlorophyll, vitamin K and tocopherols, as well as many micro and macro elements such as iron, calcium, zinc, phosphorus, magnesium and vitamin E, which strong antioxidant properties. It protects unsaturated fatty acids against oxidation reactions, thanks to which they retain their properties. In addition, it has a positive effect on the circulatory system, making blood vessels more flexible, improving blood flow and reducing the possibility of ischemic heart disease or atherosclerosis [82]. Vitamin E is otherwise known as alpha-tocopherol (5.66% of all tocopherols). Cannabis also contains gamma-tocopherol (89.11%), beta-tocopherol (0.33%) and delta-tocopherol (4.90%). These are antioxidant compounds that are involved through the interaction and active quenching of DPPH and ABTS + cationic radicals. They form metal transfer chelates with them. They also absorb oxygen radicals generated by AAPPH (ORAC) and prevent lipid peroxidation in human LDL. Thanks to such good antioxidant properties, these compounds protect proteins, lipid membranes and DNA against the harmful effects of radicals that cause oxidative damage, as mentioned in the article by Żuk-Gołaszewska et al. [80]. Hemp food improves digestion and is beneficial for the healthy digestive system, and also lowers cholesterol, reducing the risk of heart attacks, one of the main civilization diseases of the 21st century in highly developed countries [78].

2.2.3. Textile Industry

The fibers obtained from Cannabis Sativa can be used to produce high-quality fabrics that are used in the clothing industry around the world. It is worth emphasizing that the production of hemp fibers is more ecological and less water-absorbing than the widely produced and used cotton. According to Columbia History of the World, hemp fabrics have been known to man since the eighth millennium BC. From the 5th century BC up to the stage of the industrial revolution, hemp fabrics were used in the production of about 90% of sails. Until the United States of America introduced the so-called Marihuana Tax Act (1937), which also included industrial hemp, about 80% of all fabrics intended for clothes and other everyday textile products were made of hemp fabrics. According to specialists in the textile industry, hemp fabrics are more durable and three times more extensible; they are warmer, more delicate and have high water absorption than cotton fabrics. One of Ireland's exports from the decades to the 1930s was high-quality hemp-based underwear, while Italian hemp-based fabrics were considered one of the best textiles in the world. Hemp was also used to strengthen rotting and fire-resistant carpets, as opposed to artificial, flammable synthetics [69].

2.2.4. Pulp and Paper Industry

Another of the industries mentioned that uses industrial hemp is the pulp and paper industry. The first century AD saw the discovery in China that hemp paper is 50–100 times more durable than most papyrus varieties, and its production it is 100 times easier and cheaper. In the following years, this discovery spread all over the world, especially in Europe and America, where hemp paper was used to create bibles, banknotes, securities, navigation maps, logbooks, and in later years also, books and newspapers [83]. In 1776, the first declarations of independence for the United States were written on hemp paper, the popularity of which grew until the beginning of the 20th century and global industrial development. Until 1883, hemp paper accounted for most of the global paper market. Hemp has always been a significant competition in the present pulp and paper industry. Twenty to thirty percent of hemp stalks are made of hemp fiber, which is used to produce environmentally friendly paper. One hectare of hemp can produce 3–4 times more paper than the same area of trees, and the time of their growth is incomparably shorter and under favorable conditions, the harvest can take place even 3–4 times a year. Hemp paper, unlike wood pulp, does not require the clearing of long-growing, centuries-old forests that produce the oxygen necessary for life or such strong chemical processes that have a significant impact on the environment. The possibility of using recycled hemp paper is estimated at seven times, while the possibility of using wood for only three. The pulp and paper industry is one of the biggest competitors of the hemp industry, which, thanks to its political and economic influence, contributed to the introduction of the first cannabis prohibition in the United States and influenced the unfavorable perception of this plant in the world. In 1916, a method of producing hemp pulp for the production of paper was invented in the United States, using not the fibers of the stalks, as previously, but cellulose-rich fibers-shives, with four times higher efficiency, compared to wood production. The process could also use a much lower amount of sulfur and acid chemicals, and the hemp paper produced by this method does not require an environmentally harmful bleaching process. Unfortunately, no collection machines are available, and removing the outer shives from the inner fiber has not allowed this method to gain sufficient popularity. However, hemp is taking part in the production of paper again and again. As indicated in their article by Amode and Jeetah, paper production in 2018 was estimated at 400 million tons, and the annual growth each year until 2030 is forecasted at 1.1%. The data also allow the conclusion that by 2060 there will be an increase in the use of paper for printing and writing by as much as 180%. This information gives a signal that there is a need to use other sources of cellulose besides wood in the paper industry, and for environmental and ecological reasons, it will be worth increasing interest in hemp in this direction [84]. This is due to better product parameters, such as exceptional strength and mechanical and thermal resistance, resistance to abrasion and yellowing and high flexibility of the material [85].

2.2.5. Construction

Hemp is an extremely efficient and environmentally friendly building material. This is due to the fact that the increase in hemp biomass is two to four times greater than in forests managed on the same acreage. Hemp fiber is used to make furniture and decorations, partition plates are produced, thermal insulation of buildings is also carried out, or research is carried out on concrete blocks containing hemp fibers, characterized by low thermal conductivity and good acoustic barrier. From special varieties of hemp, it is easy to produce ecological bricks up to seven times stronger than concrete. Fiber hemp is also used to produce insulation material, building material for the construction of roofs, walls and floors. It is quite resistant to moisture, does not rot, is not flammable and is almost 100% recyclable. According to research by construction specialists, cellulose concrete made with hemp is resistant to fire and insects, is lighter than conventional building materials and has much better acoustic, thermal and insulating properties [86–90]. Seng et al. in their article, indicate that the thermal conductivity of hemp concrete, depending on the method, ranges from 0.103 to 0.112 $W\ m^{-1} \cdot K^{-1}$ [88]. The use of hemp concrete reduces the cost of building a residential house thanks to the simplification of the structure and the use of cheaper raw materials. Other plant derivatives, such as hemp oil, can also be an important ingredient used in the manufacture of paints and varnishes, as it dries quickly and leaves a thin, flexible film, and the use of its subsidies in petrochemicals is eliminated.

2.2.6. Automotive Industry

Hemp influenced the development of the automotive industry from the very beginning. They were used in the first cars to produce structural elements. The fibers of this plant were tested as a component in the production of car bodies by Henry Ford in 1941 and by Lotus Cars. Hemp was also used to create laminates for any type of construction. Researchers dealing with the subject of hemp-containing materials postulate that hemp-based materials are extremely durable and at the same time have a high biodegradability potential. Many specialists present hemp as a natural material that is stronger than that obtained from other sources of natural cellulose fibers such as coconut, bamboo or jute. The prospect of biomaterials that make up motor vehicles is promising due to the high cost of storing old cars and a strong impact on the natural environment. The calculation of the impact of the entire product life cycle is emphasized. Hemp fiber is used to make body parts, cockpits, seats and other interior elements. On the other hand, the obtained hemp oil can be successfully used as a pro-ecological component of paints and varnishes [91–93].

2.2.7. Cosmetics, Pharmaceutical and Medical Industries

Recently, the use of hemp derivatives in the cosmetics industry has been a very fashionable direction. Hemp oil and extracts containing regenerative, anti-aging and anti-inflammatory substances are used in the production of hemp-based cosmetics. The concentration of the four main components in industrial hemp and wild hemp varied as follows: β-caryophyllene 11–22% and 15.4–29.6%, α-humulene 4.4–7.6% and 5.3–11.9%, caryophyllene oxide 8.6–13.7% and 0.2–31.2%, and humulene epoxide 2, 2.3–5.6% and 1.2–9.5%, respectively. The concentration of CBD in the essential oil of wild hemp ranged from 6.9 to 52.4% of the total oil content, while CBD in the essential oils of registered varieties ranged from 7.1 to 25%, as described in more detail in their article by Zheljazkov et al. [94]. It is applied directly to the skin, has a protective effect, soothes inflammation, irritation and skin changes, it is recommended for people with severe allergies. Beauty salons use hemp preparations as a moisturizing and nourishing agent, reducing discoloration and evening-out skin tone. Hemp oil belongs to the so-called dry oils because it is quickly absorbed and leaves no greasy film. It can be applied directly to the skin, but today many companies produce cosmetics based on it, including care creams, lotions, massage oils, soaps, shampoos, conditioners and more. Hemp extracts largely contain cannabidiol (CBD) and resin fractions that have soothing and calming properties. The pharmaceutical and medical industries also appreciate hemp ingredients more and more. Research is being carried out on the treatment of depression, sleepiness, convulsions, degenerative diseases such as Alzheimer's disease and nutritional problems [95]. The latest reports also indicate strong antimicrobial properties, strong Gram-positive and Gram-negative effect on drug-resistant bacteria. Preliminary information also suggests possible inhibitory effects on the growth of cancer cells [94–96]. Recently, there have also been reports of the biggest problem at the moment, i.e., the SARS-CoV-2 pandemic. CBD contained in hemp was used on lung epithelial cells and in mice. Cannabidiol and its metabolite 7-OH-CBD strongly block viral replication by inhibiting gene expression and reversing the effects of infection. In this case, CBD inhibits SARS-CoV-2 replication in the early stages of the disease. This relationship is therefore indicated by Nguyen et al. as a very effective potential measure to prevent infection in the early stages of infection; however, further testing and clinical trials are needed to clearly confirm the effects of cannabidiol on this virus [97,98].

2.2.8. Polymer Industry

Currently, most branches of the economy are based on polymer products, but their negative impact significantly affects the degradation of the natural environment. For this reason, research is carried out, and newer, more environmentally friendly polymer composites are introduced to the market. Such are also composites based on hemp fibers. These fibers replace the previously commonly used glass fibers with reinforcing properties [99]. However, those used so far have been energy-consuming in the production process and difficult to utilize and non-biodegradable. On the other hand, replacing them with hemp fibers allowed for the creation of more environmentally friendly composites, which, after use, have a smaller impact on the environment during storage and are also subject to partial decomposition. The most popular biocomposites are those based on resins such as unsaturated polyester, phenolic or epoxy resins. They have found their application in the production of cars, hulls of boats and small airplanes, wind turbines and other objects made with the technology of creating laminates [43]. It is also possible to use hemp oils for the synthesis of polymers, but so far, it is a poorly developed branch. A new approach indicated in the research work of Dr. Masek's group is the use of hemp extracts and waxes in composites based on biothermoplastics and ecological vulcanizates. The hemp compounds mentioned are used as dyes, indicators of degradation processes, inhibitors and catalysts of aging processes [63,100].

2.2.9. Other Uses

Interesting and worth mentioning and emphasizing is the possibility of using hemp plants for the rehabilitation of mining excavations. These plants, due to their good adaptation to environmental conditions, high resistance to pests and diseases, are a great organism for pioneering introduction to damaged heaps and post-mining areas. They have good properties of binding heavy metals in their structure, which significantly allows the soil to be cleaned in a short and ecological way and enables the introduction of other species of fauna and flora to the reclaimed ecosystem [82,83,101].

The Figure 5 attached above gives a good indication of the fields in which interest in hemp plants has been greatest over the last 10 years. The top five with the highest number of publications on them are material science, engineering, agricultural and biological sciences, chemistry and chemical engineering. This analysis shows very interesting data, as the general opinion of the average consumer is that the greatest use of hemp is in cosmetology, pharmaceuticals and the medical industry, less so in the food industry. However, the data presented show that it is the industry, especially the materials industry, that has the greatest aspirations for the use of these plants in science and industry. As illustrated in Figure 6 below.

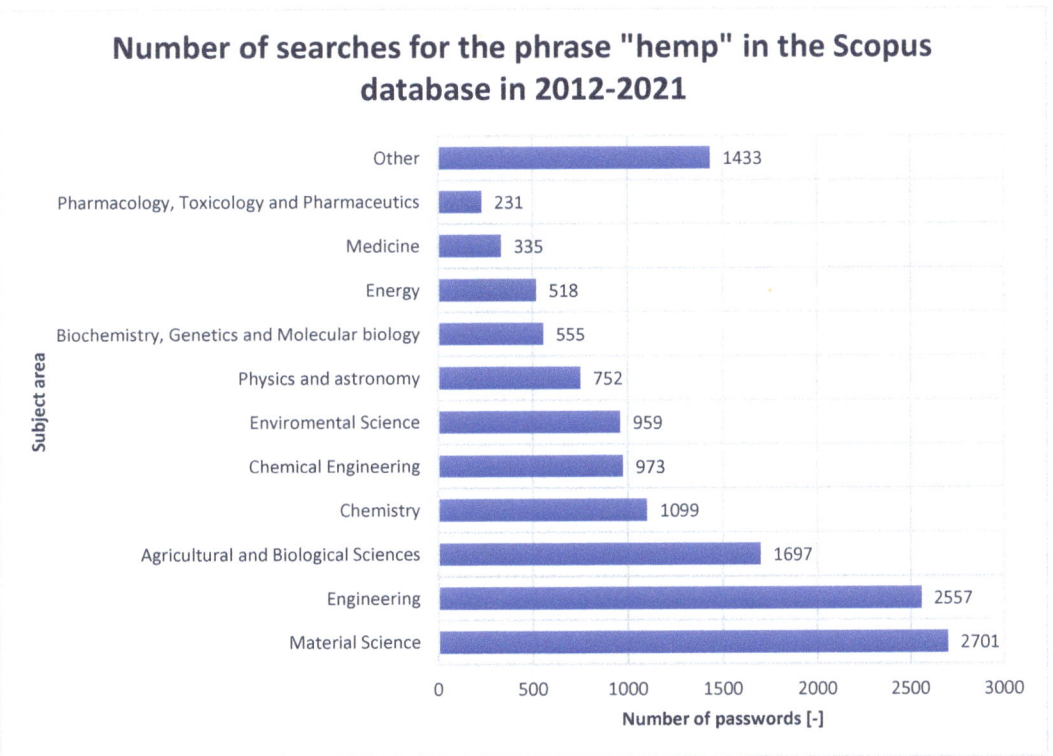

Figure 5. The number of occurrences of the phrase "hemp" in the Scopus database in the last 10 years, broken down into individual fields (as of 5 December 2021).

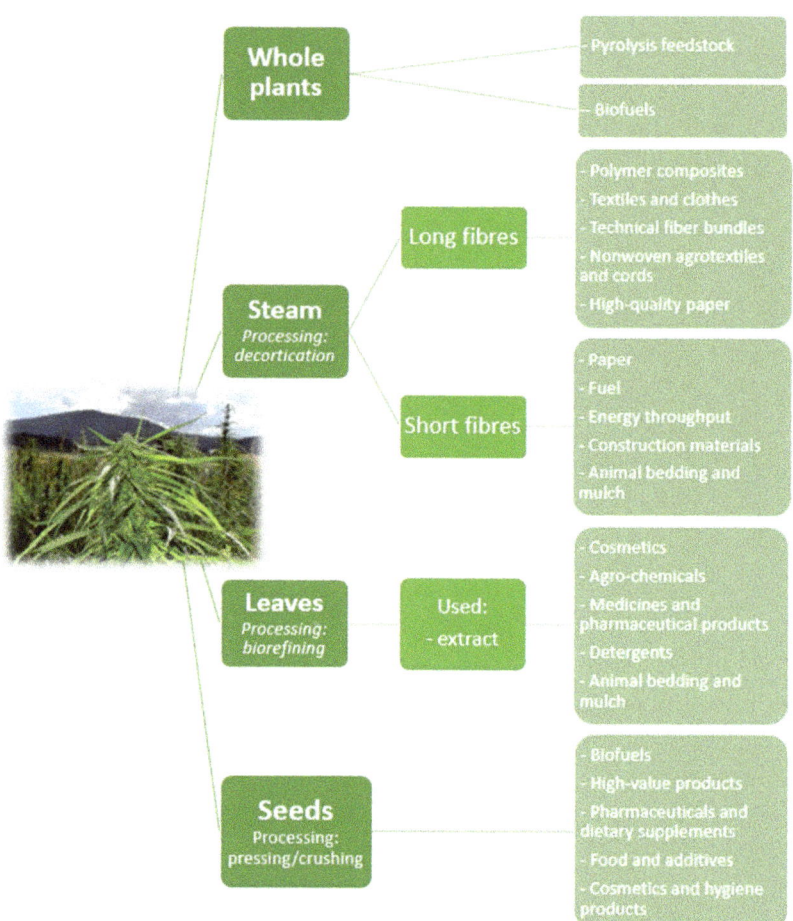

Figure 6. The use of hemp in various industries due to the division into the parts of the plant used.

3. Hemp and Derivatives in the Polymer Industry

Polymers are the most important construction material of the 20th and 21st centuries in many industries around the world. However, their cheapness and durability caused enormous havoc on the natural environment due to the deposition of huge heaps of rubbish on land and huge islands of artificial plastic seen even from space in the ratings. For this reason, the industry of biodegradable polymers and creative composites made of these polymers with natural additives has been rapidly developing in recent years. As can be seen from the above-mentioned examples of applications and the specified chemical compositions and physicochemical properties of hemp and its derivatives, these plants are an ideal candidate for a huge share in this industry sector.

3.1. Thermoplastics

Thermoplastics are currently the largest group of manufactured polymers. They have such advantages as the possibility of material, raw material or energy recycling. Over the past decades, new types of thermoplastics were developed that are made from renewable green sources and are biodegradable. Unfortunately, many of these materials do not have the required physical or chemical properties as well as conventional fossil-based thermoplastics [102]. Such conventional polymers are polypropylene (PP) and polyethylene

(PE). They are characterized by durability, stiffness, lightness, good barrier properties; they are satisfactorily chemically inert, easily processable and, above all, cheap. These features make them one of the most commonly used polymers in everyday life for the production of packaging, everyday equipment and construction materials. However, these are non-biodegradable polymers; in order to increase the potential of products made of them for biodegradation and to reduce their mass proportion in the product, natural fillers are also introduced to strengthen the mechanical properties [103]. Hemp fibers can be an excellent filler in this case. This is emphasized in their work by Sullins et al. [104–106]. The addition of modified hemp fibers for modified hemp fibers to the PP composite increased its flexural and tensile properties. The composite with up to 30% of fibers showed better properties than pure polymer or content of 15%. Hemp fibers also showed very good interfacial interactions with this polymer matrix. On the other hand, Etaati et al., in their research, indicated changes in the analysis of dynamic mechanical properties. They investigated changes in these properties of polypropylene composites with short hemp fibers at temperatures from 25 to 150 °C. They indicate in their article that the addition of fibers strengthens the composite when working at higher temperatures, above alpha relaxation. They also emphasize that when more modifications are used in the form of a coupling agent, it is also required, which confirms the earlier statement about the necessity to modify the fibers and the composites formed with them [107,108]. Researcher Oliveira et al. instead dealt with PE-based composites. In her research, she showed that the treatment of hemp fibers with alkali in order to flush out lignin and hemicellulose improved the dispersion of these fibers in the matrix and improved resistance to thermal degradation. The addition of 5% modified hemp fibers and the use of a bonding agent also improved the processability during rotational molding [109]. The introduction of hemp allows lowering or eliminating voids altogether and creates stronger connections at the fiber-polymer matrix boundary, thanks to which it does not change the strength, while the modulus of elasticity increases material. Additionally, such composites characterize a more hydrophilic surface that can affect their faster degradation and thus facilitate the recycling of polyethylene-based products [110]. The most frequently used biodegradable polymer composites with hemp fibers are those based on polymers of aliphatic polyesters such as polylactide (PLA) or polyhydroxybutyrate (PHB) [99,111–113]. In the case of these polymers, the addition of hemp fibers accelerates hydrolytic degradation, which is worth mentioning that such material decomposes faster in the environment into simple compounds such as water, carbon dioxide and biomass. According to Mazzanti et al., even such a small addition of 3% wt. causes this effect [114]. As for PP and PE, also for PLA, the addition of hemp fibers significantly enhances the mechanical and thermal properties of such refined compositions. Differential Scanning Calorimetry (DSC) studies indicated that the addition of fibers did not significantly affect the glass transition and melting temperatures [115]. At this point, it is also worth emphasizing that in the case of using such a filler as hemp fibers, it depends on the orientation of the material. Arrangement parallel to the force of the fibers leads to a strong effect of transferring stresses inside the material; therefore, an important stage in the preparation of polymer composites with them is their proper orientation through the use of appropriate unit operations in the process, such as rolling, extruding, injection or calendaring. This approach allows the best possible use of the fibers as an active filler in polymer composites [116,117]. Apart from fibers, hemp extracts are also other additives to thermoplastic polymers. Thus far, this is a supplement with antioxidant properties. In his work, Plota et al. showed the thermally stabilizing effect of CBD on polylactide and Topas. In this case, the indicative effect of this additive is also indicated, as the color of refined samples changed with the aging processes. This is one of the few works in this direction worth exploring [118]. An equally interesting approach was shown by a team of researchers led by Andriotis et al. They created water-soluble fibers produced by the electrospun method using polyvinyl (pyrrolidine) (PVP) and Eudragit L-100, in which CBD and CBG were used as active substances with therapeutic effect [119]. Thermoplastic composites using hemp materials, as you can see, are in common use and find more and more possible applications in everyday life as well as in construction materials. It is worth continuing research on this type of material in order to increase their share in green polymer composites.

3.2. Elastomers

Rubbers are another important material used by humans. Due to cross-linking, unfortunately, they are not recyclable, and their natural decomposition takes hundreds of years. For this reason, it is worth delving into and developing intensively more environmentally friendly rubber compounds. In this case, hemp materials can also help us. Previous studies indicate the use of hemp derivatives in mixtures based on natural rubber (NR) [120]. From the results obtained by Moonart et al., it follows that in order to obtain good adhesion between the fiber and the polymer matrix, they must be treated. In this case, it was proposed to prepare by treating the hemp fiber with alkali and then using a $KMnO_4$ solution and silane. Such modification resulted in an increase in the tensile strength of the fibers and a better interfacial connection of the materials [121,122]. Another study investigated NR vulcanizates with hemp fibers cross-linked with benzoyl peroxide. This arrangement exhibited increased hardness, modulus at 100% elongation, tear strength, tensile strength and elongation at break. This effect depended on the degree of fiber filling of the composite. Hemp fibers can be, in this case, a good replacement for synthetic or steel fibers due to their cheapness, biodegradability and good weight-to-strength ratio [123].

3.3. Duroplasts

Duroplasts are the last group of the polymeric materials we discuss with conspicuous additives [124]. One of the most commonly used thermosets is unsaturated epoxy resins [41,53,125–130]. It is this polymer that is one of the most modified with hemp derivatives. As a result of the addition of fibers, the tensile, compressive and bending strengths increased. It is logical because the fibers perfectly transfer stresses in materials in which they are active fillers. Modification by copolymerization involving the grafting of acrylonitrile on the surface of the fibers also allowed for a minimal improvement in thermal stability than in the case of unmodified hemp fibers [108,131,132]. In his article, Scarponi compared the use of glass and hemp fibers [133]. They showed that hemp/epoxy composites could compete with glass/epoxy composites. The covers for ultra-light airplanes produced for the purpose of the tests showed very good properties and, at the same time, greater environmental friendliness [134]. As indicated by previous research and theory, natural fibers require some modification to improve compatibility with the polymer matrix. In the case of combinations of hemp fibers with unsaturated polyester resin (UPE), esterification of the fibers is a good example. Such an operation allows to significantly improve the interfacial adhesion, as a result of which the chemical resistance but also the mechanical and thermal resistance of the obtained composites is improved. Another popular polymer matrix of hemp composites is polyurethanes. Members et al., in their publication, showed that the addition of hemp derivatives influenced such properties of polyurethane foams as morphology, mechanical, thermal and insulating properties. They showed in their work that impregnation with sunflower oil and tung oil resulted in improved thermal stability and flame retardancy of PUR foams. It reduced hydrophilicity by limiting water absorption [135]. Materials such as PUR are used in the construction industry to improve the thermal insulation properties of buildings. The introduction of the hemp filler brings us closer to a more sustainable development of this industry sector. Hemp fibers added to polyurethanes in the amount of 15% by weight will increase the tensile and bending modulus. Such an addition makes the product more environmentally friendly and reduces its cost [136,137].

4. Conclusions

This overview article shows how important cannabis has been in human history so far and what it may be in the future. The contained data illustrating the richness of the chemical composition of these plants indicates the possibility of the very wide use of active compounds in medicine, pharmacy, cosmetology and the food industry. Interest in these plants is already growing in these sectors. The use of hemp and its derivatives in the new materials sector also shows promise for the development of environmentally friendly polymer products. The polymer industry, contributing to each of the main sectors of the economy, can draw from this green source of many active phytosubstances, oils and fibers. The pro-ecological aspect of hemp cultivation, low soil and water requirements and the possibility of processing and using 100% of plants with cheap production allow us to be optimistic about the development of this production department and related science activities. There is also a lack of basic knowledge in the use of other cannabis derivatives in the polymer industry. The section dealing with elastomers is the poorest in the literature on this subject. It is a signal for researchers, technologists and entrepreneurs with a possible niche to research and use this valuable source of substances, not only fibers as a strengthening additive but also extracts and waxes as an antioxidant, antimicrobial substances, plasticizers and aging time indicators. In our opinion, scientists from around the world should intensify research on environmentally friendly materials such as hemp, which is the material of the future.

Author Contributions: Conceptualization, formal analysis, data analysis, investigation, methodology, review and editing, A.M.; data analysis, investigation, methodology and writing, K.T.; description of results, methodology, data analysis, investigation and writing K.T.; supervisor A.M. All authors have read and agreed to the published version of the manuscript.

Funding: No external funding provided.

Institutional Review Board Statement: Not applicable.

Informed Consent Statement: Not applicable.

Data Availability Statement: No data are available while the first author was a doctoral candidate in the Interdisciplinary Doctoral School at the Lodz University of Technology, Poland.

Acknowledgments: This work was completed while the second author was a doctoral candidate in the Interdisciplinary Doctoral School at the Lodz University of Technology, Poland.

Conflicts of Interest: The authors declare no conflict of interest.

Abbreviations

AAPPH (ORAC)	Oxygen Radical Absorbance Capacity
ABTS	2,2-azinobis-(3-ethylbenzothiazoline-6-sulfonate)
CBC	Cannabichromene
CBD	Cannabidiol
CBDA	Cannabinoid acid
CBG	Cannabigerol
CBN	Cannabinol
C-NMR	Carbon-13 (C13) nuclear magnetic resonance
DPPH	2,2-diphenyl-1-picrylhydrazyl.
FT-IR	Infrared Spectroscopy with Fourier Transformation
NaOH	Sodium hydroxide
THC	Delta-9-tetrahydrocannabinol
THCV	Tetrahydrocannabivarin

References

1. Lambert, S.; Sinclair, C.; Boxall, A. Occurrence, Degradation, and Effect of Polymer-Based Materials in the Environment. In *Reviews of Environmental Contamination and Toxicology*; Whitacre, D.M., Ed.; Springer International Publishing: Cham, Switzerland, 2014; Volume 227, pp. 1–53. ISBN 978-3-319-01327-5.
2. Leja, K.; Lewandowicz, G. Polymer biodegradation and biodegradable polymers—A review. *Pol. J. Environ. Stud.* **2010**, *19*, 255–266.

3. Dominguiano, O.; Chandra, R.; Rustgi, R. Related papers Polyet hylene and biodegradable mulches for agricult ural applicat ions: A review Nguyen Van Biological degradat ion of plast ics: A comprehensive review biodegradable polymers. *Prog. Polym. Sci.* **1998**, *23*, 1273–1335.
4. Luckachan, G.E.; Pillai, C.K.S. Biodegradable Polymers—A Review on Recent Trends and Emerging Perspectives. *J. Polym. Environ.* **2011**, *19*, 637–676. [CrossRef]
5. Iwata, T. Biodegradable and Bio-Based Polymers: Future Prospects of Eco-Friendly Plastics. *Angew. Chem. Int. Ed.* **2015**, *54*, 3210–3215. [CrossRef] [PubMed]
6. Gallo-Molina, A.C.; Castro-Vargas, H.I.; Garzón-Méndez, W.F.; Martínez Ramírez, J.A.; Rivera Monroy, Z.J.; King, J.W.; Parada-Alfonso, F. Extraction, isolation and purification of tetrahydrocannabinol from the Cannabis sativa L. plant using supercritical fluid extraction and solid phase extraction. *J. Supercrit. Fluids* **2019**, *146*, 208–216. [CrossRef]
7. Long, T.; Wagner, M.; Demske, D.; Leipe, C.; Tarasov, P.E. Cannabis in Eurasia: Origin of human use and Bronze Age transcontinental connections. *Veg. Hist. Archaeobot.* **2017**, *26*, 245–258. [CrossRef]
8. Fike, J. The history of hemp. In *Industrial Hemp as a Modern Commodity Crop*; Wiley: Hoboken, NJ, USA, 2019; pp. 1–25. [CrossRef]
9. Fike, J. Industrial Hemp: Renewed Opportunities for an Ancient Crop. *CRC Crit. Rev. Plant Sci.* **2016**, *35*, 406–424. [CrossRef]
10. Shibata, T. Cellulose and its derivatives in medical use. In *Renewable Resources for Functional Polymers and Biomaterials*; RCS Royal Chemistry Society: London, UK, 2011; pp. 48–87. [CrossRef]
11. Szymański, Ł.; Grabowska, B.; Kurleto, Ż.; Kaczmarska, K. Celuloza i jej pochodne—Zastosowanie w przemyśle. *Arch. Foundry Eng.* **2015**, *15*, 129–132.
12. Samyn, P.; Nordell, P. Photosynthesis—Gigantic factory of cellulose. *Przegląd Pap.* **2015**, *48*, 6455–6498. [CrossRef]
13. Poletto, M.; Ornaghi Júnior, H.L.; Zattera, A.J. Native cellulose: Structure, characterization and thermal properties. *Materials* **2014**, *7*, 6105–6119. [CrossRef] [PubMed]
14. Saito, T.; Isogai, A. TEMPO-mediated oxidation of native cellulose. The effect of oxidation conditions on chemical and crystal structures of the water-insoluble fractions. *Biomacromolecules* **2004**, *5*, 1983–1989. [CrossRef] [PubMed]
15. Keshk, S.M.A.S.; Gouda, M. Natural biodegradable medical polymers: Cellulose. In *Science and Principles of Biodegradable and Bioresorbable Medical Polymers*; Woodhead Publishing: Sawston, UK, 2017; pp. 279–294. [CrossRef]
16. Penczek, S.; Pretula, J.; Lewiński, P. Polimery z odnawialnych surowców, polimery biodegradowalne. *Polimery* **2013**, *58*, 835–846. [CrossRef]
17. Kaczmarek, H.; Bajer, K. Badanie przebiegu biodegradacji kompozytów poli(chlorek winylu)/celuloza. *Polimery* **2008**, *53*, 631–638. [CrossRef]
18. Kubiak, K.; Kalinowska, H.; Peplińska, M.; Bielecki, S. Celuloza bakteryjna jako nanobiomateriał. *Postępy Biol. Komórki* **2009**, *36*, 85–98.
19. Wasilewska, A.W.; Pietruszka, D.B.L. Materiały naturalne w ekobudownictwie. *Przegląd Bud.* **2017**, *88*, 50–53.
20. Antczak, T.; Wietecha, J.; Kazimierczak, J.; Bloda, A.; Ciechańska, D. Nanowłókna celulozowe wytwarzane z biomasy roślinnej. *Chemik* **2014**, *68*, 755–760.
21. Lee, Y.-H.; Fan, L.T. Properties and mode of action of cellulase. *Adv. Biochem. Eng.* **1980**, *17*, 101–129.
22. Seidl, P.R.; Freire, E.; Borschiver, S. Non-fuel applications of sugars in Brazil. In *Biomass Sugars for Non-Fuel Applications*; Royal Society of Chemistry: London, UK, 2016; pp. 228–257. [CrossRef]
23. Henriksson, G.; Lennholm, H. Cellulose and carbohydrate chemistry. *Wood Chem. Wood Biotechnol.* **2009**, 71–100. [CrossRef]
24. Nishino, T. Natural fibre sources. *Green Compos.* **2005**, 49–80. [CrossRef]
25. Bledzki, A.K.; Urbaniak, M.; Jaszkiewicz, A.; Feldmann, M. Cellulose fibres as an alternative for glass fibres in polymer composites. *Polimery* **2014**, *59*, 372–382. [CrossRef]
26. Gardner, K.H.; Blackwell, J. The hydrogen bonding in native cellulose. *BBA Gen. Subj.* **1974**, *343*, 232–237. [CrossRef]
27. Cousins, S.K.; Brown, R.M. Cellulose I microfibril assembly: Computational molecular mechanics energy analysis favours bonding by van der Waals forces as the initial step in crystallization. *Polymer* **1995**, *36*, 3885–3888. [CrossRef]
28. Tanasă, F.; Zănoagă, M.; Teacă, C.A.; Nechifor, M.; Shahzad, A. Modified hemp fibers intended for fiber-reinforced polymer composites used in structural applications—A review. I. Methods of modification. *Polym. Compos.* **2020**, *41*, 5–31. [CrossRef]
29. Sołowski, G. *Wybrane Zagadnienia z Zakresu Ochrony Środowiska i Energii Odnawialnej*; Wydawnictwo Uniwersytetu Przyrodniczego w Lublinie: Lublin, Poland, 2016; ISBN 9788365598165.
30. Heinze, T.; Liebert, T. Unconventional methods in cellulose functionalization. *Prog. Polym. Sci.* **2001**, *26*, 1689–1762. [CrossRef]
31. Maréchal, Y.; Chanzy, H. The hydrogen bond network in I(β) cellulose as observed by infrared spectrometry. *J. Mol. Struct.* **2000**, *523*, 183–196. [CrossRef]
32. Chundawat, S.P.S.; Bellesia, G.; Uppugundla, N.; Da Costa Sousa, L.; Gao, D.; Cheh, A.M.; Agarwal, U.P.; Bianchetti, C.M.; Phillips, G.N.; Langan, P.; et al. Restructuring the crystalline cellulose hydrogen bond network enhances its depolymerization rate. *J. Am. Chem. Soc.* **2011**, *133*, 11163–11174. [CrossRef]
33. Kondo, T.; Sawatari, C. A Fourier transform infra-red spectroscopic analysis of the character of hydrogen bonds in amorphous cellulose. *Polymer (Guildf.)* **1996**, *37*, 393–399. [CrossRef]
34. French, A.D.; Miller, D.P.; Aabloo, A. Miniature crystal models of cellulose polymorphs and other carbohydrates. *Int. J. Biol. Macromol.* **1993**, *15*, 30–36. [CrossRef]

35. Marrinan, H.J.; Mann, J. A study by infra-red spectroscopy of hydrogen bonding in cellulose. *J. Appl. Chem.* **2007**, *4*, 204–211. [CrossRef]
36. Dri, F.L.; Hector, L.G., Jr.; Moon, R.J.; Zavattieri, P.D. Anisotropy of the elastic properties of crystalline cellulose Iβ from first principles density functional theory with Van der Waals interactions. *Cellulose* **2013**, *20*, 2703–2718. [CrossRef]
37. Rees, D.A.; Skerrett, R.J. Conformational analysis of cellobiose, cellulose, and xylan. *Carbohydr. Res.* **1968**, *7*, 334–348. [CrossRef]
38. Pizzi, A.; Eaton, N. The Structure of Cellulose by Conformational Analysis. Part 4. Crystalline Cellulose II. *J. Macromol. Sci. -Chem.* **1984**, *24*, 901–918. [CrossRef]
39. Kroon-Batenburg, L.M.; Kroon, J. The crystal and molecular structures of cellulose I and II. *Glycoconj. J.* **1997**, *14*, 677–690. [CrossRef]
40. Manaia, J.P.; Manaia, A.T.; Rodrges, L. Industrial Hemp Fibres: An Overview. *Fibers* **2019**, *7*, 106. [CrossRef]
41. Väisänen, T.; Batello, P.; Lappalainen, R.; Tomppo, L. Modification of hemp fibers (Cannabis Sativa L.) for composite applications. *Ind. Crops Prod.* **2018**, *111*, 422–429. [CrossRef]
42. Marrot, L.; Lefeuvre, A.; Pontoire, B.; Bourmaud, A.; Baley, C. Analysis of the hemp fiber mechanical properties and their scattering (Fedora 17). *Ind. Crops Prod.* **2013**, *51*, 317–327. [CrossRef]
43. Müssig, J.; Amaducci, S.; Bourmaud, A.; Beaugrand, J.; Shah, D.U. Transdisciplinary top-down review of hemp fibre composites: From an advanced product design to crop variety selection. *Compos. Part C Open Access* **2020**, *2*, 100010. [CrossRef]
44. Duval, A.; Bourmaud, A.; Augier, L.; Baley, C. Influence of the sampling area of the stem on the mechanical properties of hemp fibers. *Mater. Lett.* **2011**, *65*, 797–800. [CrossRef]
45. Kaczmar, J.W.; Pach, J.; Burgstaller, C. The chemically treated hemp fibres to reinforce polymers. *Polimery* **2011**, *56*, 817–822. [CrossRef]
46. Yang, H.; Yan, R.; Chen, H.; Lee, D.H.; Zheng, C. Characteristics of hemicellulose, cellulose and lignin pyrolysis. *Fuel* **2007**, *86*, 1781–1788. [CrossRef]
47. Patel, H.A.; Somani, R.S.; Bajaj, H.C.; Jasra, R.V. Preparation and characterization of phosphonium montmorillonite with enhanced thermal stability. *Appl. Clay Sci.* **2007**, *35*, 194–200. [CrossRef]
48. Salmén, L.; Bergström, E. Cellulose structural arrangement in relation to spectral changes in tensile loading FTIR. *Cellulose* **2009**, *16*, 975–982. [CrossRef]
49. Ołdak, D.; Kaczmarek, H.; Buffeteau, T.; Sourisseau, C. Photo- and bio-degradation processes in polyethylene, cellulose and their blends studied by ATR-FTIR and raman spectroscopies. *J. Mater. Sci.* **2005**, *40*, 4189–4198. [CrossRef]
50. Chwanninger, M.; Rodrigues, J.C.; Pereira, H.; Hinterstoisser, B. Effects of short-time vibratory ball milling on the shape of FT-IR spectra of wood and cellulose. *Vib. Spectrosc.* **2004**, *36*, 23–40. [CrossRef]
51. Chen, C.; Luo, J.; Qin, W.; Tong, Z. Elemental analysis, chemical composition, cellulose crystallinity, and FT-IR spectra of Toona sinensis wood. *Mon. Für Chem. -Chem. Mon.* **2014**, *145*, 175–185. [CrossRef]
52. Cichosz, S.; Masek, A. Cellulose fibers hydrophobization via a hybrid chemical modification. *Polymers* **2019**, *11*, 1174. [CrossRef]
53. Pandey, K.K.; Pitman, A.J. FTIR studies of the changes in wood chemistry following decay by brown-rot and white-rot fungi. *Int. Biodeterior. Biodegrad.* **2003**, *52*, 151–160. [CrossRef]
54. Gulmine, J.V.; Janissek, P.R.; Heise, H.M.; Akcelrud, L. Polyethylene characterization by FTIR. *Polym. Test.* **2002**, *21*, 557–563. [CrossRef]
55. Owen, N.L.; Thomas, D.W. Infrared studies of "hard" and "soft" woods. *Appl. Spectrosc.* **1989**, *43*, 451–455. [CrossRef]
56. Sarkanen, K.; Ludwig, C. *Liguins. Occurrence, Formation, Structure, and Reactions*; Wiley-Interscience: New York, NY, USA, 1971.
57. Müller, U.; Rätzsch, M.; Schwanninger, M.; Steiner, M.; Zöbl, H. Yellowing and IR-changes of spruce wood as result of UV-irradiation. *J. Photochem. Photobiol. B Biol.* **2003**, *69*, 97–105. [CrossRef]
58. Oh, S.Y.; Yoo, D.I.; Shin, Y.; Seo, G. FTIR analysis of cellulose treated with sodium hydroxide and carbon dioxide. *Carbohydr. Res.* **2005**, *340*, 417–428. [CrossRef]
59. Lebo, S.E.; Lonsky, W.; McDonough, T.; Medvecz, P. The Occurrence and Light Induced Formation of Ortho Quinonoid Lignin Structures in White Spruce Refiner Mechanical Pulp. In Proceedings of the International Pulp Bleaching Conference in Orlando, Orlando, FL, USA, 5–9 June 1988.
60. Anderson, E.L.; Owen, N.L.; Feist, W.C.; Pawlak, Z. Infrared Studies of Wood Weathering. Part I: Softwoods. *Appl. Spectrosc.* **1991**, *45*, 641–647. [CrossRef]
61. Morán, J.I.; Alvarez, V.A.; Cyras, V.P.; Vázquez, A. Extraction of cellulose and preparation of nanocellulose from sisal fibers. *Cellulose* **2008**, *15*, 149–159. [CrossRef]
62. Łojewska, J.; Miśkowiec, P.; Łojewski, T.; Proniewicz, L.M. Cellulose oxidative and hydrolytic degradation: In situ FTIR approach. *Polym. Degrad. Stab.* **2005**, *88*, 512–520. [CrossRef]
63. Ramirez, C.L.; Fanovich, M.A.; Churio, M.S. *Cannabinoids: Extraction Methods, Analysis, and Physicochemical Characterization*, 1st ed.; Elsevier B.V.: Amsterdam, The Netherlands, 2018; Volume 61, ISBN 9780444641830.
64. Horanin, A.; Bryndal, I. Hemp—Active Ingredients, Medicinal Properties and Using. *Pr. Nauk. Uniw. Ekon. we Wrocławiu* **2017**, 76–84. [CrossRef]
65. Ahmed, S.A.; Ross, S.A.; Slade, D.; Radwan, M.M.; Zulfiqar, F.; ElSohly, M.A. Cannabinoid ester constituents from high-potency Cannabis sativa. *J. Nat. Prod.* **2008**, *71*, 536–542. [CrossRef]

66. Radwan, M.M.; Chandra, S.; Gul, S.; Elsohly, M.A. Cannabinoids, phenolics, terpenes and alkaloids of cannabis. *Molecules* **2021**, *26*, 2774. [CrossRef] [PubMed]
67. Datta, S.; Ramamurthy, P.C.; Anand, U.; Singh, S.; Singh, A.; Dhanjal, D.S.; Dhaka, V.; Kumar, S.; Kapoor, D.; Nandy, S.; et al. Wonder or evil?: Multifaceted health hazards and health benefits of Cannabis sativa and its phytochemicals. *Saudi J. Biol. Sci.* **2021**, *28*, 7290–7313. [CrossRef] [PubMed]
68. Filipiuc, L.E.; Ababei, D.C.; Alexa-Stratulat, T.; Pricope, C.V.; Bild, V.; Stefanescu, R.; Stanciu, G.D.; Tamba, B.-I. Major Phytocannabinoids and Their Related Compounds: Should We Only Search for Drugs That Act on Cannabinoid Receptors? *Pharmaceutics* **2021**, *13*, 1823. [CrossRef] [PubMed]
69. Filer, C.N. Acidic Cannabinoid Decarboxylation. *Cannabis Cannabinoid Res.* **2021**. [CrossRef] [PubMed]
70. Attard, T.M.; Bainier, C.; Reinaud, M.; Lanot, A.; McQueen-Mason, S.J.; Hunt, A.J. Utilisation of supercritical fluids for the effective extraction of waxes and Cannabidiol (CBD) from hemp wastes. *Ind. Crops Prod.* **2018**, *112*, 38–46. [CrossRef]
71. Rock, E.M.; Parker, L.A. Constituents of Cannabis Sativa. *Adv. Exp. Med. Biol.* **2021**, *1264*, 1–13. [CrossRef]
72. Atalay, S.; Jarocka-karpowicz, I.; Skrzydlewskas, E. Antioxidative and anti-inflammatory properties of cannabidiol. *Antioxidants* **2020**, *9*, 21. [CrossRef] [PubMed]
73. Farinon, B.; Molinari, R.; Costantini, L.; Merendino, N. The seed of industrial hemp (Cannabis sativa l.): Nutritional quality and potential functionality for human health and nutrition. *Nutrients* **2020**, *12*, 1935. [CrossRef]
74. Lewis, M.M.; Yang, Y.; Wasilewski, E.; Clarke, H.A.; Kotra, L.P. Chemical Profiling of Medical Cannabis Extracts. *ACS Omega* **2017**, *2*, 6091–6103. [CrossRef] [PubMed]
75. Radwan, M.M.; ElSohly, M.A.; El-Alfy, A.T.; Ahmed, S.A.; Slade, D.; Husni, A.S.; Manly, S.P.; Wilson, L.; Seale, S.; Cutler, S.J.; et al. Isolation and Pharmacological Evaluation of Minor Cannabinoids from High-Potency Cannabis sativa. *J. Nat. Prod.* **2015**, *78*, 1271–1276. [CrossRef] [PubMed]
76. Radwan, M.M.; ElSohly, M.A.; Slade, D.; Ahmed, S.A.; Khan, I.A.; Ross, S.A. Biologically active cannabinoids from high-potency Cannabis sativa. *J. Nat. Prod.* **2009**, *72*, 906–911. [CrossRef]
77. Pugazhendhi, A.; Suganthy, N.; Chau, T.P.; Sharma, A.; Unpaprom, Y.; Ramaraj, R.; Karuppusamy, I.; Brindhadevi, K. Cannabinoids as anticancer and neuroprotective drugs: Structural insights and pharmacological interactions—A review. *Process Biochem.* **2021**, *111*, 9–31. [CrossRef]
78. Apostol, L. Studies on using hemp seed as a functional ingredient in the production of functional food products. *J. EcoAgri Tour.* **2017**, *13*, 12–17.
79. Leyva-Gutierrez, F.M.A.; Munafo, J.P.; Wang, T. Characterization of By-Products from Commercial Cannabidiol Production. *J. Agric. Food Chem.* **2020**, *68*, 7648–7659. [CrossRef] [PubMed]
80. Żuk-Gołaszewska, K.; Gołaszewski, J. Cannabis sativa L.—Cultivation and quality of raw material. *J. Elem.* **2018**, *23*, 971–984. [CrossRef]
81. Kraszkiewicz, A.; Kachel, M.; Parafiniuk, S.; Zając, G.; Niedziółka, I.; Sprawka, M. Assessment of the Possibility of Using Hemp Biomass (Cannabis Sativa L.) for Energy Purposes. *Appl. Sci.* **2019**, *9*, 4437. [CrossRef]
82. Kaniewski, R.; Pniewska, I.; Kubacki, A.; Strzelczyk, M.; Chudy, M.; Oleszak, G. Konopie siewne (Cannabis sativa L.)—wartościowa roślina użytkowa i lecznicza. *Postępy Fitoter.* **2017**, *18*, 139–144. [CrossRef]
83. Crini, G.; Lichtfouse, E.; Chanet, G.; Morin-Crini, N. Applications of hemp in textiles, paper industry, insulation and building materials, horticulture, animal nutrition, food and beverages, nutraceuticals, cosmetics and hygiene, medicine, agrochemistry, energy production and environment: A review. *Environ. Chem. Lett.* **2020**, *18*, 1451–1476. [CrossRef]
84. Shamreen Amode, N.; Jeetah, P. Paper Production from Mauritian Hemp Fibres. *Waste Biomass Valoriz.* **2021**, *12*, 1781–1802. [CrossRef]
85. van der Werf, H.M.G.; Harsveld van der Veen, J.E.; Bouma, A.T.M.; ten Cate, M. Quality of hemp (*Cannabis sativa* L.) stems as a raw material for paper. *Ind. Crops Prod.* **1994**, *2*, 219–227. [CrossRef]
86. Stevulova, N.; Kidalova, L.; Junak, J.; Cigasova, J.; Terpakova, E. Effect of hemp shive sizes on mechanical properties of lightweight fibrous composites. *Procedia Eng.* **2012**, *42*, 496–500. [CrossRef]
87. Hussain, A.; Calabria-Holley, J.; Lawrence, M.; Ansell, M.P.; Jiang, Y.; Schorr, D.; Blanchet, P. Development of novel building composites based on hemp and multi-functional silica matrix. *Compos. Part B Eng.* **2019**, *156*, 266–273. [CrossRef]
88. Seng, B.; Magniont, C.; Lorente, S. Characterization of a precast hemp concrete. Part I: Physical and thermal properties. *J. Build. Eng.* **2019**, *24*, 100540. [CrossRef]
89. Barnat-Hunek, D.; Smarzewski, P.; Brzyski, P. Properties of Hemp–Flax Composites for Use in the Building Industry. *J. Nat. Fibers* **2017**, *14*, 410–425. [CrossRef]
90. Lekavicius, V.; Shipkovs, P.; Ivanovs, S.; Rucins, A. Thermo-insulation properties of hemp-based products. *Latv. J. Phys. Tech. Sci.* **2015**, *52*, 38–51. [CrossRef]
91. La Rosa, A.D.; Cozzo, G.; Latteri, A.; Mancini, G.; Recca, A.; Cicala, G. A comparative life cycle assessment of a composite component for automotive. *Chem. Eng. Trans.* **2013**, *32*, 1723–1728. [CrossRef]
92. Murugu Nachippan, N.; Alphonse, M.; Bupesh Raja, V.K.; Shasidhar, S.; Varun Teja, G.; Harinath Reddy, R. Experimental investigation of hemp fiber hybrid composite material for automotive application. *Mater. Today Proc.* **2021**, *44*, 3666–3672. [CrossRef]

93. Mastura, M.T.; Sapuan, S.M.; Mansor, M.R.; Nuraini, A.A. Materials selection of thermoplastic matrices for 'green' natural fibre composites for automotive anti-roll bar with particular emphasis on the environment. *Int. J. Precis. Eng. Manuf. Green Technol.* **2018**, *5*, 111–119. [CrossRef]
94. Zheljazkov, V.D.; Sikora, V.; Dincheva, I.; Kačániová, M.; Astatkie, T.; Semerdjieva, I.B.; Latkovic, D. Industrial, CBD, and Wild Hemp: How Different Are Their Essential Oil Profile and Antimicrobial Activity? *Molecules* **2020**, *25*, 4631. [CrossRef] [PubMed]
95. Casiraghi, A.; Roda, G.; Casagni, E.; Cristina, C.; Musazzi, U.M.; Franzè, S.; Rocco, P.; Giuliani, C.; Fico, G.; Minghetti, P.; et al. Extraction Method and Analysis of Cannabinoids in Cannabis Olive Oil Preparations. *Planta Med.* **2018**, *84*, 242–249. [CrossRef]
96. Blaskovich, M.A.T.; Kavanagh, A.M.; Elliott, A.G.; Zhang, B.; Ramu, S.; Amado, M.; Lowe, G.J.; Hinton, A.O.; Pham, D.M.T.; Zuegg, J.; et al. The antimicrobial potential of cannabidiol. *Commun. Biol.* **2021**, *4*, 7. [CrossRef]
97. Chi Nguyen, L.; Yang, D.; Nicolaescu, V.; Best, T.J.; Gula, H.; Saxena, D.; Gabbard, J.D.; Chen, S.-N.; Ohtsuki, T.; Brent Friesen, J.; et al. Cannabidiol inhibits SARS-CoV-2 replication through induction of the host ER stress and innate immune responses. *Sci. Adv.* **2022**, *8*, 6110. [CrossRef] [PubMed]
98. Raj, V.; Park, J.G.; Cho, K.H.; Choi, P.; Kim, T.; Ham, J.; Lee, J. Assessment of antiviral potencies of cannabinoids against SARS-CoV-2 using computational and in vitro approaches. *Int. J. Biol. Macromol.* **2021**, *168*, 474–485. [CrossRef]
99. Pappu, A.; Pickering, K.L.; Thakur, V.K. Manufacturing and characterization of sustainable hybrid composites using sisal and hemp fibres as reinforcement of poly (lactic acid) via injection moulding. *Ind. Crops Prod.* **2019**, *137*, 260–269. [CrossRef]
100. Rovetto, L.J.; Aieta, N.V. Supercritical carbon dioxide extraction of cannabinoids from *Cannabis sativa* L. *J. Supercrit. Fluids* **2017**, *129*, 16–27. [CrossRef]
101. Mańkowski, J.; Kubacki, A.; Kołodziej, J.; Pniewska, I.; Teren, Z.; Konin, K.W.B. *Rekultywacja Terenów Zdegradowanych w Wyniku Działania Kopalni Odkrywkowych*; Biuletyn Informacyjny Polskiej Izby Lnu i Konopi: Poznań, Poland, 2013.
102. Terzopoulou, Z.N.; Papageorgiou, G.Z.; Papadopoulou, E.; Athanassiadou, E.; Alexopoulou, E.; Bikiaris, D.N. Green composites prepared from aliphatic polyesters and bast fibers. *Ind. Crops Prod.* **2015**, *68*, 60–79. [CrossRef]
103. Sergi, C.; Tirillò, J.; Seghini, M.C.; Sarasini, F.; Fiore, V.; Scalici, T. Durability of basalt/hemp hybrid thermoplastic composites. *Polymers* **2019**, *11*, 603. [CrossRef]
104. Sullins, T.; Pillay, S.; Komus, A.; Ning, H. Hemp fiber reinforced polypropylene composites: The effects of material treatments. *Compos. Part B Eng.* **2017**, *114*, 15–22. [CrossRef]
105. Panaitescu, D.M.; Fierascu, R.C.; Gabor, A.R.; Nicolae, C.A. Effect of hemp fiber length on the mechanical and thermal properties of polypropylene/SEBS/hemp fiber composites. *J. Mater. Res. Technol.* **2020**, *9*, 10768–10781. [CrossRef]
106. Yallew, T.B.; Kumar, P.; Singh, I. Sliding behaviour of woven industrial hemp fabric reinforced thermoplastic polymer composites. *Int. J. Plast. Technol.* **2015**, *19*, 347–362. [CrossRef]
107. Etaati, A.; Pather, S.; Fang, Z.; Wang, H. The study of fibre/matrix bond strength in short hemp polypropylene composites from dynamic mechanical analysis. *Compos. Part B Eng.* **2014**, *62*, 19–28. [CrossRef]
108. Shahzad, A. Hemp fiber and its composites—A review. *J. Compos. Mater.* **2012**, *46*, 973–986. [CrossRef]
109. Oliveira, M.A.S.; Pickering, K.L.; Sunny, T.; Lin, R.J.T. Treatment of hemp fibres for use in rotational moulding. *J. Polym. Res.* **2021**, *28*, 53. [CrossRef]
110. Ziąbka, M.; Szaraniec, B. Polymeric composites with natural fiber additives. *Composites* **2010**, *10*, 138–142.
111. Xiao, X.; Chevali, V.S.; Song, P.; He, D.; Wang, H. Polylactide/hemp hurd biocomposites as sustainable 3D printing feedstock. *Compos. Sci. Technol.* **2019**, *184*, 107887. [CrossRef]
112. Khattab, M.M.; Dahman, Y. Production and recovery of poly-3-hydroxybutyrate bioplastics using agro-industrial residues of hemp hurd biomass. *Bioprocess Biosyst. Eng.* **2019**, *42*, 1115–1127. [CrossRef] [PubMed]
113. Oza, S.; Ning, H.; Ferguson, I.; Lu, N. Effect of surface treatment on thermal stability of the hemp-PLA composites: Correlation of activation energy with thermal degradation. *Compos. Part B Eng.* **2014**, *67*, 227–232. [CrossRef]
114. Mazzanti, V.; Salzano de Luna, M.; Pariante, R.; Mollica, F.; Filippone, G. Natural fiber-induced degradation in PLA-hemp biocomposites in the molten state. *Compos. Part A Appl. Sci. Manuf.* **2020**, *137*, 105990. [CrossRef]
115. Baghaei, B.; Skrifvars, M.; Berglin, L. Manufacture and characterisation of thermoplastic composites made from PLA/hemp co-wrapped hybrid yarn prepregs. *Compos. Part A Appl. Sci. Manuf.* **2013**, *50*, 93–101. [CrossRef]
116. Mazzanti, V.; Pariante, R.; Bonanno, A.; Ruiz de Ballesteros, O.; Mollica, F.; Filippone, G. Reinforcing mechanisms of natural fibers in green composites: Role of fibers morphology in a PLA/hemp model system. *Compos. Sci. Technol.* **2019**, *180*, 51–59. [CrossRef]
117. Marrot, L.; Alao, P.F.; Mikli, V.; Kers, J. Properties of Frost-Retted Hemp Fibers for the Reinforcement of Composites. *J. Nat. Fibers* **2021**, *18*, 1–12. [CrossRef]
118. Plota, A.; Masek, A. Plant-Origin Stabilizer as an Alternative of Natural Additive to Polymers Used in Packaging Materials. *Int. J. Mol. Sci.* **2021**, *22*, 4012. [CrossRef] [PubMed]
119. Andriotis, E.G.; Chachlioutaki, K.; Monou, P.K.; Bouropoulos, N.; Tzetzis, D.; Barmpalexis, P.; Chang, M.W.; Ahmad, Z.; Fatouros, D.G. Development of Water-Soluble Electrospun Fibers for the Oral Delivery of Cannabinoids. *AAPS PharmSciTech* **2021**, *22*, 1–14. [CrossRef] [PubMed]
120. Stelescu, M.D.; Manaila, E.; Craciun, G.; Dumitrascu, M. New green polymeric composites based on hemp and natural rubber processed by electron beam irradiation. *Sci. World J.* **2014**, *2014*, 684047. [CrossRef] [PubMed]
121. Moonart, U.; Utara, S. Effect of surface treatments and filler loading on the properties of hemp fiber/natural rubber composites. *Cellulose* **2019**, *26*, 7271–7295. [CrossRef]

122. Koushki, P.; Kwok, T.H.; Hof, L.; Wuthrich, R. Reinforcing silicone with hemp fiber for additive manufacturing. *Compos. Sci. Technol.* **2020**, *194*, 108139. [CrossRef]
123. Manaila, E.; Stelescu, M.D.; Craciun, G.; Surdu, L. Effects of benzoyl peroxide on some properties of composites based on hemp and natural rubber. *Polym. Bull.* **2014**, *71*, 2001–2022. [CrossRef]
124. Dayo, A.Q.; Gao, B.; Wang, J.; Liu, W.; Derradji, M.; Shah, A.H.; Babar, A.A. Natural hemp fiber reinforced polybenzoxazine composites: Curing behavior, mechanical and thermal properties. *Compos. Sci. Technol.* **2017**, *144*, 114–124. [CrossRef]
125. del Borrello, M.; Mele, M.; Campana, G.; Secchi, M. Manufacturing and characterization of hemp-reinforced epoxy composites. *Polym. Compos.* **2020**, *41*, 2316–2329. [CrossRef]
126. Di Landro, L.; Janszen, G. Composites with hemp reinforcement and bio-based epoxy matrix. *Compos. Part B Eng.* **2014**, *67*, 220–226. [CrossRef]
127. Ribeiro, M.P.; de, M. Neuba, L.; da Silveira, P.H.P.M.; da Luz, F.S.; da S. Figueiredo, A.B.-H.; Monteiro, S.N.; Moreira, M.O. Mechanical, thermal and ballistic performance of epoxy composites reinforced with Cannabis sativa hemp fabric. *J. Mater. Res. Technol.* **2021**, *12*, 221–233. [CrossRef]
128. Khdier, H.M.; Ali, A.H.; Salih, W.M. Manufacturing of Thermal and Acoustic Insulation From (Polymer Blend/Recycled Natural Fibers). *Eng. Technol. J.* **2020**, *38*, 1801–1807. [CrossRef]
129. Gupta, M.K.; Gond, R.K.; Bharti, A. Effects of treatments on the properties of polyester based hemp composite. *Indian J. Fibre Text. Res.* **2018**, *43*, 313–319.
130. Caprino, G.; Carrino, L.; Durante, M.; Langella, A.; Lopresto, V. Low impact behaviour of hemp fibre reinforced epoxy composites. *Compos. Struct.* **2015**, *133*, 892–901. [CrossRef]
131. Singha, A.S.; Rana, A.K. Preparation and characterization of graft copolymerized Cannabis indica L. fiber-reinforced unsaturated polyester matrix-based biocomposites. *J. Reinf. Plast. Compos.* **2012**, *31*, 1538–1553. [CrossRef]
132. Inbakumar, J.P.; Ramesh, S. Mechanical, wear and thermal behaviour of hemp fibre/egg shell particle reinforced epoxy resin bio composite. *Trans. Can. Soc. Mech. Eng.* **2018**, *42*, 280–285. [CrossRef]
133. Scarponi, C. Hemp fiber composites for the design of a Naca cowling for ultra-light aviation. *Compos. Part B Eng.* **2015**, *81*, 53–63. [CrossRef]
134. Vinod, B.; Sudev, L.J. Investigation on Effect of Cryogenic Temperature on Mechanical Behavior of Jute and Hemp Fibers Reinforced Polymer Composites. *Appl. Mech. Mater.* **2019**, *895*, 76–82. [CrossRef]
135. Członka, S.; Strąkowska, A.; Kairytė, A. The impact of hemp shives impregnated with selected plant oils on mechanical, thermal, and insulating properties of polyurethane composite foams. *Materials* **2020**, *13*, 4709. [CrossRef]
136. Sair, S.; Oushabi, A.; Kammouni, A.; Tanane, O.; Abboud, Y.; Oudrhiri Hassani, F.; Laachachi, A.; El Bouari, A. Effect of surface modification on morphological, mechanical and thermal conductivity of hemp fiber: Characterization of the interface of hemp-Polyurethane composite. *Case Stud. Therm. Eng.* **2017**, *10*, 550–559. [CrossRef]
137. Sair, S.; Oushabi, A.; Kammouni, A.; Tanane, O.; Abboud, Y.; El Bouari, A. Mechanical and thermal conductivity properties of hemp fiber reinforced polyurethane composites. *Case Stud. Constr. Mater.* **2018**, *8*, 203–212. [CrossRef]

MDPI
St. Alban-Anlage 66
4052 Basel
Switzerland
Tel. +41 61 683 77 34
Fax +41 61 302 89 18
www.mdpi.com

Materials Editorial Office
E-mail: materials@mdpi.com
www.mdpi.com/journal/materials